PURE AND APPLIED MATHEMATICS

A Series of Texts and Monographs

Edited by: R. COURANT · L. BERS · J. J. STOKER

Vol. I: **Supersonic Flow and Shock Waves**
By R. Courant and K. O. Friedrichs

Vol. II: **Nonlinear Vibrations in Mechanical and Electrical Systems**
By J. J. Stoker

Vol. III: **Dirichlet's Principle, Conformal Mapping, and Minimal Surfaces**
By R. Courant

Vol. IV: **Water Waves**
By J. J. Stoker

Vol. V: **Integral Equations**
By F. G. Tricomi

Vol. VI: **Differential Equations: Geometric Theory**
By Solomon Lefschetz

Vol. VII: **Linear Operators—Parts I and II**
By Nelson Dunford and Jacob T. Schwartz

Vol. VIII: **Modern Geometrical Optics**
By Max Herzberger

Vol. IX: **Orthogonal Functions**
By G. Sansone

Vol. X: **Lectures on Differential and Integral Equations**
By K. Yosida

Vol. XI: **Representation Theory of Finite Groups and Associative Algebras**
By C. W. Curtis and I. Reiner

Vol. XII: **Electromagnetic Theory and Geometrical Optics**
By Morris Kline and Irvin W. Kay

Vol. XIII: **Combinatorial Group Theory**
By W. Magnus, A. Karrass, and D. Solitar

Vol. XIV: **Asymptotic Expansions for Ordinary Differential Equations**
By Wolfgang Wasow

Vol. XV: **Tchebycheff Systems: With Applications in Analysis and Statistics**
By Samuel Karlin and William J. Studden

Vol. XVI: **Convex Polytopes**
By Branko Grünbaum

Vol. XVII: **Fourier Analysis in Several Complex Variables**
By Leon Ehrenpreis (*in preparation*)

Vol. XVIII: **Generalized Integral Transformations with Applications**
By Armen H. Zemanian (*in preparation*)

Vol. XIX: **Introduction to the Theory of Categories and Functors**
By Ion Bucur and Aristide Deleanu

Additional volumes in preparation

PURE AND APPLIED MATHEMATICS

A Series of Texts and Monographs

Edited by: R. COURANT · L. BERS · J. J. STOKER

VOLUME XIX

INTRODUCTION TO THE THEORY OF CATEGORIES AND FUNCTORS

ION BUCUR and ARISTIDE DELEANU

with the collaboration of
Peter J. Hilton

A WILEY-INTERSCIENCE PUBLICATION

John Wiley & Sons Ltd

London · New York · Sydney

Library of Congress Catalog Card Number 68–56285

SBN 470 11651 X

Set on Monophoto Filmsetter and printed by J. W. Arrowsmith Ltd., Bristol, England

PREFACE

The theory of categories has arisen in the last twenty-five years and now constitutes an autonomous branch of mathematics. It owes its origin and early inspiration to developments in algebraic topology. When the basic concepts of *category*, *functor*, *natural transformation* and *natural equivalence* were first formulated by Eilenberg and MacLane they served immediately to provide the appropriate framework for describing the way in which algebraic tools were used, and could be used, in the study of topology. It was surely evident from the outset, to the inventors of these fundamental notions and to others, that their domain of application certainly extended far beyond that of algebraic topology. Indeed there were clearly many applications within algebra itself, and *homological algebra* began to emerge as a mathematical discipline in its own right concerned with *abelian* categories and their specializations to categories of modules. However, it was not clear in the early stages that there was a "pure" theory latent within the domain of categories and functors which was capable of assuming substantial proportions within the body of mathematics. Of course, the original basic concepts came to be reinforced by auxiliary notions suggested by applications of those concepts; and many arguments traditionally carried out in a more specialized setting were seen to fit naturally into the more abstract framework of category theory. Nevertheless, it is only in the last ten years, or less, that the source of inspiration for advances in category theory has come to any considerable extent from within the theory itself. Once this process had begun it accelerated rapidly, so that now the corpus of knowledge has increased enormously and with this advance has come a great increase in the scope of application of the theory, to include such widely scattered parts of mathematics as functional analysis and mathematical logic.

It seems therefore that the time is ripe for a book devoted to category theory, and suitable both for those wishing to work within the theory itself and those wishing to use the theory—or at least its basic aspects—in other mathematical disciplines such as algebra, topology, algebraic geometry, logic etc. The present volume is intended to serve such a double purpose and, hopefully, to be suitable not only as a reference source but also as a text for a graduate course. The mathematical background required is very slight, being certainly no more than that available to a student on completion of his first degree; but it should be said from the outset that some sophistication is called for from the reader if he is to appreciate the rather abstract viewpoint and arguments of category theory.

The purpose of this book is not the same as that of the valuable texts by Freyd and Mitchell (to which reference is made in the bibliography). Freyd's book is devoted more or less exclusively to abelian categories and Mitchell's book is also rather different in scope, emphasis and direction from the present one. In order to render their text as useful as possible to specialists in various fields and to indicate as clearly as possible the flavour of the theory, the authors have included, as well as basic topics such as the representability and prorepresentability of functors and embedding theorems for categories, specialized applications such as the study of Grothendieck topologies and the theory of sheaves. The authors have taken the theory right up to (but excluding) the modern theory of triples, algebras and equationally-defined categories but have provided an extensive bibliography to guide further reading.

It is a pleasure to preface this book by my two colleagues and friends Ion Bucur and Aristide Deleanu.

Courant Institute of Mathematical Sciences,
New York University, and Cornell University,
August 1968 PETER HILTON

ACKNOWLEDGEMENTS

It is a pleasure for the authors to record their gratitude to Dr. Nicolae Popescu who has written paragraph 10 of chapter V and paragraphs 5, 6 and 8 of chapter VI.

They also wish to express their thanks to Mrs. Simona Pascu who has done a fine job in typing the manuscript.

ION BUCUR
ARISTIDE DELEANU

July 1968

CONTENTS

CHAPTER 1

Basic Concepts

1. The Notion of a Category. Duality. Subcategories. Examples

The notions of category and functor were introduced to mathematics by S. Eilenberg and S. MacLane in 1944, in connexion with the problem of axiomatizing the theory of homology and cohomology groups of topological spaces. These notions have gradually found applications in other branches of mathematics as well.

Definition. We shall say that a *category* $\mathscr{C}$ is given if a class $\mathrm{Ob}\mathscr{C}$ of elements called objects is given such that:

1. for each pair (A, B) of objects of $\mathscr{C}$ a set $\mathrm{Hom}_{\mathscr{C}}(A, B)$, called the set of morphisms of A into B, is given (we shall frequently write $u: A \longrightarrow B$ or $A \xrightarrow{u} B$ instead of $u \in \mathrm{Hom}_{\mathscr{C}}(A, B)$).

2. for each triple (A, B, C) of objects of $\mathscr{C}$ a map

$$\mu: \mathrm{Hom}_{\mathscr{C}}(A, B) \times \mathrm{Hom}_{\mathscr{C}}(B, C) \longrightarrow \mathrm{Hom}_{\mathscr{C}}(A, C)$$

is given (if $u \in \mathrm{Hom}_{\mathscr{C}}(A, B)$, $v \in \mathrm{Hom}_{\mathscr{C}}(B, C)$, we shall denote by vu or $v \circ u$ the image $\mu(u, v)$ of the pair (u, v) under μ; we shall say that μ is the composition map of morphisms).

3. The sets $\mathrm{Hom}_{\mathscr{C}}(A, B)$ and the composition maps of morphisms satisfy the following axioms:

(α) The composition map is associative: if $A \xrightarrow{u} B \xrightarrow{v} C \xrightarrow{w} D$, then

$$w(vu) = (wv)u.$$

(β) For each object A of $\mathscr{C}$ there is a morphism $1_A: A \longrightarrow A$ (called the identity morphism of A) such that if $B \xrightarrow{u} A$ and $A \xrightarrow{v} C$ then $1_A u = u$, $v 1_A = v$.

(γ) If the pairs (A, B), (A', B') are distinct, then the intersection of the sets $\mathrm{Hom}_{\mathscr{C}}(A, B)$, $\mathrm{Hom}_{\mathscr{C}}(A', B')$ is void.

Examples of categories

1. *The category* $\mathscr{E}ns$. The objects of the category $\mathscr{E}ns$ are the sets. If A and B are two arbitrary sets we shall define $\mathrm{Hom}_{\mathscr{E}ns}(A, B)$ to be the set

of all maps defined on the set A and taking values in the set B. The composition map $(f, g) \longrightarrow g f$ is the usual composition of maps.

2. The category $\mathscr{T}op$. The objects of the category $\mathscr{T}op$ are the topological spaces. If X and Y are two arbitrary topological spaces we shall define the $\mathrm{Hom}_{\mathscr{T}op}(X, Y)$ to be the sets of all continuous maps defined on the space X and taking their values in the space Y. The composition map is the usual composition of maps (the composite of two continuous maps is also continuous).

3. The category $\mathscr{G}$. The objects of the category $\mathscr{G}$ are the groups. If A and B are two arbitrary groups we shall define $\mathrm{Hom}_{\mathscr{G}}(A, B)$ to be the set of all homomorphisms of the group A into the group B. The composition map is the usual composition of homomorphisms.

4. The category $\mathscr{A}b$. The objects of the category $\mathscr{A}b$ are the Abelian groups, and the morphisms are again the homomorphisms.

5. The category ${}_\Lambda\mathscr{C}$. Let Λ be a ring with a unit element. The objects of the category ${}_\Lambda\mathscr{C}$ are the left Λ-modules unitary over the ring Λ. The morphisms are homomorphisms of Λ-modules, and the composition map is again the usual composition of homomorphisms. There is a similar definition for $\mathscr{C}_\Lambda$.

Definition. If $(\mathscr{C}_i)_{i\in I}$ is a family of categories we define the product category $\prod_{i\in I} \mathscr{C}_i$ as follows:

The objects of $\prod_{i\in I} \mathscr{C}_i$ are all the families of the form $(X_i)_{i\in I}$ where for each $i \in I$, X_i is an object of $\mathscr{C}_i$.

$$\mathrm{Hom}_{\prod_{i\in I}\mathscr{C}_i}((X_i)_{i\in I}, (Y_i)_{i\in I}) = \prod_{i\in I} \mathrm{Hom}_{\mathscr{C}_i}(X_i, Y_i)$$

If

$$(u_i)_{i\in I} \in \mathrm{Hom}_{\prod_{i\in I}\mathscr{C}_i}((X_i)_{i\in I}, (Y_i)_{i\in I})$$

and

$$(v_i)_{i\in I} \in \mathrm{Hom}_{\prod_{i\in I}\mathscr{C}_i}((Y_i)_{i\in I}, (Z_i)_{i\in I}),$$

then

$$(v_i)_{i\in I} \circ (u_i)_{i\in I} = (v_i \circ u_i)_{i\in I}.$$

Definition. Given a category $\mathscr{C}$, a new category $\mathscr{C}^0$, called the *dual category of* $\mathscr{C}$, is obtained in the following manner:

(i) The objects of the category $\mathscr{C}^0$ coincide with the objects of the category $\mathscr{C}$, in other words $\mathrm{Ob}\,\mathscr{C} = \mathrm{Ob}\,\mathscr{C}^0$.

(ii) The set of morphisms $\mathrm{Hom}_{\mathscr{C}^0}(A, B)$ is by definition identical with $\mathrm{Hom}_{\mathscr{C}}(B, A)$.

(iii) The composition map

$$\mathrm{Hom}_{\mathscr{C}^0}(A, B) \times \mathrm{Hom}_{\mathscr{C}^0}(B, C) \longrightarrow \mathrm{Hom}_{\mathscr{C}^0}(A, C)$$

is defined as follows:

$$\text{if } u \in \mathrm{Hom}_{\mathscr{C}^0}(A, B) \quad \text{and} \quad v \in \mathrm{Hom}_{\mathscr{C}^0}(B, C),$$

then

$$v \circ_{\mathscr{C}^0} u = u \circ_{\mathscr{C}} v.$$

Clearly $(\mathscr{C}^0)^0 = \mathscr{C}$.

The possibility of associating with each category $\mathscr{C}$ its dual category $\mathscr{C}^0$ enables one to dualize each notion and each statement with respect to a category $\mathscr{C}$ into a notion and a statement with respect to the category $\mathscr{C}^0$ (from a practical point of view, this is the procedure of 'reversing the arrows').

Definition. Let $\mathscr{C}$ be a category. By a *subcategory of* $\mathscr{C}$ we shall mean a category $\mathscr{C}'$ which satisfies the following conditions:

(i) $\mathrm{Ob}\,\mathscr{C}' \subset \mathrm{Ob}\,\mathscr{C}$.

(ii) $\mathrm{Hom}_{\mathscr{C}'}(A, B) \subset \mathrm{Hom}_{\mathscr{C}}(A, B)$.

(iii) The composition of morphisms in $\mathscr{C}'$ is induced by the composition of morphisms in $\mathscr{C}$.

(iv) The identity morphisms in $\mathscr{C}'$ are identity morphisms in $\mathscr{C}$.

Definition. A subcategory $\mathscr{C}'$ of a category $\mathscr{C}$ is said to be *full* if, for each pair (A, B) of objects of $\mathscr{C}'$, we have

$$\mathrm{Hom}_{\mathscr{C}'}(A, B) = \mathrm{Hom}_{\mathscr{C}}(A, B).$$

2. Monomorphisms, epimorphisms, isomorphisms

Let $\mathscr{C}$ be a category and $u: A \longrightarrow B$ a morphism in $\mathscr{C}$. For each object X of the category $\mathscr{C}$ we consider the map

$$\mathrm{Hom}_{\mathscr{C}}(X, A) \longrightarrow \mathrm{Hom}_{\mathscr{C}}(X, B) \tag{1}$$

which sends the element $v \in \mathrm{Hom}_{\mathscr{C}}(X, A)$ onto $uv \in \mathrm{Hom}_{\mathscr{C}}(X, B)$.

Definition. The morphism u is said to be a *monomorphism* or an *injection* if the map (1) is univalent for any object X of the category $\mathscr{C}$.

Examples. 1. Let $\mathscr{C}$ be one of the categories $\mathscr{E}ns$ or ${}_\Lambda\mathscr{C}$, where Λ is an arbitrary ring, and let $u: A \longrightarrow B$ be a morphism in $\mathscr{C}$. Then the following two propositions are equivalent:

(a) u is a monomorphism.

(b) for any $x_1, x_2 \in A$, $x_1 \neq x_2$ implies $u(x_1) \neq u(x_2)$, that is, u is univalent.

PROOF. We shall content ourselves with giving the proof for the category ${}_\Lambda\mathscr{C}$, since the case of the category $\mathscr{E}ns$ is analogous. We prove (a) $\Rightarrow$ (b). Let $N = u^{-1}(0)$. Assume $N \neq \{0\}$; then the map $\mathrm{Hom}_\Lambda(N, A) \longrightarrow \mathrm{Hom}_\Lambda(N, B)$ is not univalent. For, consider the elements $v_1, v_2 \in \mathrm{Hom}_\Lambda(N, A)$ defined as follows:

$$v_1(\xi) = 0 \quad \text{for any} \quad \xi \in N,$$
$$v_2(\xi) = \xi \quad \text{for any} \quad \xi \in N.$$

From the hypothesis it follows that $v_1 \neq v_2$, but their images under the map $\mathrm{Hom}_\Lambda(N, A) \longrightarrow \mathrm{Hom}_\Lambda(N, B)$ coincide.

We now prove the implication (b) $\Rightarrow$ (a). Assume that there exists a left Λ-module M and $v \in \mathrm{Hom}_\Lambda(M, A)$ such that $v \neq 0$ and the image of v under the map $\mathrm{Hom}_\Lambda(M, A) \longrightarrow \mathrm{Hom}_\Lambda(M, B)$ is the null homomorphism. It follows from this that there exists $\xi \in M$ such that $v(\xi) \neq 0$ and $u(v(\xi)) = 0$, which contradicts (b).

2. For any object A of an arbitrary category $\mathscr{C}$, $1_A: A \longrightarrow A$ is a monomorphism.

PROPOSITION 1.1. *The composition of two monomorphisms is a monomorphism.*

The proof follows immediately from the fact that the composition of two univalent maps is a univalent map.

PROPOSITION 1.2. *If $u: A \longrightarrow B$, $v: B \longrightarrow C$, and vu is a monomorphism, then u is a monomorphism.*

The proof is straightforward.

With the notations above, we now define a map

$$\mathrm{Hom}_\mathscr{C}(B, X) \longrightarrow \mathrm{Hom}_\mathscr{C}(A, X) \tag{2}$$

for any morphism $u: A \longrightarrow B$ and any object X of the category $\mathscr{C}$ as follows:

The map (2) sends the element $v \in \mathrm{Hom}_\mathscr{C}(B, X)$ onto $vu \in \mathrm{Hom}_\mathscr{C}(A, X)$.

Definition. The morphism u is said to be an *epimorphism* or a *surjection* if the map (2) is univalent for any object X of the category $\mathscr{C}$.

It is obvious that u is an epimorphism if and only if it is a monomorphism when considered as an element in $\mathrm{Hom}_{\mathscr{C}^0}(B, A)$. In other words, the notion of epimorphism is obtained by dualizing the notion of monomorphism.

PROPOSITION 1.1′. *The composition of two epimorphisms is an epimorphism.*

PROPOSITION 1.2′. *If vu is an epimorphism then v is an epimorphism.*

Examples. 1. In the categories $\mathscr{E}ns$ and ${}_\Lambda\mathscr{C}$ the notions epimorphism and map *onto* or homomorphism *onto* coincide.

We shall content ourselves with indicating the proof for the category ${}_\Lambda\mathscr{C}$. Assume $u : A \longrightarrow B$ is a homomorphism onto, where A, B are Λ-modules. If the homomorphisms $v, w : B \longrightarrow C$ are distinct, then there exists at least one element $x \in B$ such that $v(x) \neq w(x)$. But there exists at least one element $y \in A$ such that $u(y) = x$. We infer

$$(vu)(y) = v(u(y)) = v(x) \neq w(x) = w(u(y)) = (wu)(y),$$

i.e. $vu \neq wu$, hence u is an epimorphism.

Conversely, assume $u : A \longrightarrow B$ is an epimorphism. Suppose that $u(A)$ is a proper submodule of B. Let C be the direct sum of two Λ-modules B_1 and B_2, each being isomorphic with B, and let $v_i : B \longrightarrow C$ $(i = 1, 2)$ be the isomorphism of B onto B_i followed by the canonical injection. Let D be the factor module of C with respect to its sub-group generated by all the elements of the form $v_1(x) - v_2(x)$ where $x \in u(A)$. Finally, let $w : C \longrightarrow D$ be the canonical projection. Then $wv_1 \neq wv_2$ and $(wv_1)u = (wv_2)u$, which is absurd. Consequently $u(A) = B$, i.e. u is a homomorphism onto.

2. Let $\mathscr{C}$ be the category whose objects are Abelian topological groups (separated in the sense of Hausdorff) whose morphisms are the continuous homomorphisms. If $u : A \longrightarrow B$ is a morphism in $\mathscr{C}$ the following two propositions are equivalent:

(a) u is an epimorphism.

(b) $\overline{u(A)} = B$.

PROOF. (a) $\Rightarrow$ (b). Suppose $\overline{u(A)} = B_1 \neq B$. Then B/B_1 is an object of $\mathscr{C}$. Let $v_1, v_2 : B \longrightarrow B/B_1$ where v_1 is the canonical homomorphism and v_2 is the identically zero homomorphism. The hypothesis $B_1 \neq B$ implies $v_1 \neq v_2$; but $v_1 u = v_2 u$, which contradicts (a).

(b) $\Rightarrow$ (a). Let C be an arbitrary object of the category $\mathscr{C}$ and $v_1, v_2 : B \longrightarrow C$ such that $v_1 u = v_2 u$. Hence the continuous maps v_1, v_2 coincide on the subset $u(A)$, which is everywhere dense in B, and therefore $v_1 = v_2$.

3. For any category $\mathscr{C}$ and any object A of $\mathscr{C}$ the morphism $1_A : A \longrightarrow A$ is an epimorphism.

Definition. In an arbitrary category $\mathscr{C}$, the morphism $u : A \longrightarrow B$ is said to be a *bijection* if u is simultaneously an injection and a surjection.

Definition. The morphism $u : A \longrightarrow B$ is said to be an *isomorphism* if there exists $v : B \longrightarrow A$ such that

$$vu = 1_A, \quad uv = 1_B.$$

PROPOSITION 1.3. *Any isomorphism is a bijection.*

PROOF. From the relation $vu = 1_A$ and from the fact that 1_A is a monomorphism it follows, according to proposition 1.2, that u is a monomorphism. From the relation $uv = 1_B$ and from the fact that 1_B is an epimorphism it follows, according to proposition 1.2′, that u is an epimorphism.

Remark. There exist bijections which fail to be isomorphisms. For example, consider in the category of Abelian Hausdorff topological groups the following objects and morphisms:

R is the additive topological group of real numbers, Q is the additive topological group of rational numbers, $u : Q \longrightarrow R$ is the inclusion of Q into R.

It is easily checked using the example 2 above that u is a bijection; on the other hand, u fails to be an isomorphism, since it is not *onto*.

PROPOSITION 1.4. *The composition of two isomorphisms is an isomorphism.*

The proof is obvious.

Let $\mathscr{C}$ be a category and A an object of $\mathscr{C}$. Consider two monomorphisms $u : U \longrightarrow A$ and $v : V \longrightarrow A$. We shall say that the monomorphism u *majorates* the monomorphism v, and we shall write $u \geq v$ if there exists a morphism $v_1 : V \longrightarrow U$ such that we have $uv_1 = v$.

If one such v_1 exists, then, from the fact that v is a monomorphism and from proposition 1.2, it follows that v_1 is a monomorphism, and from the fact that u is a monomorphism, it follows directly from the definition of a monomorphism that v_1 is uniquely determined.

The relation '$\geq$' so defined is reflexive and transitive.

Suppose now we have also $v \geq u$. In this case there exists a unique isomorphism $u_1 : U \longrightarrow V$ and a unique isomorphism $v_1 : V \longrightarrow U$ such that $v = uv_1$, $u = vu_1$. For, from the definition results the existence of monomorphisms v_1, u_1 such that $v = uv_1$, $u = vu_1$, and from the relations $v = v(u_1 v_1)$, $u = u(v_1 u_1)$ and from the fact that v and u are monomorphisms, it results that $u_1 v_1 = 1_V$, $v_1 u_1 = 1_U$.

If two monomorphisms $u: U \longrightarrow A$, $v: V \longrightarrow A$ are such that $u \geq v$ and $v \geq u$ we shall say that they are equivalent.

In each class of equivalent monomorphisms, we choose once for all a representative (by using, for example, the axiom of choice). The monomorphisms obtained in this way are called *subobjects* of the object A.

Hence it follows that the subobjects of A are not simply objects of the category $\mathscr{C}$, but certain pairs of the form (U, u), where $u: U \longrightarrow A$ is a monomorphism which is called the canonical injection of U into A. In order not to complicate the writing and the language, we shall write simply U instead of (U, u) whenever there can be no possibility of confusion.

The notion dual to that of subobject is called a *quotient object*, and the dual of the canonical injection is called a canonical surjection.

We note that the relation '$\geq$' between subobjects or quotient objects is in fact a partial order relation.

Example. Let A be an object of the category $\mathscr{Ab}$. We may assume that the choice occurring in the definition of the subobject is so performed that any subobject of A is of the form (A_1, α), where A_1 is a subset of A and $\alpha(x) = x$ for any $x \in A_1$. In other words, a subobject in $\mathscr{Ab}$ is thus identified with what is currently understood by a subgroup of A.

3. Functors

Definition. Let $\mathscr{C}_1$ and $\mathscr{C}_2$ be two categories. A covariant (respectively contravariant) functor F from $\mathscr{C}_1$ into $\mathscr{C}_2$ consists of:

(a) a map $A \longrightarrow F(A)$ which associates to each object A of $\mathscr{C}_1$ an object $F(A)$ of $\mathscr{C}_2$.

(b) for each pair (A, B) of objects of $\mathscr{C}_1$, a map

$$F(A, B): \mathrm{Hom}_{\mathscr{C}_1}(A, B) \longrightarrow \mathrm{Hom}_{\mathscr{C}_2}(F(A), F(B))$$

$$(\text{respectively } F(A, B): \mathrm{Hom}_{\mathscr{C}_1}(A, B) \longrightarrow \mathrm{Hom}_{\mathscr{C}_2}(F,(B), F(A)))$$

such that, if we write simply $F(u)$ instead of $F(A, B)$, $(u): F(1_A) = 1_{F(A)}$ and $F(vu) = F(v)F(u)$ (respectively $F(vu) = F(u)F(v)$).

Clearly to each contravariant functor F from $\mathscr{C}_1$ into $\mathscr{C}_2$ a covariant functor from $\mathscr{C}_1^0$ into $\mathscr{C}_2$ and a covariant functor from $\mathscr{C}_1$ into $\mathscr{C}_2^0$ are associated in a natural manner, and vice versa, so that the general study of contravariant functors can be reduced to the study of covariant functors.

Examples. Modern mathematics reveals a large number of functors. It is reasonable to assume that this rich display of functors made it possible to develop the concept of a category. We shall limit ourselves to present here only a few examples which will be used in the sequel.

1. Let $\mathscr{C}$ be an arbitrary category and $\mathscr{E}ns$ the category of sets. Consider the product category $\mathscr{C}^0 \times \mathscr{C}$ (see definition on p. 2). We define a covariant functor from the category $\mathscr{C}^0 \times \mathscr{C}$ to the category $\mathscr{E}ns$ if we set:

$$F(A, B) = \mathrm{Hom}_{\mathscr{C}}(A, B)$$

$$[F(u, v)](f) = vfu \quad \text{for any} \quad (u, v):(A, B) \longrightarrow (A_1, B_1)$$

and

$$f \in \mathrm{Hom}_{\mathscr{C}}(A, B).$$

2. Let $\mathscr{M}$ be the category whose objects are ordered triples (A, L, M) where A is a ring, L is a right A-module and M is a left A-module. A morphism of (A, L, M) into (A_1, L_1, M_1) is, by definition, an ordered triple (φ, u, v) where (φ, u) is a bihomomorphism of the pair (A, L) into the pair (A_1, L_1) and (φ, v) is a bihomomorphism of (A, M) into (A_1, M_1). We obtain a covariant functor T from the category $\mathscr{M}$ into the category $\mathscr{A}b$ of abelian groups if we set:

$$T(A, L, M) = L \bigotimes_A M$$

$T(\varphi, u, v) =$ the homomorphism induced by (φ, u, v) from $L \bigotimes_A M$ into $L_1 \bigotimes_{A_1} M_1$.

3. Let X be a fixed object of the category $\mathscr{C}$. We define a contravariant functor h_X from the category $\mathscr{C}$ into the category $\mathscr{E}ns$ if we set:

$$h_X(Y) = \mathrm{Hom}_{\mathscr{C}}(Y, X)$$

for any morphism $u: Y \longrightarrow Y'$ in $\mathscr{C}$, $h_X(u)$ is the map of $\mathrm{Hom}_{\mathscr{C}}(Y', X)$ into $\mathrm{Hom}_{\mathscr{C}}(Y, X)$ which sends v onto vu.

Similarly, we define a covariant functor $\bar{h}_X$ from the category $\mathscr{C}$ into the category $\mathscr{E}ns$ if we set

$$\bar{h}_X(Y) = \mathrm{Hom}_{\mathscr{C}}(X, Y)$$

for any morphism $u: Y \longrightarrow Y'$ in $\mathscr{C}$, $\bar{h}_X(u)$ is the map of $\mathrm{Hom}_{\mathscr{C}}(X, Y)$ into $\mathrm{Hom}_{\mathscr{C}}(X, Y')$ which sends v onto uv.

4. Let ${}_A\mathscr{C}$ be the category of left A-modules and X a fixed right A-module. We obtain a covariant functor T_X from the category ${}_A\mathscr{C}$ into the category $\mathscr{Ab}$ by setting

$$T_X(Y) = Y \bigotimes_A X$$

for any homomorphism $u: Y \longrightarrow Y'$ of the A-module Y into the A-module Y', $T_X(u)$ is the map from $Y \bigotimes_A X$ into $Y' \bigotimes_A X$ induced by u.

Definition. Let $\mathscr{C}_1$ and $\mathscr{C}_2$ be two fixed categories and let F, G be two covariant functors from $\mathscr{C}_1$ into $\mathscr{C}_2$. We say that a *functorial morphism* f from the functor F into the functor G is given, if for each object A in $\mathscr{C}_1$ there is given a morphism $f(A): F(A) \longrightarrow G(A)$ in $\mathscr{C}_2$ such that the diagram

$$\begin{array}{ccc} F(A) & \xrightarrow{f(A)} & G(A) \\ {\scriptstyle F(u)}\downarrow & & \downarrow{\scriptstyle G(u)} \\ F(B) & \xrightarrow[f(B)]{} & G(B) \end{array}$$

is commutative for any morphism $u: A \longrightarrow B$ in $\mathscr{C}_1$.

If $f(A)$ is an isomorphism for each A in $\mathscr{C}_1$, then f is called a *functorial isomorphism.*

If the functors F, G are contravariant, then in the definition of the functorial morphism the above diagram has of course to be replaced by the following diagram:

$$\begin{array}{ccc} F(A) & \xrightarrow{f(A)} & G(A) \\ {\scriptstyle F(u)}\uparrow & & \uparrow{\scriptstyle G(u)} \\ F(B) & \xrightarrow[f(B)]{} & G(B) \end{array}$$

Example. Let $w: X \longrightarrow X'$ be a morphism in the category $\mathscr{C}$. We associate to this morphism a functorial morphism from the functor h_X into the functor $h_{X'}$ in the following way.

Let Y be an arbitrary object in $\mathscr{C}$. We denote by $h_w(Y)$ the map from the set $\mathrm{Hom}_{\mathscr{C}}(Y, X)$ into the set $\mathrm{Hom}_{\mathscr{C}}(Y, X')$ which sends $v \in \mathrm{Hom}_{\mathscr{C}}(Y, X) = h_X(Y)$ onto $wv \in \mathrm{Hom}_{\mathscr{C}}(Y, X') = h_{X'}(Y)$. To conclude that we have thus defined a functorial morphism from h_X into $h_{X'}$ we have to check that the diagram

$$\begin{array}{ccc} h_X(Y) & \xrightarrow{h_w(Y)} & h_{X'}(Y) \\ {\scriptstyle h_X(u)}\uparrow & & \uparrow{\scriptstyle h_{X'}(u)} \\ h_X(Y') & \xrightarrow[h_w(Y')]{} & h_{X'}(Y') \end{array}$$

is commutative for any morphism $u: Y \longrightarrow Y'$ in $\mathscr{C}$. But this follows immediately, since, if $f \in h_X(Y') = \mathrm{Hom}_{\mathscr{C}}(Y', X)$, we have

$$[h_w(Y) \cdot h_X(u)](f) = (h_w(Y))(fu) = w(fu)$$

$$[h_{X'}(u) \cdot h_w(Y')](f) = h_{X'}(u)(wf) = (wf)u.$$

It is obvious that if w is an isomorphism then h_w is a functorial isomorphism.

If f is a functorial morphism from F into G, and g is a functorial morphism from G into H, then a functorial morphism h from F into H can be defined by setting for each object A in $\mathscr{C}_1$, $h(A) = g(A)f(A)$. The morphism thus defined is called the composition of the functorial morphisms f and g and is denoted by $h = gf$. The operation of composition thus defined is associative. If the objects of the category $\mathscr{C}_1$ form a *set*, then one can define in a natural way a new category, $\mathrm{Hom}(\mathscr{C}_1, \mathscr{C}_2)$, whose objects are the covariant functors from $\mathscr{C}_1$ into $\mathscr{C}_2$, and the set of morphisms from the covariant functor F into the covariant functor G, $\mathrm{Hom}(F, G)$ in the sense of this new category, is by definition the set of all functorial morphisms from F to G.

If F and G are two covariant or contravariant functors from the category $\mathscr{C}_1$ into the category $\mathscr{C}_2$, we shall always denote by $\mathrm{Hom}(F, G)$ the class of all functorial morphisms from F to G (not only in the case where $\mathscr{C}_1$ is a set !). We shall also use the notation $f: F \longrightarrow G$ to mean that f is a functorial morphism from the functor F to the functor G, both for covariant and for contravariant functors. Also, 1_F will denote the functorial morphism $1_F: F \longrightarrow F$ defined by $1_F(Y) = 1_{F(Y)}$ for any Y.

The logical difficulties involved in the construction of the category $\mathrm{Hom}(\mathscr{C}_1, \mathscr{C}_2)$ can be avoided by using the concept of a Grothendieck universe.

A *universe* $\mathscr{U}$ is a set such that the following axioms are satisfied:

U_1: If $(X_i)_{i \in I}$ is a family of sets belonging to $\mathscr{U}$ and if I is an element of $\mathscr{U}$, then the union $\bigcup_{i \in I} X_i$ is an element of $\mathscr{U}$.

U_2: If x belongs to $\mathscr{U}$, then the set $\{x\}$ consisting of a single element belongs to $\mathscr{U}$.

U_3: If x belongs to X and if X belongs to $\mathscr{U}$, then x belongs to $\mathscr{U}$.

U_4: If X is a set belonging to $\mathscr{U}$, the set $\mathscr{P}(X)$ of subsets of X is an element of $\mathscr{U}$.

U_5: The pair (x, y) is an element of $\mathscr{U}$ if and only if x and y are elements of $\mathscr{U}$.

A category $\mathscr{C}$ is said to be a $\mathscr{U}$-category, if:

1. The objects of $\mathscr{C}$ form a set which is an element of $\mathscr{U}$, and
2. The set $\mathrm{Hom}_{\mathscr{C}}(X, Y)$ is an element of $\mathscr{U}$ for each pair (X, Y) of objects of $\mathscr{C}$.

If $\mathscr{C}_1$ and $\mathscr{C}_2$ are $\mathscr{U}$-categories, then $\mathrm{Hom}(\mathscr{C}_1, \mathscr{C}_2)$ is also a $\mathscr{U}$-category, because the objects of $\mathrm{Hom}(\mathscr{C}_1, \mathscr{C}_2)$ form a subset of the set of maps of $\bigcup_{X,Y \in \mathscr{C}_1} \mathrm{Hom}_{\mathscr{C}_1}(X, Y)$ into $\bigcup_{Z,V \in \mathscr{C}_2} \mathrm{Hom}_{\mathscr{C}_2}(Z, V)$, and for any $F, G \in \mathrm{Hom}(\mathscr{C}_1, \mathscr{C}_2)$, $\mathrm{Hom}(F, G)$ is a subset of the set $\prod_{X \in \mathscr{C}_1} \mathrm{Hom}_{\mathscr{C}_2}(F(X), G(X))$.

PROPOSITION 1.5. If $f: F \longrightarrow G$ *is a functorial isomorphism from the contravariant functor* F *to the contravariant functor* G, *there exists a unique functorial morphism* $g: G \longrightarrow F$ *such that* $gf = 1_F, fg = 1_G$.

(Similar statement for the case of covariant functors).

PROOF. The uniqueness being obvious, it remains to prove the existence. To do this, let Y be an object of the category $\mathscr{C}_1$. There exists a unique morphism $g(Y): G(Y) \longrightarrow F(Y)$ in the category $\mathscr{C}_2$ such that $f(Y)g(Y) = 1_{G(Y)}$, $g(Y)f(Y) = 1_{F(Y)}$. We claim that we have thus defined a functorial morphism $g: G \longrightarrow F$; for, the diagram

$$\begin{array}{ccc} G(Y) & \xrightarrow{g(Y)} & F(Y) \\ {\scriptstyle G(u)}\uparrow & & \uparrow{\scriptstyle F(u)} \\ G(Y') & \xrightarrow[g(Y')]{} & F(Y') \end{array}$$

is commutative for any $u: Y \longrightarrow Y'$. To see this, we prove that

$$f(Y)g(Y)G(u) = f(Y)F(u)g(Y'),$$

taking into account that f is a functorial morphism and the diagram

$$\begin{array}{ccc} F(Y) & \xrightarrow{f(Y)} & G(Y) \\ {\scriptstyle F(u)}\uparrow & & \uparrow{\scriptstyle G(u)} \\ F(Y') & \xrightarrow[f(Y')]{} & G(Y') \end{array}$$

is therefore commutative. Hence we have

$$f(Y)g(Y)G(u) = G(u)$$

$$f(Y)F(u)g(Y') = G(u)f(Y')g(Y') = G(u).$$

Let F be a contravariant functor from a category $\mathscr{C}$ into the category $\mathscr{E}ns$. We propose to determine the class $\mathrm{Hom}(h_X, F)$ of all functorial morphisms from the functor h_X to the functor F. We obtain the following:

THEOREM 1.6. *There exists a canonical map* $\alpha:\mathrm{Hom}(h_X, F) \longrightarrow F(X)$ *which is one-to-one and onto.*

PROOF. Let f be a functorial morphism from h_X to F. For any object Y of the category $\mathscr{C}$ there exists a morphism $f(Y):h_X(Y) \longrightarrow F(Y)$ such that for any morphism $u:Y \longrightarrow Y'$ in $\mathscr{C}$ the following diagram is commutative:

$$\begin{array}{ccc} h_X(Y) & \xrightarrow{f(Y)} & F(Y) \\ {\scriptstyle h_X(u)}\uparrow & & \uparrow{\scriptstyle F(u)} \\ h_X(Y') & \xrightarrow[f(Y')]{} & F(Y') \end{array}$$

In particular, there exists a morphism $f(X):h_X(X) = \mathrm{Hom}_{\mathscr{C}}(X, X) \longrightarrow F(X)$. We set by definition $\alpha(f) = (f(X))(1_X)$.

We now define a map $\beta:F(X) \longrightarrow \mathrm{Hom}(h_X, F)$ as follows:

Let $\xi \in F(X)$ and let Y be an arbitrary object of the category $\mathscr{C}$. We define a morphism $(\beta(\xi))(Y):h_X(Y) = \mathrm{Hom}_{\mathscr{C}}(Y, X) \longrightarrow F(Y)$ by the equation

$$[(\beta(\xi))(Y)](u) = [F(u)](\xi).$$

To conclude that we have thus defined a functorial morphism from h_X to F we have to check that the diagram

$$\begin{array}{ccc} h_X(Y) & \xrightarrow{(\beta(\xi))(Y)} & F(Y) \\ {\scriptstyle h_X(v)}\uparrow & & \uparrow{\scriptstyle F(v)} \\ h_X(Y') & \xrightarrow[(\beta(\xi))(Y')]{} & F(Y') \end{array}$$

is commutative for any morphism $v:Y \longrightarrow Y'$ in $\mathscr{C}$. To do this, let $w:Y \longrightarrow X$. We have:

$$[F(v)(\beta(\xi))(Y')](w) = (F(v))(F(w)(\xi)) = (F(v)F(w))(\xi) = (F(wv))(\xi).$$

$$[(\beta(\xi))(Y) \cdot h_X(v)](w) = [(\beta(\xi))(Y)](wv) = (F(wv))(\xi).$$

We now show that α and β are bijections inverse to each other.

(a) $\beta(\alpha(f)) = f$. Let Y be an object of $\mathscr{C}$. We shall show that $[\beta(\alpha(f))](Y) = f(Y)$. By definition we have for any $u \in h_X(Y) = \mathrm{Hom}_{\mathscr{C}}(Y, X)$:

$$\{[\beta(\alpha(f))](Y)\}(u) = (F(u))(\alpha(f)) = (F(u))(f(X)(1_X)) = (F(u)f(X))(1_X)$$

But, since f is a functorial morphism the following diagram is commutative

$$\begin{array}{ccc} h_X(X) & \xrightarrow{f(X)} & F(X) \\ {\scriptstyle h_X(u)}\downarrow & & \downarrow{\scriptstyle F(u)} \\ h_X(Y) & \xrightarrow[f(Y)]{} & F(Y) \end{array}$$

which means that

$$(F(u) \cdot f(X))(1_X) = (f(Y)h_X(u))(1_X) = (f(Y))(u).$$

(b) $\alpha(\beta(\xi)) = \xi$. Indeed,

$$\alpha(\beta(\xi)) = [(\beta(\xi))(X)](1_X) = (F(1_X))(\xi) = 1_{F(X)}(\xi) = \xi.$$

COROLLARY 1.7. *If X and X' are two objects of the category $\mathscr{C}$ then for any functorial morphism $f: h_X \longrightarrow h_{X'}$ there exists a unique morphism $\xi: X \longrightarrow X'$ in the category $\mathscr{C}$ such that*

$$(f(Y))(u) = \xi u$$

for any object X in $\mathscr{C}$. If f is a functorial isomorphism then ξ is an isomorphism and conversely.

PROOF. Set $\xi = \alpha(f)$. It follows that $\beta(\xi) = f$, whence the desired relation.

If f is a functorial isomorphism, then there exists according to proposition 1.5 a functorial morphism $g: h_{X'} \longrightarrow h_X$ such that $gf = 1_{h_X}$, $fg = 1_{h_{X'}}$. Let $\xi' = \alpha(g)$.

We have:

$$\begin{aligned} 1_X = \alpha(gf) &= [(gf)(X)](1_X) = [g(X)f(X)](1_X) \\ &= (g(X))(f(X)(1_X)) = g(X)(\xi) = \xi'\xi. \end{aligned}$$

Similarly $1_{X'} = \xi\xi'$.

4. Representable functors

Definition. Let F be a contravariant functor from a category $\mathscr{C}$ to the category $\mathscr{Ens}$. F is said to be *representable* if there exists an object X in $\mathscr{C}$ such that there exists a functorial isomorphism f from h_X to F. Using the above notations, we shall say that the pair $(X, \alpha(f))$ represents the functor F.

By a 'Abus de langage' we shall sometimes say that X represents the functor F.

PROPOSITION 1.8. *If the pairs* (X, ξ), (X', ξ') *represent the functor* F, *then there exists a unique isomorphism* $u: X \longrightarrow X'$ *such that* $(F(u))(\xi') = \xi$.

PROOF. Existence. The hypothesis implies that there exists a functorial isomorphism $g: h_X \longrightarrow h_{X'}$ such that the diagram

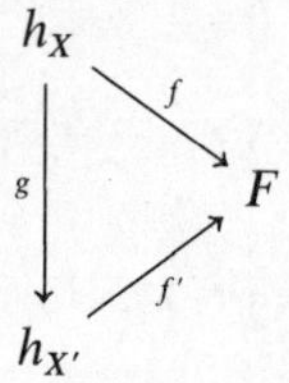

is commutative, where $\alpha(f) = \xi$, $\alpha(f') = \xi'$. By corollary 1.7 there exists $u: X \longrightarrow X'$, such that $(g(Y))(w) = uw$ for any $w: Y \longrightarrow X$. We have $\xi = (f(X))(1_X) = (f'(X)g(X))(1_X) = f'(X)(u) = (F(u))(\xi')$.

Uniqueness. If $u, v: X \longrightarrow X'$ are such that $(F(u))(\xi') = (F(v))(\xi') = \xi$, then for both induced functorial morphisms $g_1, g_2: h_X \longrightarrow h_{X'}$ the above diagram is commutative, which implies that $g_1 = g_2$, and $u = v$. Indeed, $(f'(Y)g_1(Y))(w) = (f'(Y))(uw) = (F(uw))(\xi') = (F(w) \cdot F(u))(\xi') = [F(w)](F(u)(\xi')) = F(w)\xi = (F(Y))(w)$.

Similarly $(f'(Y)g_2(Y))(w) = (F(Y))(w)$. But this implies $u = v$.

A great many problems of mathematics reduce to that of deciding whether or not a given functor is representable. For example the classification of fibre bundles with a fixed structural group, homology and cohomology theories, the theory of Picard varieties, and so on.

Unfortunately, we do not dispose at present of convenient general criteria whereby we could decide whether or not a given functor F is representable. A necessary condition of representability is exposed in chapter 3 (corollary 3.8).

We now give some classical examples of representable functors.

The tensor product of two modules. Let Λ be a commutative ring with a unit element and let A and B be two Λ-modules. We denote by $\mathscr{B}(X)$ the set of all the bilinear maps defined on $A \times B$ and taking their values in the Λ-module X. We thus obtain a covariant functor $\mathscr{B}: {}_\Lambda\mathscr{C} \longrightarrow {}_\Lambda\mathscr{C}$. $\mathscr{B}$ is representable by $A \underset{\Lambda}{\otimes} B$.

Rings of fractions. Let $\mathscr{A}$ be the category of commutative rings with unit element and of ring homomorphism which preserve the unit elements.

Let A be an object of the category $\mathscr{A}$ and S a multiplicatively closed subset of A. For each ring B we shall denote by $Q(B)$ the set of all homomorphisms f of A into B with the property that $f(s)$ is an invertible element of B for any $s \in S$. We thus obtain a covariant functor $Q: \mathscr{A} \longrightarrow \mathscr{E}ns$. This functor is representable: $Q(B) = \operatorname{Hom}_{\mathscr{A}}(A[S^{-1}], B)$.

The completion of a metric space. Let $\mathscr{T}$ be the category of metric topological spaces and of uniformly continuous maps. We distinguish in this category the full subcategory $\overline{\mathscr{T}}$ of complete metric spaces. Let A be an arbitrary metric topological space. Consider the covariant functor $F: \overline{\mathscr{T}} \longrightarrow \mathscr{E}ns$ defined as follows:

$$F(X) = \{f: A \longrightarrow X \mid f \text{ uniformly continuous}\}$$

This functor is representable: $F(X) = \operatorname{Hom}_{\overline{\mathscr{T}}}(\bar{A}, X)$, where $\bar{A}$ is the completion of the space A.

The Alexandroff compactification of a locally compact space. Let $\mathscr{T}_{lc}$ be the category of locally compact topological spaces and of proper continuous maps. We distinguish in this category the full subcategory $\mathscr{T}_c$ of compact spaces. Consider the covariant functor $F: \mathscr{T}_c \longrightarrow \mathscr{E}ns$ defined as follows:

$$F(X) = \operatorname{Hom}_{\mathscr{T}_{lc}}(A, X).$$

This functor is representable: $F(X) = \operatorname{Hom}_{\mathscr{T}_c}(A^*, X)$; where A^* is the Alexandroff compactification of the space A.

PROPOSITION 1.9. *Let F, G be two representable covariant functors and let (A, ξ), respectively (B, η) be the pairs which represent them. Suppose there exists a functorial morphism $\mu: F \longrightarrow G$. Under these conditions, there exists a unique morphism $\mu: B \longrightarrow A$ such that $[\mu(A)](\xi) = [G(u)](\eta)$. Moreover, if H is a third representable functor and $v: G \longrightarrow H$ a functorial morphism, then the morphism associated in the manner described above to the functorial morphism $v\mu: F \longrightarrow H$ coincides with uv.*

PROOF. For any object X, there exist morphisms $\varphi(X)$, $\gamma(X)$, $\lambda(X)$ such that the diagram

$$\begin{array}{ccc} F(X) & \xrightarrow{\mu(X)} & G(X) \\ \uparrow{\scriptstyle \varphi(X)} & & \uparrow{\scriptstyle \gamma(X)} \\ \operatorname{Hom}_{\mathscr{C}}(A, X) & \xrightarrow[\lambda(X)]{} & \operatorname{Hom}_{\mathscr{C}}(B, X) \end{array} \qquad (3)$$

is commutative. Let $\mu = (\lambda(A))(1_A)$. We shall show that u satisfies the condition in the statement of the proposition. The commutativity of diagram (3) implies the relation

$$[\mu(A)](\xi) = [\mu(A)][(\varphi(A))(1_A)] = [\gamma(A)\lambda(A)](1_A) = (\gamma(A))(u) = (G(u))(\eta),$$

so that the existence of u is proved. The unicity follows from the fact that $\gamma(A)$ is a bijection.

To prove the second part of the proposition, it is sufficient to verify the relation

$$[(\nu\mu)(A)](\xi) = [H(uv)](\zeta)$$

This is a consequence of the following relations:

$$\begin{aligned}[(\nu u)(A)](\xi) &= \nu(A)(u(A)\xi) = \nu(A)(G(u)\eta) = (H(u)\nu(B))(\eta)\\ &= H(u)(\nu(B)\eta) = (H(u)H(v))(\zeta).\end{aligned}$$

To write the third equality we have used the commutativity of the diagram

$$\begin{array}{ccc} G(A) & \xrightarrow{\nu(A)} & H(A) \\ {\scriptstyle G(u)}\uparrow & & \uparrow{\scriptstyle H(u)} \\ G(B) & \xrightarrow[\nu(B)]{} & H(B) \end{array}$$

This completes the proof of the proposition. As a corollary we obtain again the proposition: If the functor F is represented simultaneously by the pairs (A, ξ), (B, η) then there exists a unique isomorphism $u: B \longrightarrow A$ such that $[F(u)](\eta) = \xi$.

5. Adjoint functors

Let $\mathscr{C}_1$ and $\mathscr{C}_2$ be two categories and let F be a covariant functor from the category $\mathscr{C}_1$ into the category $\mathscr{C}_2$, and G a covariant functor from the category $\mathscr{C}_2$ into the category $\mathscr{C}_1$. We shall associate to these two functors another two functors from the product category $\mathscr{C}_1^0 \times \mathscr{C}_2$ (see Definition on p. 2) into the category $\mathscr{E}ns$, denoted by $\mathrm{Hom}_{\mathscr{C}_1}(G)$, $\mathrm{Hom}_{\mathscr{C}_2}(F)$, which are defined as follows (we shall limit ourselves here and elsewhere to give the explicit form of the maps defined by the respective functors only for the case of objects):

$\mathrm{Hom}_{\mathscr{C}_1}(G)$ associates to the object (A, B) of $\mathscr{C}_1^0 \times \mathscr{C}_2$ the object $\mathrm{Hom}_{\mathscr{C}_1}(A, G(B))$ and $\mathrm{Hom}_{\mathscr{C}_2}(F)$ associates to the object (A, B) the object

$\mathrm{Hom}_{\mathscr{C}_2}(F(A), B)$. We say that the functor G is an *adjoint* functor of the functor F if there exists a functorial isomorphism

$$\varphi : \mathrm{Hom}_{\mathscr{C}_2}(F) \longrightarrow \mathrm{Hom}_{\mathscr{C}_1}(G).$$

PROPOSITION 1.10. *If G is an adjoint functor of the functor F and G_1 is an adjoint functor of the functor F_1 and if $f : F \longrightarrow F_1$ is a functorial morphism from F to F_1, then there exists a unique functorial morphism $g : G_1 \longrightarrow G$ such that the diagram*

$$\begin{array}{ccc} \mathrm{Hom}_{\mathscr{C}_2}(F(A), B) & \xrightarrow{\varphi(A,B)} & \mathrm{Hom}_{\mathscr{C}_1}(A, G(B)) \\ {\scriptstyle \mathrm{Hom}_{\mathscr{C}_2}(f(A), B)} \uparrow & & \uparrow {\scriptstyle \mathrm{Hom}_{\mathscr{C}_1}(A, g(B))} \\ \mathrm{Hom}_{\mathscr{C}_2}(F_1(A), B) & \xrightarrow[\varphi_1(A,B)]{} & \mathrm{Hom}_{\mathscr{C}_1}(A, G_1(B)) \end{array} \tag{4}$$

is commutative for any object A of $\mathscr{C}_1$ and any object B of $\mathscr{C}_2$. (The significance of the morphisms $\mathrm{Hom}_{\mathscr{C}_2}(f(A), B)$, $\mathrm{Hom}_{\mathscr{C}_1}(A, g(B))$, is clear). Moreover, if G_2 is an adjoint functor of F_2 and if $f_1 : F_1 \longrightarrow F_2$ and $g_1 : G_2 \longrightarrow G_1$ is the functorial morphism associated in this manner, then the morphism associated to $f_1 f$ coincides with $g g_1$.

PROOF. Let B be an arbitrary element of the category $\mathscr{C}_2$. Consider the functors $h_{G(B)}$, $h_{G_1(B)}$ from the category $\mathscr{C}_1$ into the category $\mathscr{Ens}$. From the hypothesis we infer that there exists a functorial morphism from $h_{G_1(B)}$ to $h_{G(B)}$. By corollary 1.7 there exists a unique morphism $g(B) : G_1(B) \longrightarrow G(B)$ such that diagram (4) is commutative. It remains to prove that in this manner a functorial morphism $g : G_1 \longrightarrow G$ is defined. Let $u : B \longrightarrow B_1$ be a morphism in the category $\mathscr{C}_1$. We have to show that the diagram

$$\begin{array}{ccc} G(B_1) & \xleftarrow{G(u)} & G(B) \\ {\scriptstyle g(B_1)} \uparrow & & \uparrow {\scriptstyle g(B)} \\ G_1(B_1) & \xleftarrow[G_1(u)]{} & G_1(B) \end{array} \tag{5}$$

is commutative. By using again corollary 1.7 it will be sufficient to prove that the morphisms

$$\mathrm{Hom}_{\mathscr{C}_1}(A, G(u))\, \mathrm{Hom}_{\mathscr{C}_1}(A, g(B)),$$

$$\mathrm{Hom}_{\mathscr{C}_1}(A, g(B_1))\, \mathrm{Hom}_{\mathscr{C}_1}(A, G_1(u)) : \mathrm{Hom}_{\mathscr{C}_1}(A, G_1(B)) \longrightarrow \mathrm{Hom}_{\mathscr{C}_1}(A, G(B_1))$$

coincide or, equivalently, the same thing about the morphisms obtained by composing these two with $\varphi_1(A, B)$. Taking into account the fact that the following diagrams are commutative

$$\begin{array}{ccc}
\mathrm{Hom}_{\mathscr{C}_2}(F_1(A), B) & \xrightarrow{\varphi_1(A, B)} & \mathrm{Hom}_{\mathscr{C}_1}(A, G_1(B)) \\
\Big\downarrow{\scriptstyle \mathrm{Hom}_{\mathscr{C}_2}(F_1(A), u)} & & \Big\downarrow{\scriptstyle \mathrm{Hom}_{\mathscr{C}_1}(A, G_1(u))} \\
\mathrm{Hom}_{\mathscr{C}_2}(F_1(A), B_1) & \xrightarrow[\varphi_1(A, B_1)]{} & \mathrm{Hom}_{\mathscr{C}_1}(A, G_1(B_1))
\end{array}$$

(φ_1 is a functorial morphism),

$$\begin{array}{ccc}
\mathrm{Hom}_{\mathscr{C}_2}(F_1(A), B_1) & \xrightarrow{\varphi_1(A, B_1)} & \mathrm{Hom}_{\mathscr{C}_1}(A, G_1(B_1)) \\
\Big\downarrow{\scriptstyle \mathrm{Hom}_{\mathscr{C}_2}(f(A), B_1)} & & \Big\downarrow{\scriptstyle \mathrm{Hom}_{\mathscr{C}_1}(A, g(B_1))} \\
\mathrm{Hom}_{\mathscr{C}_2}(F(A), B_1) & \xrightarrow[\varphi(A, B_1)]{} & \mathrm{Hom}_{\mathscr{C}_1}(A, G(B_1))
\end{array}$$

(the definition of $g(B_1)$, in other words diagram (4) in which B is replaced by B_1)

we obtain the relations

$$\begin{aligned}
&\mathrm{Hom}_{\mathscr{C}_1}(A, g(B_1))\, \mathrm{Hom}_{\mathscr{C}_1}(A, G_1(u))\varphi_1(A, B) \\
&\quad = \mathrm{Hom}_{\mathscr{C}_1}(A, g(B_1))(\varphi_1(A, B_1))\, \mathrm{Hom}_{\mathscr{C}_2}(F_1(A), u) \\
&\quad = \varphi(A, B_1)\, \mathrm{Hom}_{\mathscr{C}_2}(f(A), B_1)\, \mathrm{Hom}_{\mathscr{C}_2}(F_1(A), u).
\end{aligned}$$

In the same way, from the commutativity of diagram (4) and of the diagram

$$\begin{array}{ccc}
\mathrm{Hom}_{\mathscr{C}_2}(F(A), B_1) & \xrightarrow{\varphi(A, B_1)} & \mathrm{Hom}_{\mathscr{C}_1}(A, G(B_1)) \\
\Big\uparrow{\scriptstyle \mathrm{Hom}_{\mathscr{C}_2}(F(A), u)} & & \Big\uparrow{\scriptstyle \mathrm{Hom}_{\mathscr{C}_1}(A, G(u))} \\
\mathrm{Hom}_{\mathscr{C}_2}(F(A), B) & \xrightarrow[\varphi(A, B)]{} & \mathrm{Hom}_{\mathscr{C}_1}(A, G(B))
\end{array}$$

(φ is a functorial morphism)

we infer:

$$\begin{aligned}
&\mathrm{Hom}_{\mathscr{C}_1}(A, G(u))\, \mathrm{Hom}_{\mathscr{C}_1}(A, g(B))\varphi_1(A, B) \\
&\quad = \mathrm{Hom}_{\mathscr{C}_1}(A, G(u))(\varphi(A, B))\, \mathrm{Hom}_{\mathscr{C}_2}(f(A), B) \\
&\quad = \varphi(A, B_1)\, \mathrm{Hom}_{\mathscr{C}_2}(F(A), u)\, \mathrm{Hom}_{\mathscr{C}_2}(f(A), B).
\end{aligned}$$

Hence the equality of the two mentioned morphisms follows by using the commutativity of the diagram

$$\begin{array}{ccc}
\mathrm{Hom}_{\mathscr{C}_2}(F(A), B) & \xrightarrow{\mathrm{Hom}_{\mathscr{C}_2}(F(A), u)} & \mathrm{Hom}_{\mathscr{C}_2}(F(A), B_1) \\
\Big\uparrow{\scriptstyle \mathrm{Hom}_{\mathscr{C}_2}(f(A), B)} & & \Big\uparrow{\scriptstyle \mathrm{Hom}_{\mathscr{C}_2}(f(A), B_1)} \\
\mathrm{Hom}_{\mathscr{C}_2}(F_1(A), B) & \xrightarrow[\mathrm{Hom}_{\mathscr{C}_2}(F_1(A), u)]{} & \mathrm{Hom}_{\mathscr{C}_2}(F_1(A), B_1)
\end{array}$$

(this diagram is commutative because the composition of morphisms is associative).

As for the second part of the proposition, it follows immediately by a simple computation.

COROLLARY 1.11. *If G_1 and G_2 are two adjoint functors of the functor F, then there exists a functorial isomorphism from G_1 to G_2.*

Classical examples of adjoint functors

1. Let Λ and Γ be two rings with unit element and let B be a left Γ-module which is simultaneously a right Λ-module. We denote as usually by ${}_\Lambda\mathscr{C}$ (respectively $\mathscr{C}_\Gamma$) the category of left Λ-modules (respectively of right Γ-modules). If A is a left Λ-module, then $A \bigotimes_\Lambda B$ can be provided in a natural way with a structure of left Γ-module such that for any $\gamma \in \Gamma$ we have $\gamma(a \otimes b) = a \otimes \gamma b$.

We thus obtain a functor $F: {}_\Lambda\mathscr{C} \longrightarrow {}_\Gamma\mathscr{C}$ by setting $F(A) = A \bigotimes_\Lambda B$ considered as a Γ-module in the manner just described.

If C is a left Γ-module, then a structure of left Λ-module may be introduced in a natural way into the Abelian group $\mathrm{Hom}_\Gamma(B, C)$ as follows: if $f: B \longrightarrow C$ and $\lambda \in \Lambda$, we set $(\lambda f)(x) = f(x\lambda)$.

We thus obtain a functor $G: {}_\Gamma\mathscr{C} \longrightarrow {}_\Lambda\mathscr{C}$ by setting $G(C) = \mathrm{Hom}_\Gamma(B, C)$ considered as a Λ-module in the manner just described.

The functor G is an adjoint functor of the functor F.

2. Let $\mathscr{S}$ be the category whose objects are pairs (X, x), where X is an arcwise connected topological space and x is a base-point; a morphism from the object (X, x) into (Y, y) is by definition the class of all continuous base-point preserving maps homotopical with a continuous map $f: X \longrightarrow Y$.

Let $\mathscr{G}$ be the category of groups and homomorphisms (see example 3 on p. 2). We then have the following functors:

$$\pi: \mathscr{S} \longrightarrow \mathscr{G}, \qquad \varepsilon: \mathscr{G} \longrightarrow \mathscr{S}.$$

The former associates to the pair (X, x) the fundamental group $\pi_1(X, x)$, whereas the latter associates to the group G the Eilenberg–MacLane space $K(G, 1)$ together with a base-point e_0. It is proved in homotopy theory as a result of certain theorems of Hurewicz and Eilenberg–MacLane that the functor ε is an adjoint of the functor π.

3. Let $\mathscr{W}$ be the category of spaces with base-point having the homotopy type of a CW-complex and of homotopy classes of base-point preserving continuous maps. We define two covariant functors:

$$\Omega:\mathscr{W} \longrightarrow \mathscr{W}, \qquad \Sigma:\mathscr{W} \longrightarrow \mathscr{W}$$

as follows:

Ω associates to the pair (X, x) of $\mathscr{W}$ its loop-space ΩX, i.e. the space of all continuous maps $f:[0, 1] \longrightarrow X$ with $f(0) = f(1) = x$ with the compact-open topology; the base-point of ΩX is the map f with $f([0, 1]) = x$.

Σ associates to the pair (X, x) its suspension ΣX, i.e. the space obtained from the cylinder $[0, 1] \times X$ by pinching to a point the subspace $\{0\} \times X \cup \{1\} \times X \cup [0, 1] \times \{x\}$. This point is the base-point of ΣX.

It is proved in topology that the functor Σ is an adjoint functor of the functor Ω.

4. Let $\mathscr{Top}$ be the category of topological spaces and of continuous maps (see example 2, on p. 2) and let $\mathscr{K}$ be the category of semi-simplicial complexes and semi-simplicial maps. We define two covariant functors

$$S:\mathscr{Top} \longrightarrow \mathscr{K}, \qquad R:\mathscr{K} \longrightarrow \mathscr{Top}$$

as follows:

S associates to each topological space X its total singular complex SX; R associates to each semi-simplicial complex K its geometric realization in the sense of Milnor RK.

It is proved in algebraic topology that the functor R is an adjoint functor of the functor S.

5. Consider the category $\mathscr{Ens}$ of sets and let B be a fixed object of the category $\mathscr{Ens}$. We define two covariant functors

$$F, G:\mathscr{Ens} \longrightarrow \mathscr{Ens}$$

as follows:

F associates to each set A the Cartesian product $F(A) = A \times B$; G associates to each set C the set of all maps defined on B and taking their values in $C:G(C) = \mathrm{Hom}_{\mathscr{Ens}}(B, C)$.

The functor G is an adjoint functor of the functor F, since it is easy to see that

$$\mathrm{Hom}_{\mathscr{Ens}}(A \times B, C) = \mathrm{Hom}_{\mathscr{Ens}}(A, \mathrm{Hom}_{\mathscr{Ens}}(B, C)).$$

PROPOSITION 1.12. *Let* $F:\mathscr{C}_1 \longrightarrow \mathscr{C}_2$ *and* $G:\mathscr{C}_2 \longrightarrow \mathscr{C}_1$ *be covariant functors, and let* $\psi:\mathrm{Hom}_{\mathscr{C}_2}(F) \longrightarrow \mathrm{Hom}_{\mathscr{C}_1}(G)$ *be a functorial morphism.*

For each object A of the category $\mathscr{C}_1$ let $\varepsilon_\psi(A)$ denote the morphism $[\psi(A, F(A))](1_{F(A)}): A \longrightarrow (GF)(A)$ in the category $\mathscr{C}_1$. This defines a functorial morphism $\varepsilon_\psi: 1_{\mathscr{C}_1} \longrightarrow GF$. The map $\psi \longrightarrow \varepsilon_\psi$ from the class of all functorial morphisms from $\mathrm{Hom}_{\mathscr{C}_2}(F)$ to $\mathrm{Hom}_{\mathscr{C}_1}(G)$ into the class of all functorial morphisms from $1_{\mathscr{C}_1}$ to GF is one-to-one and onto.

PROOF. Let $\mu: A \longrightarrow A_1$. We have to show that the diagram

$$\begin{array}{ccc} A & \xrightarrow{\varepsilon_\psi(A)} & (GF)(A) \\ {\scriptstyle \mu}\downarrow & & \downarrow{\scriptstyle (GF)(\mu)} \\ A_1 & \xrightarrow[\varepsilon_\psi(A_1)]{} & (GF)(A_1) \end{array}$$

is commutative. But from the fact that ψ is a functorial morphism it follows that the diagram

$$\begin{array}{ccc} \mathrm{Hom}_{\mathscr{C}_2}(F(A), F(A)) & \xrightarrow{\psi(A, F(A))} & \mathrm{Hom}_{\mathscr{C}_1}(A, (GF)(A)) \\ \downarrow{\scriptstyle \cdot F(\mu)} & & \downarrow{\scriptstyle \cdot (GF)(\mu)} \\ \mathrm{Hom}_{\mathscr{C}_2}(F(A), F(A_1)) & \xrightarrow[\psi(A, F(A_1))]{} & \mathrm{Hom}_{\mathscr{C}_1}(A, (GF)(A_1)) \end{array}$$

is commutative, where the vertical arrows denote the morphisms obtained by composing at right with $F(\mu)$ respectively $(GF)(\mu)$. In other words we have:

$$(GF)(\mu)\varepsilon_\psi(A) = (GF)(\mu)(\psi(A, F(A))(1_{F(A)})) = \psi(A, F(A_1))F(\mu). \tag{6}$$

By the same reason as above, the diagram

$$\begin{array}{ccc} \mathrm{Hom}_{\mathscr{C}_2}(F(A), F(A_1)) & \xrightarrow{\psi(A, F(A_1))} & \mathrm{Hom}_{\mathscr{C}_1}(A, (GF)(A_1)) \\ {\scriptstyle F(\mu)\cdot}\uparrow & & \uparrow{\scriptstyle \mu\cdot} \\ \mathrm{Hom}_{\mathscr{C}_2}(F(A_1), F(A_1)) & \xrightarrow[\psi(A_1, F(A_1))]{} & \mathrm{Hom}_{\mathscr{C}_1}(A_1, (GF)(A_1)) \end{array}$$

is commutative, whence:

$$\psi(A, F(A_1))F(\mu) = \mu[\psi(A_1, F(A_1))](1_{F(A_1)}), \tag{7}$$

i.e. the desired commutativity relation.

We now show that we can associate to each functorial morphism $\varphi: 1_{\mathscr{C}_1} \longrightarrow GF$ a morphism $\delta_\varphi: \mathrm{Hom}_{\mathscr{C}_2}(F) \longrightarrow \mathrm{Hom}_{\mathscr{C}_1}(G)$ such that $\varepsilon_{\delta_\varphi} = \varphi$, $\delta_{\varepsilon_\psi} = \psi$.

The definition of δ_φ is as follows:

If A is an object of $\mathscr{C}_1$, B an object of $\mathscr{C}_2$ and $u: F(A) \longrightarrow B$ in the category $\mathscr{C}_2$, then $[\delta_\varphi(A, B)](u) = G(u)\varphi(A)$. The relations $\varepsilon_{\delta_\varphi} = \varphi$, $\delta_{\varepsilon_\psi} = \psi$ are readily checked.

It remains only to show that δ_φ is a functorial morphism. To this end, let $(\mu, v):(A, B) \longrightarrow (A_1, B_1)$ be a morphism in the category $\mathscr{C}_1^0 \times \mathscr{C}_2$. We must prove that the diagram

$$\begin{array}{ccc} \mathrm{Hom}_{\mathscr{C}_2}(F(A), B) & \xrightarrow{\delta_\varphi(A, B)} & \mathrm{Hom}_{\mathscr{C}_1}(A, G(B)) \\ \downarrow & & \downarrow \\ \mathrm{Hom}_{\mathscr{C}_2}(F(A_1), B_1) & \xrightarrow[\delta_\varphi(A_1, B_1)]{} & \mathrm{Hom}_{\mathscr{C}_1}(A_1, G(B_1)) \end{array}$$

is commutative (the vertical arrows represent the morphisms induced by μ and v). We have

$$[\delta_\varphi(A_1, B_1)][vuF(\mu)] = G(vuF(\mu))\varphi(A_1) = G(v)G(u)(GF)(\mu)\varphi(A_1) \quad (8)$$

$$G(v)[\delta_\varphi(A, B)(u)]\mu = G(v)G(u)\varphi(A)\mu = (G(v)G(u))\varphi(A)\mu. \quad (9)$$

But φ is a functorial morphism from $1_{\mathscr{C}_1}$ to GF, and therefore the following diagram is commutative

$$\begin{array}{ccc} A_1 & \xrightarrow{\varphi(A_1)} & (GF)(A_1) \\ \mu \downarrow & & \downarrow (GF)(\mu) \\ A & \xrightarrow[\varphi(A)]{} & (GF)(A) \end{array}$$

i.e. $\varphi(A)\mu = (GF)(\mu)\varphi(A_1)$, which, according to equations (8), (9), proves precisely the desired commutativity relation.

The following proposition is proved in a similar way, the notations being the same as in proposition 1.12.

PROPOSITION 1.12′. *Let $\Phi: FG \longrightarrow 1_{\mathscr{C}_2}$ be a functorial morphism. For each object (A, B) of the category $\mathscr{C}_1^0 \times \mathscr{C}_2$ we define a morphism*

$$\varphi(A, B): \mathrm{Hom}_{\mathscr{C}_1}(A, G(B)) \longrightarrow \mathrm{Hom}_{\mathscr{C}_2}(F(A), B)$$

by the formula

$$(\varphi(A, B))(v) = \Phi(B)F(v) \qquad \text{for any } v: A \longrightarrow G(B).$$

We thus obtain a functorial morphism $\varphi: \mathrm{Hom}_{\mathscr{C}_1}(G) \longrightarrow \mathrm{Hom}_{\mathscr{C}_2}(F)$. The map $\Phi \longrightarrow \varphi$ thus defined is a bijection from the class of functorial morphisms of FG into $1_{\mathscr{C}_2}$ to the class of functorial morphisms of $\mathrm{Hom}_{\mathscr{C}_1}(G)$ into $\mathrm{Hom}_{\mathscr{C}_2}(F)$.

PROPOSITION 1.13. *Let $\mathscr{C}_1$, $\mathscr{C}_2$ be two categories and $F: \mathscr{C}_1 \longrightarrow \mathscr{C}_2$, $G: \mathscr{C}_2 \longrightarrow \mathscr{C}_1$ be two functors. In order that the functor G be adjoint to the functor F it is necessary and sufficient that there exist two functorial* *morphisms* $\Phi : 1_{\mathscr{C}_1} \longrightarrow GF$, $\Psi : FG \longrightarrow 1_{\mathscr{C}_2}$ *such that the following conditions are satisfied*:

(a) *for any object* A *of the category* $\mathscr{C}_1$ *we have*

$$\Psi(F(A))F(\Phi(A)) = 1_{F(A)}.$$

(b) *For any object* B *of the category* $\mathscr{C}_2$ *we have*

$$G(\Psi(B)) \cdot \Phi(G(B)) = 1_{G(B)}.$$

PROOF. It is immediate by using proposition 1.12. We content ourselves with indicating in detail the sufficiency of the condition. Let φ $\mathrm{Hom}_{\mathscr{C}_2}(F) \longrightarrow \mathrm{Hom}_{\mathscr{C}_1}(G)$ and $\psi : \mathrm{Hom}_{\mathscr{C}_1}(G) \longrightarrow \mathrm{Hom}_{\mathscr{C}_2}(F)$ the functorial morphisms associated to Φ respectively to Ψ in the sense of propositions 1.12, 1.12′. We show that $\psi\varphi$ = identity, $\varphi\psi$ = identity. From the definitions of φ and ψ we have

$$(\varphi(A, B))(u) = G(u)\Phi(A) \qquad \text{for any } u : F(A) \longrightarrow B$$

$$(\psi(A, B))(v) = \Psi(B)F(v) \qquad \text{for any } v : A \longrightarrow G(B)$$

Hence

$$(\psi(A, B))((\varphi(A, B))(u)) = (\psi(A, B))(G(u)\Phi(A)) = \Psi(B)(G(u)\Phi(A))$$
$$= \Psi(B)(FG)(u)F(\Phi(A)).$$

But, since ψ is a functorial morphism, the diagram

$$\begin{array}{ccc} (FG)(B) & \xrightarrow{\Psi(B)} & B \\ \uparrow{\scriptstyle (FG)(u)} & & \uparrow{\scriptstyle u} \\ (FG)(F(A)) & \xrightarrow[\Psi(F(A))]{} & F(A) \end{array}$$

is commutative, so that, according to property (a):

$$(\psi(A, B))((\varphi(A, B))(u)) = u\Psi(F(A))F(\Phi(A)) = u1_{F(A)} = u.$$

The relation $\varphi\psi$ = identity is proved in the same way.

PROPOSITION 1.13′. *Let* $\mathscr{C}$, $\mathscr{C}'$ *be two categories*, $T : \mathscr{C} \longrightarrow \mathscr{C}'$ *a functor and* $S : \mathscr{C}' \longrightarrow \mathscr{C}$ *an adjoint functor of* T. *Let* $\Phi : 1_{\mathscr{C}} \longrightarrow S\Psi : TS \longrightarrow 1_{\mathscr{C}'}$ *be as in proposition* 1.13. *Then* Φ *is an isomorphism if and only if* S *is fully faithful, i.e. the map* $\operatorname{Hom}_{\mathscr{C}'}(X, Y) \longrightarrow \operatorname{Hom}_{\mathscr{C}}(SX, SY)$ *defined by* $u \longrightarrow Su$ *is a bijection for any objects* X, Y *of* $\mathscr{C}'$.

PROOF. Let X, Y be two objects of $\mathscr{C}'$. We have the commutative diagram

$$\begin{array}{ccc} \operatorname{Hom}_{\mathscr{C}'}(X, Y) & \xrightarrow{S(X,Y)} & \operatorname{Hom}_{\mathscr{C}}(SX, SY) \\ & {\scriptstyle \operatorname{Hom}_{\mathscr{C}'}(\Phi(X), Y)} \searrow \quad \swarrow {\scriptstyle \approx} & \\ & \operatorname{Hom}_{\mathscr{C}'}(TSX, Y) & \end{array}$$

which represents the decomposition of the adjunction isomorphism according to proposition 1.12′.

It is evident that $S(X, Y)$ is a bijection if and only if $\operatorname{Hom}_{\mathscr{C}'}(\Phi(X), Y)$ is a bijection.

But the fact that $\operatorname{Hom}_{\mathscr{C}}(\Phi(X), Y)$ is a bijection for any Y is equivalent by proposition 1.7 to the fact that $\Phi(X)$ is an isomorphism.

PROPOSITION 1.14. *Let* $\mathscr{C}_1$ *and* $\mathscr{C}_2$ *be two categories and* $G: \mathscr{C}_2 \longrightarrow \mathscr{C}_1$ *be a covariant functor. Assume that there are given:*

(i) *a function which associates to each object* A *of the category* $\mathscr{C}_1$ *an object* $F(A)$ *of the category* $\mathscr{C}_2$.

(ii) *a morphism* $\rho_A: A \longrightarrow G(F(A))$ *in* $\mathscr{C}_1$, *for each object* A *of the category* $\mathscr{C}_1$.

For each object B *of the category* $\mathscr{C}_2$, *this gives rise to a map*

$$\varphi(A, B): \operatorname{Hom}_{\mathscr{C}_2}(F(A), B) \longrightarrow \operatorname{Hom}_{\mathscr{C}_2}(A, G(B))$$

defined as follows

$$[\varphi(A, B)](v) = G(v)\rho_A. \tag{10}$$

$\varphi(A, B)$ *is supposed to be a bijection for each object* A *of* $\mathscr{C}_1$, *and each object* B *of* $\mathscr{C}_2$.

Under these conditions, there exists a covariant functor $F: \mathscr{C}_1 \longrightarrow \mathscr{C}_2$ *such that* G *is its adjoint functor.*

PROOF. The definition of the functor $F: \mathscr{C}_1 \longrightarrow \mathscr{C}_2$. Let M, N be two arbitrary objects of the category $\mathscr{C}_1$ and $u: M \longrightarrow N$. We have to define $F(u): F(M) \longrightarrow F(N)$, and to verify the usual properties of functors. Consider the morphism $\rho_N u: M \longrightarrow G(F(N))$. If follows from the hypothesis that there exists a unique element $F(u): F(M) \longrightarrow F(N)$ such that

$$[\varphi(M, F(N))][F(u)] = \rho_N u. \tag{11}$$

It is obvious that we have according to this definition $F(1_M) = 1_{F(M)}$. It remains to prove the relation:

$$F(vu) = F(v)\,F(u), \tag{12}$$

where $u:M \longrightarrow N$, $v:N \longrightarrow P$. To this end, it is sufficient to prove the following relation:

$$[\varphi(M, F(P))][F(v)\,F(u)] = [\varphi(M, F(P))][F(vu)]. \tag{13}$$

But we infer from the definition

$$[\varphi(M, F(P))][F(vu)] = \rho_P(vu).$$

$$G(F(v))\rho_N = [\varphi(N, F(P))][F(v)] = \rho_P v \tag{14}$$

$$G(F(u))\rho_M = [\varphi(M, F(N))][F(u)] = \rho_N u$$

hence

$$\begin{aligned}[\varphi(M, F(P))][F(vu)] &= (\rho_P v)u = G(F(v))\rho_N u \\ &= G(F(v))\,G(F(u))\rho_M = G(F(v)F(u))\rho_M. \end{aligned}\tag{15}$$

On the other hand, by definition

$$[\varphi(M, F(P))][F(v)F(u)] = G(F(v)F(u))\rho_M. \tag{16}$$

By comparing equations (15) and (16), we obtain equation (13), whence the desired equation (12).

Remark. The notion of adjoint functor we have introduced can be sharpened in the following manner:

We shall say that the functor G is a *right-adjoint* functor for F, if G is an adjoint functor for F in the sense introduced above. If we start from a functor G, we shall say that a functor F is a *left-adjoint* functor for G if the same condition is fulfilled.

One obtains the dual notion of left- and right-adjoint functor of a contravariant functor. For instance, G is a right-adjoint functor for F, if we have a functorial isomorphism

$$\mathrm{Hom}_{\mathscr{C}_2}(B, F(A)) \xrightarrow{\approx} \mathrm{Hom}_{\mathscr{C}_1}(A, G(B)).$$

Note the symmetry which occurs in this case.

All the propositions about adjoint functors given in section 5 are valid also for the other cases considered.

6. The Notion of Equivalence Between Categories

Let $\mathscr{C}_1, \mathscr{C}_2, \mathscr{C}_3$ be three categories and let $F, G: \mathscr{C}_1 \longrightarrow \mathscr{C}_2$, $H: \mathscr{C}_2 \longrightarrow \mathscr{C}_3$ be functors. Let $\varphi: F \longrightarrow G$ be a functorial morphism from the functor F to the functor G (all functors are assumed to be covariant).

We can consider the functors $H \circ F$, $H \circ G: \mathscr{C}_1 \longrightarrow \mathscr{C}_3$. If we consider for each object A of $\mathscr{C}_1$ the morphisms $H(\varphi(A)): (H \circ F)(A) \longrightarrow (H \circ G)(A)$, it is easily seen that we obtain in this way a functorial morphism from the functor $H \circ F$ to the functor $H \circ G$, which we denote by $H\varphi$ or $H * \varphi$. In other words we have by definition

$$(H\varphi)(A) = H(\varphi(A)).$$

Similarly, consider the functors $F_1, F_2: \mathscr{C}_2 \longrightarrow \mathscr{C}_3$, $F_3: \mathscr{C}_1 \longrightarrow \mathscr{C}_2$, and assume that a functorial morphism $\psi: F_1 \longrightarrow F_2$ has been given. We define a functorial morphism from the functor $F_1 \circ F_3$ to the functor $F_2 \circ F_3$, denoted by ψF_3 or $\psi * F_3$, by setting

$$(\psi F_3)(A) = \psi(F_3(A)),$$

for each object A of the category $\mathscr{C}_1$.

Definition. Let $\mathscr{C}_1$ and $\mathscr{C}_2$ be two categories. The covariant functor $F: \mathscr{C}_1 \longrightarrow \mathscr{C}_2$ is said to define an *equivalence* of the category $\mathscr{C}_1$ with the category $\mathscr{C}_2$ if there exists a functor $G: \mathscr{C}_2 \longrightarrow \mathscr{C}_1$ and two functorial isomorphisms $\varphi: 1_{\mathscr{C}_1} \longrightarrow GF$, $\psi: 1_{\mathscr{C}_2} \longrightarrow FG$ such that

$$F\varphi = \psi F. \tag{17}$$

PROPOSITION 1.15. *Assume that the functor $F: \mathscr{C}_1 \longrightarrow \mathscr{C}_2$ defines an equivalence of the category $\mathscr{C}_1$ with the category $\mathscr{C}_2$. Then the following assertions are true:*

(a) *For any pair (A, B) of objects of $\mathscr{C}_1$, the map $F(A, B): \mathrm{Hom}_{\mathscr{C}_1}(A, B) \longrightarrow \mathrm{Hom}_{\mathscr{C}_2}(F(A), F(B))$ which associates to each morphism $u: A \longrightarrow B$ in $\mathscr{C}_1$ the morphism $F(u): F(A) \longrightarrow F(B)$ in $\mathscr{C}_2$ is one-to-one and onto.*

(b) *Any object M of $\mathscr{C}_2$ is isomorphic with an object of the form $F(A)$.*

PROOF. We prove that the morphism obtained by composing the following morphisms

$$\mathrm{Hom}_{\mathscr{C}_2}(F(A), F(B)) \xrightarrow{G(F(A), F(B))} \mathrm{Hom}_{\mathscr{C}_1}(GF(A), GF(B)) \xrightarrow{\varphi_*} \mathrm{Hom}_{\mathscr{C}_1}(A, B) \tag{18}$$

is an inverse for the morphism $F(A, B)$. The map φ_* is defined by using the functorial isomorphism φ as follows:

$$\varphi_*(w) = (\varphi(B))^{-1}w\varphi(A) \qquad \text{for any } w: GF(A) \longrightarrow GF(B).$$

Now, let $u: A \longrightarrow B$. We have to show that the morphism $F(u)$ composed with the morphism in the sequence (18) yields precisely the morphism u. This morphism is:

$$(\varphi(B))^{-1}GF(u)\varphi(A) = \alpha : A \longrightarrow B.$$

It follows that the diagram

$$\begin{array}{ccc} A & \xrightarrow{\alpha} & B \\ {\scriptstyle \varphi(A)}\downarrow & & \downarrow{\scriptstyle \varphi(B)} \\ GF(A) & \xrightarrow[GF(u)]{} & GF(B) \end{array}$$

is commutative.

But from the definition of functorial morphisms it follows that the diagram

$$\begin{array}{ccc} A & \xrightarrow{u} & B \\ {\scriptstyle \varphi(A)}\downarrow & & \downarrow{\scriptstyle \varphi(B)} \\ GF(A) & \xrightarrow[GF(u)]{} & GF(B) \end{array}$$

is also commutative, whence it follows that $\alpha = u$, since φ is a functorial isomorphism.

Now, let $v: F(A) \longrightarrow F(B)$. We have to show that $F[\varphi_*(G(v))] = v$. But

$$\begin{aligned} F[\varphi_*(G(v))] &= F[(\varphi(B))^{-1}G(v)\varphi(A)] = F((\varphi(B))^{-1})(FG)(v)F(\varphi(A)) \\ &= F((\varphi(B))^{-1})(FG)(v)(\psi F)(A) = F((\varphi(B))^{-1})(FG)(v)\psi(F(A)). \end{aligned}$$

By using the commutativity of the diagram

$$\begin{array}{ccc} F(A) & \xrightarrow{v} & F(B) \\ {\scriptstyle \psi(F(A))}\downarrow & & \downarrow{\scriptstyle \psi(F(B))} \\ (FG)(F(A)) & \xrightarrow[(FG)(v)]{} & (FG)(F(B)) \end{array}$$

we get

$$\begin{aligned} F[\varphi_*(G(v))] &= F((\varphi(B))^{-1})\psi(F(B))v = F((\varphi(B))^{-1})(F\varphi)(B)v \\ &= F((\varphi(B))^{-1})F(\varphi(B))v = F((\varphi(B))^{-1}\varphi(B))v \\ &= F(1_B)v = 1_{F(B)}v = v. \end{aligned}$$

We have obviously used in an essential manner equation (17). This completes the proof of assertion (a).

The proof of assertion (b) follows immediately, for the object M of $\mathscr{C}_2$ is isomorphic with $F(G(M))$, for example by the isomorphism $\psi(M):M \longrightarrow FG(M)$.

PROPOSITION 1.16. *Let* $F:\mathscr{C}_1 \longrightarrow \mathscr{C}_2$ *be an equivalence of the category* $\mathscr{C}_1$ *with the category* $\mathscr{C}_2$, *and let* $G:\mathscr{C}_2 \longrightarrow \mathscr{C}_1$, $\varphi:1_{\mathscr{C}_1} \longrightarrow GF$, $\psi:1_{\mathscr{C}_2} \longrightarrow FG$ *be such that the conditions in the definition on p. 26 are satisfied. Under these conditions, the following assertions are true*:

(a) *For each pair* (M, N) *of objects of* $\mathscr{C}_2$, *the map* $G(M, N):\mathrm{Hom}_{\mathscr{C}_2}(M, N) \longrightarrow \mathrm{Hom}_{\mathscr{C}_1}(G(M), G(N))$ *is one-to-one and onto.*

(b) *The functor G is adjoint to the functor F and the functor F is adjoint to the functor G.*

(c) *The following relation is valid*

$$G\psi = \varphi G.$$

PROOF. (a) Let $v:M \longrightarrow N$. By taking into account the commutativity of the diagram

$$\begin{array}{ccc} GFG(M) & \xrightarrow{GFG(v)} & GFG(N) \\ {\scriptstyle \varphi(G(M))}\uparrow & & \uparrow{\scriptstyle \varphi(G(N))} \\ G(M) & \xrightarrow[G(v)]{} & G(N) \end{array}$$

it is clear that the map

$$\mathrm{Hom}_{\mathscr{C}_2}(M, N) \xrightarrow{G(M, N)} \mathrm{Hom}_{\mathscr{C}_1}(G(M), G(N))$$

coincides with the composition of the maps in the sequence

$$\begin{aligned} &\mathrm{Hom}_{\mathscr{C}_2}(M, N) \xrightarrow{FG(M, N)} \mathrm{Hom}_{\mathscr{C}_2}(FG(M), FG(N)) \\ &\quad = \mathrm{Hom}_{\mathscr{C}_2}(F(G(M)), F(G(N))) \longrightarrow \mathrm{Hom}_{\mathscr{C}_1}(G(M), G(N)), \qquad (19) \end{aligned}$$

the last arrow representing the inverse map of the map

$$\mathrm{Hom}_{\mathscr{C}_1}(G(M), G(N)) \xrightarrow{F(G(M), G(N))} \mathrm{Hom}_{\mathscr{C}_2}(F(G(M)), F(G(N))),$$

as it was defined in proposition 1.15.

Indeed, let $v:M \longrightarrow N$. By applying successively the morphisms in the sequence of equation (19) we get the morphism

$$\varphi(G(N))^{-1}GF(G(v))\varphi(G(M)) = \beta:G(M) \longrightarrow G(N),$$

so that, as readily seen, the diagram

$$\begin{array}{ccc} GF(G(N)) & \xrightarrow{GF(G(v))} & GF(G(N)) \\ {\scriptstyle \varphi(G(M))}\uparrow & & \uparrow{\scriptstyle \varphi(G(N))} \\ G(M) & \xrightarrow[\beta]{} & G(N) \end{array}$$

is commutative. By using the commutativity of the diagram mentioned at the beginning, it follows that $\beta = G(v)$. Since $\psi : 1_{\mathscr{C}_2} \longrightarrow FG$ is an isomorphism, $FG(M, N)$ is a bijection. As according to proposition 1.15 the last isomorphism in equation (19) is also a bijection, we infer that their composition is a bijection, in other words $G(M, N)$ is one-to-one and onto.

(b) Let A be an object of $\mathscr{C}_1$ and B an object of $\mathscr{C}_2$. We have the following sequence of bijections:

$$\mathrm{Hom}_{\mathscr{C}_2}(F(A), B) \xrightarrow{G(F(A), B)} \mathrm{Hom}_{\mathscr{C}_1}(GF(A), G(B)) \longrightarrow \mathrm{Hom}_{\mathscr{C}_1}(A, G(B)).$$

The last bijection is induced by the isomorphism $A \xrightarrow{\varphi(A)} GF(A)$.

The bijection $\mathrm{Hom}_{\mathscr{C}_2}(F(A), B) \longrightarrow \mathrm{Hom}_{\mathscr{C}_1}(A, G(B))$ obtained by composition has a functorial character, hence G is an adjoint functor of the functor F.

One proves in a similar way that F is an adjoint functor of the functor G.

(c) From the fact that F is an adjoint functor of the functor G and from proposition 1.13, we infer that there exist two functorial morphisms $1_{\mathscr{C}_2} \xrightarrow{\beta} FG$, $GF \xrightarrow{\alpha} 1_{\mathscr{C}_1}$ such that

$$\alpha G \cdot G\beta = 1_G. \tag{20}$$

The explicit form of the functorial morphisms α, β in this case is readily obtained:

$$\alpha = \varphi^{-1}, \beta = \psi \tag{21}$$

Now, let M be an arbitrary object of the category $\mathscr{C}_2$. Relation (20) yields:

$$(\alpha G)(M) \cdot (G\beta)(M) = 1_{G(M)},$$

i.e., in view of equation (21):

$$1_{G(M)} = \varphi^{-1}(G(M))G(\psi(M)).$$

By composing with $\varphi(G(M))$ we obtain:

$$\varphi(G(M)) = G(\psi(M)),$$

that is precisely the desired relation.

COROLLARY 1.17. *If* $F:\mathscr{C}_1 \longrightarrow \mathscr{C}_2$ *defines an equivalence of the category* $\mathscr{C}_1$ *with the category* $\mathscr{C}_2$ *and if the functors* G'', $G':\mathscr{C}_2 \longrightarrow \mathscr{C}_1$ *satisfy the conditions stated for* G *in the definition on p. 26, then* G', G'' *are canonically isomorphic.*

COROLLARY 1.18. *If* $F:\mathscr{C}_1 \longrightarrow \mathscr{C}_2$ *defines an equivalence of the category* $\mathscr{C}_1$ *with the category* $\mathscr{C}_2$ *and if the functor* $G:\mathscr{C}_2 \longrightarrow \mathscr{C}_1$ *satisfies the conditions in the definition on p. 26, then* G *defines an equivalence of the category* $\mathscr{C}_2$ *with the category* $\mathscr{C}_1$.

PROPOSITION 1.19. *In order that the covariant functor* $F:\mathscr{C}_1 \longrightarrow \mathscr{C}_2$ *define an equivalence of the category* $\mathscr{C}_1$ *with the category* $\mathscr{C}_2$, *it is necessary and sufficient that the following conditions be satisfied:*

(a) *For each pair* (A, B) *of objects of* $\mathscr{C}_1$ *the map* $F(A, B):\mathrm{Hom}_{\mathscr{C}_1}(A, B) \longrightarrow \mathrm{Hom}_{\mathscr{C}_2}(F(A), F(B))$ *is one-to-one and onto.*

(b) *Any object* M *of* $\mathscr{C}_2$ *is isomorphic with an object of the form* $F(A)$.

PROOF. The necessity follows from proposition 1.15. To prove the sufficiency, we choose for each object M of the category $\mathscr{C}_2$ an object A in the category $\mathscr{C}_1$ and an isomorphism

$$v_M:M \longrightarrow F(A).$$

This is possible by property (b). We define a functor $G:\mathscr{C}_2 \longrightarrow \mathscr{C}_1$ as follows: For each object M of the category $\mathscr{C}_2$ we set $G(M) = A$. Let $w:M \longrightarrow N$ and let $w':F(A) \longrightarrow F(B)$ be such that the diagram

$$\begin{array}{ccc} M & \xrightarrow{v_M} & F(A) \\ {\scriptstyle w}\downarrow & & \downarrow{\scriptstyle w'} \\ N & \xrightarrow[v_N]{} & F(B) \end{array}$$

is commutative. Then there exists by property (a) a morphism $w_1:A \longrightarrow B$ such that $F(w_1) = w'$. We set $G(w) = w_1$.

We now define the functorial isomorphisms φ, ψ in the Definition on p. 26.

To do this, let A be an arbitrary object of $\mathscr{C}_1$ and $B = G(F(A))$. From the definition of the functor G it results that we have the isomorphism $v_{F(A)}:F(A) \longrightarrow F(B)$. According to property (a) there exists a unique isomorphism $\varphi(A):A \longrightarrow B = GF(A)$ such that

$$(F\varphi)(A) = v_{F(A)}. \tag{22}$$

The isomorphisms $\varphi(A):A \longrightarrow GF(A)$ define a functorial isomorphism $\varphi:1_{\mathscr{C}_1} \longrightarrow GF$.

Let now M be an arbitrary object of $\mathscr{C}_2$. If $A = G(M)$ we have the isomorphism:

$$v_M : M \longrightarrow F(A) = FG(M)$$

which defines a functorial isomorphism $\psi : 1_{\mathscr{C}_2} \longrightarrow FG$. Hence we have from the definition

$$\psi(M) = v_M.$$

Therefore the relation

$$\psi(F(A)) = v_{F(A)}$$

follows, which yields, in view of equation (22) precisely the relation $\psi F = F\varphi$.

Definition. A covariant functor $F : \mathscr{C}_1 \longrightarrow \mathscr{C}_2$ is said to define an *isomorphism* of the category $\mathscr{C}_1$ with the category $\mathscr{C}_2$ if:

1. For any object Z of $\mathscr{C}_2$ there exists a unique object X of $\mathscr{C}_1$ such that $F(X) = Z$, and

2. for any pair (X, Y) of objects of $\mathscr{C}_1$, the map $F(X, Y) : \mathrm{Hom}_{\mathscr{C}_1}(X, Y) \longrightarrow \mathrm{Hom}_{\mathscr{C}_2}(F(X), F(Y))$ which associates to each morphism $u : X \longrightarrow Y$ the morphism $F(u) : F(X) \longrightarrow F(Y)$ is a bijection.

The two categories $\mathscr{C}_1$ and $\mathscr{C}_2$ are then said to be isomorphic.

Definition. A category is said to be *concrete*, if it is isomorphic with a subcategory of the category $\mathscr{E}ns$ of all sets.

Definition. An object I of a category $\mathscr{C}$ is called *integral*, if

1. for any object X of $\mathscr{C}$, the set $\mathrm{Hom}_{\mathscr{C}}(I, X)$ is non-void, and

2. for any pair X, Y of objects of $\mathscr{C}$ and any pair of distinct morphisms $u, v : X \longrightarrow Y$, there exists a morphism $w : I \longrightarrow X$ such that $uw \neq vw$.

Dually, an object I of a category $\mathscr{C}$ is called *cointegral*, if it is an integral object of the dual category $\mathscr{C}^0$.

As examples, each object of the category $\mathscr{E}ns$ is integral; in the category of all groups, the infinite cyclic groups are integral objects.

In the category $\mathscr{E}ns$, each set containing at least two elements is a cointegral object.

PROPOSITION 1.20. *Any category $\mathscr{C}$ which has integral objects is concrete.*

PROOF. Let I be an integral object of $\mathscr{C}$. We define a functor $F : \mathscr{C} \longrightarrow \mathscr{E}ns$ as follows:

$$F(X) = \mathrm{Hom}_{\mathscr{C}}(I, X), \qquad \text{for any object } X \text{ of } \mathscr{C},$$

$$[F(u)](v) = uv \qquad \text{for any morphism } u : X \longrightarrow Y$$

and $v : I \longrightarrow X$.

Now, if $X \neq Y$, then $F(X) \neq F(Y)$ by condition (γ) in the definition of a category.

If $u_1, u_2 : X \longrightarrow Y$, $u_1 \neq u_2$, then there exists by the definition of integral objects a morphism $v : I \longrightarrow X$ such that $u_1 v \neq u_2 v$, whence $F(u_1) \neq F(u_2)$.

COROLLARY 1.21. *The dual of $\mathcal{E}ns$ is a concrete category. Hence, the dual of any concrete category is a concrete category.*

CHAPTER 2

Sums and Products

1. Direct Sums and Products

Let A be an object of the category $\mathscr{C}$ and I an arbitrary set of indices. We assume that there are given a family of objects $(A_i)_{i \in I}$ of the category $\mathscr{C}$ and a family of morphisms $u_i : A \longrightarrow A_i$. We shall say that the object A together with the family of morphisms $(u_i)_{i \in I}$ is a *direct product* of the family of objects $(A_i)_{i \in I}$ if the following condition is satisfied: For any object B of the category $\mathscr{C}$ and for any family $(v_i)_{i \in I}$ of morphisms $v_i : B \longrightarrow A_i$, there exists a unique morphism $v : B \longrightarrow A$ such that the diagram

$$\begin{array}{ccc} B & \xrightarrow{v} & A \\ & {\scriptstyle v_i}\searrow \quad \swarrow{\scriptstyle u_i} & \\ & A_i & \end{array}$$

is commutative for any $i \in I$. We shall also say sometimes that the family $(A_i)_{i \in I}$ possesses a direct product. The morphisms u_i are called canonical projections.

The condition in this definition may be stated equivalently as follows: Consider the set-theoretic cartesian product $\prod_{i \in I} \mathrm{Hom}_{\mathscr{C}}(X, A_i)$, where X is an arbitrary object of the category $\mathscr{C}$. There exists obviously a map

$$\mathrm{Hom}_{\mathscr{C}}(X, A) \longrightarrow \prod_{i \in I} \mathrm{Hom}_{\mathscr{C}}(X, A_i) \qquad (*)$$

defined as follows: to each $u : X \longrightarrow A$, we associate the family $(u_i u)_{i \in I}$.

With these preparations, the above definition is equivalent with the following one:

The object A together with the family of morphisms $(u_i)_{i \in I}$ is a direct product of the family of objects $(A_i)_{i \in I}$ if and only if the map (*) is one-to-one and onto.

Proposition 2.1. *Let $(A_i)_{i \in I}$ be a family of objects of the category $\mathscr{C}$ and let A together with the morphisms $(u_i)_{i \in I}$ be a direct product of the* *family of objects* $(A_i)_{i \in I}$. *Suppose that the object* A' *together with the family of morphisms* $(u_i')_{i \in I}$ *is also a direct product of the same family of objects. Then there exists a canonical isomorphism* $A \xrightarrow{\alpha} A'$ *such that the diagram*

$$\begin{array}{ccc} A & \xrightarrow{\alpha} & A' \\ & {\scriptstyle u_i}\searrow \quad \swarrow{\scriptstyle u_i'} & \\ & A_i & \end{array}$$

is commutative for any $i \in I$.

The proof is immediate.

This proposition justifies the following choice:

Let $(A_i)_{i \in I}$ be a family of objects of the category $\mathscr{C}$ such that there exists at least one direct product of it. In the class of all direct products of this family we choose once for all a copy (e.g. by using the axiom of choice) which we denote by $\prod_{i \in I} A_i$ or $\bigtimes_{i \in I} A_i$. It should be emphasized that $\prod_{i \in I} A_i$ is not just an object of the category $\mathscr{C}$, but an object *together* with the canonical morphisms u_i. However, we shall sometimes use notations of the following type: $u_i : \prod_{i \in I} A_i \longrightarrow A_i$, u_i = canonical projections

If the family $(B_i)_{i \in I}$ possesses also a direct product, then to any family of morphisms $w_i : A_i \longrightarrow B_i$ is associated in a unique way a morphism

$$w : \prod_{i \in I} A_i \longrightarrow \prod_{i \in I} B_i$$

such that the diagram

$$\begin{array}{ccc} \prod_{j \in I} A_j & \xrightarrow{w} & \prod_{j \in I} B_j \\ {\scriptstyle u_i}\downarrow & & \downarrow \\ A_i & \xrightarrow{w_i} & B_i \end{array}$$

is commutative for any $i \in I$. This morphism is denoted by $\prod_{i \in I} w_i$ (or $\bigtimes_{i \in I} w_i$) and is called the product of the morphisms w_i or the extension to products of the morphisms w_i.

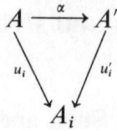

As an example, let $I = \{1, 2\}$ and $A_1 = A_2 = A$. Assume that the direct product of this family exists. We denote it by $A \times A$ and the canonical projections by $\mathrm{pr}_1 : A \times A \longrightarrow A$, $\mathrm{pr}_2 : A \times A \longrightarrow A$. Then there exist two morphisms $\Delta_A : A \longrightarrow A \times A$ and $S_A : A \times A \longrightarrow A \times A$ such that the diagrams

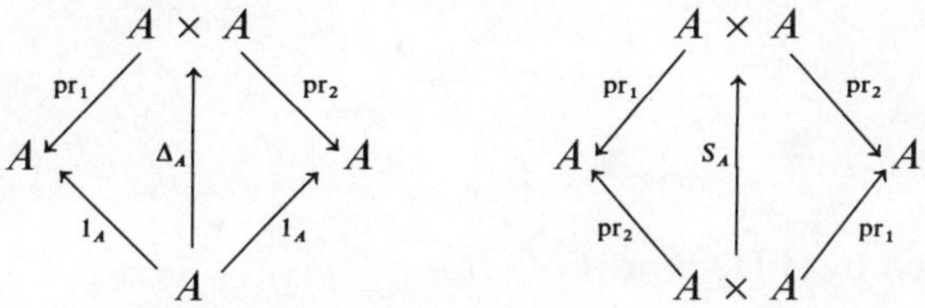

are commutative. The morphism Δ_A is called 'the diagonal imbedding,' whereas S_A is called 'the symmetry of factors' morphism.

Definition. The category $\mathscr{C}$ is said to be a *category with direct products* if any finite family of objects of $\mathscr{C}$ possesses at least one direct product. If any family of objects of the category $\mathscr{C}$ possesses at least one direct product, then $\mathscr{C}$ is said to be a *category with infinite direct products.*

Examples. The categories $\mathscr{E}ns$, $\mathscr{A}b$, $\mathscr{T}op$, ${}_\Lambda\mathscr{C}$ possess infinite direct products. The direct products and canonical projections obtained by the classical standard constructions (Cartesian product, direct product of Abelian groups, and so on) are denoted by $\prod_{i \in I} A_i$ or $\bigtimes_{i \in I} A_i$.

There exist categories in which there are finite families which possess no direct product, for example the category of ordered abelian groups.

Let $\mathscr{C}$ be a category and S an object of $\mathscr{C}$. We consider the category $\mathscr{C}/S$ defined as follows:

The objects of the category $\mathscr{C}/S$ are pairs of the form (A, α) where $\alpha : A \longrightarrow S$ is a morphism in $\mathscr{C}$.

The morphisms from the object (A, α) to the object (B, β) are the morphisms $\gamma : A \longrightarrow B$ in the category $\mathscr{C}$ such that the diagram

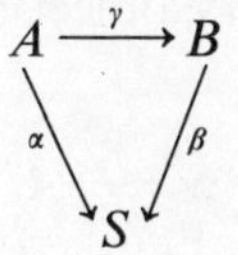

is commutative.

Let (A, α), (B, β) be two objects in the category $\mathscr{C}/S$. A direct product of these two objects is called *fibered product* of the objects A, B associated to the scheme

$$\begin{array}{ccc} A & & \\ & \searrow^{\alpha} & \\ & & S \\ & \nearrow_{\beta} & \\ B & & \end{array}$$

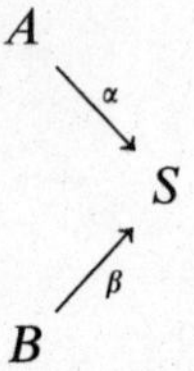

and is denoted by $A \Pi_S B$ or $A \times_S B$.

PROPOSITION 2.2. *Let $(A \Pi_S S, p_A, p_S)$ be a fibered product associated to the scheme*

$$\begin{array}{ccc} A & & \\ & \searrow^{\xi} & \\ & & S \\ & \nearrow_{1_S} & \\ S & & \end{array}$$

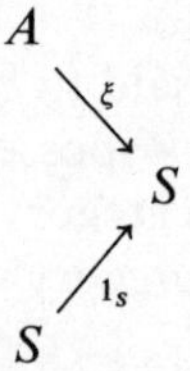

Then there exists a unique isomorphism $\mu : A \longrightarrow A \Pi_S S$ such that the diagram

$$\begin{array}{ccc} & & S \\ & \nearrow^{\xi} & \uparrow p_S \\ A & \xrightarrow{\mu} & A \Pi_S S \\ & \searrow_{1_A} & \downarrow p_A \\ & & A \end{array}$$

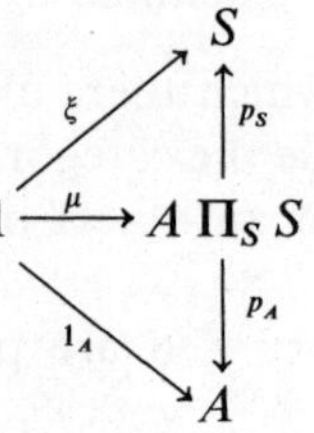

is commutative.

PROOF. Let U be an arbitrary object and $\mu_A : U \longrightarrow A$, $\mu_S : U \longrightarrow S$ be such that the diagram

$$\begin{array}{ccc} S & \xleftarrow{\mu_S} & U \\ {\scriptstyle 1_S}\downarrow & & \downarrow{\scriptstyle \mu_A} \\ S & \xleftarrow[\xi]{} & A \end{array}$$

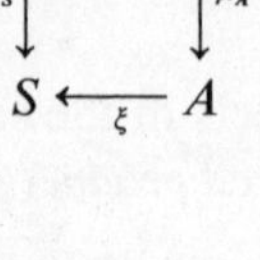

is commutative.

By virtue of proposition 2.1, it will be sufficient to show that there exists a unique morphism $u:U \longrightarrow A$ such that the diagram

$$\begin{array}{ccc} S & \xleftarrow{\mu_S} & U \\ {\scriptstyle\xi}\uparrow & \swarrow{\scriptstyle\mu} & \downarrow{\scriptstyle\mu_A} \\ A & \xrightarrow[1_A]{} & A \end{array}$$

is commutative, and this is evident.

PROPOSITION 2.3. *Consider the morphisms* $X \xrightarrow{\xi} S$, $S' \xrightarrow{\varphi} S$, $S'' \xrightarrow{\varphi'} S'$, *and denote by* $X_\varphi = X \Pi_S S'$, $X_{\varphi\varphi'} = X \Pi_S S''$, $(X_\varphi)_{\varphi'} = X_\varphi \Pi_{S'} S''$ *the fibered products associated to the schemes*

$$\begin{array}{ccc} X & & \\ & \searrow{\scriptstyle\xi} & \\ & & S, \\ & \nearrow{\scriptstyle\varphi} & \\ S' & & \end{array} \qquad \begin{array}{ccc} X & & \\ & \searrow{\scriptstyle\xi} & \\ & & S, \\ & \nearrow{\scriptstyle\varphi\varphi'} & \\ S'' & & \end{array} \qquad \begin{array}{ccc} X & & \\ & \searrow & \\ & & S'. \\ & \nearrow{\scriptstyle\varphi'} & \\ S'' & & \end{array}$$

Under these conditions, there exists a unique isomorphism $\alpha:(X_\varphi)_{\varphi'} \longrightarrow X_{\varphi\varphi'}$ *such that the diagram*

$$\begin{array}{ccc} (X_\varphi)_{\varphi'} & \xrightarrow{\alpha} & X_{\varphi\varphi'} \\ & \searrow \quad \swarrow & \\ & X_\varphi & \end{array}$$

is commutative.

PROOF. Let $\mu:U \longrightarrow X_\varphi$, $\nu:U \longrightarrow S''$ be such that the diagram

$$\begin{array}{ccc} U & \xrightarrow{\mu} & X_\varphi \\ {\scriptstyle\nu}\downarrow & & \downarrow \\ S'' & \xrightarrow[\varphi']{} & S' \end{array}$$

is commutative. According to proposition 2.1, it is sufficient to prove that there exists a unique morphism $\rho:U \longrightarrow X_{\varphi\varphi'}$ such that the diagram

$$\begin{array}{ccc} U & \xrightarrow{\mu} & X_\varphi \\ {\scriptstyle\nu}\downarrow & \searrow{\scriptstyle\rho} & \uparrow \\ S'' & \longleftarrow & X_{\varphi\varphi'} \end{array}$$

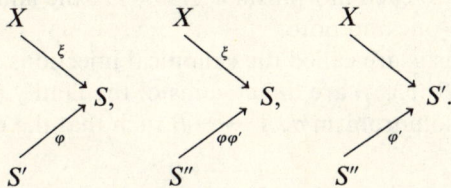

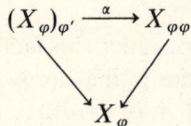

is commutative. But this follows immediately from the commutativity of the diagram

$$\begin{array}{ccccc} X_{\varphi\varphi'} & \longrightarrow & X_{\varphi} & \longrightarrow & X \\ \downarrow & & \downarrow & & \downarrow \xi \\ S'' & \xrightarrow[\varphi']{} & S' & \xrightarrow[\varphi]{} & S . \end{array}$$

Definition. We say that the object A together with the family of morphisms $u_i : A_i \longrightarrow A (i \in I)$ is a *direct sum* of the family of objects $(A_i)_{i \in I}$, if the map

$$\mathrm{Hom}_{\mathscr{C}}(A, X) \longrightarrow \prod_{i \in I} \mathrm{Hom}_{\mathscr{C}}(A_i, X) \qquad (**)$$

which associates to each morphism $u : A \longrightarrow X$ the family of morphisms $(uu_i)_{i \in I}$, is one-to-one and onto.

The morphisms u_i are called the canonical injections.

If $(A, (u_i)_{i \in I})$, $(B, (v_i)_{i \in I})$ are direct sums of the family $(A_i)_{i \in I}$, then there exists a unique isomorphism $\alpha : A \longrightarrow B$ such that the diagram

$$\begin{array}{ccc} A & \xrightarrow{\alpha} & B \\ {\scriptstyle u_i}\nwarrow & & \nearrow{\scriptstyle v_i} \\ & A_i & \end{array}$$

is commutative for any $i \in I$.

As in the case of the direct product, this fact suggests the idea of choosing from the class of all direct sums of the family of objects $(A_i)_{i \in I}$ a representative which we denote by $\coprod_{i \in I} A_i$ or $\bigoplus_{i \in I} A_i$.

Despite their name, the canonical injections $u_i : A_i \longrightarrow \bigoplus_{j \in I} A_j$ are not in general injections (monomorphisms). Similarly, the canonical projections $\prod_{j \in I} A_j \longrightarrow A_i$ are not in general surjections.

This fact can be seen on the following

Example. Consider the category $\mathscr{A}$ defined as follows:

(i) the objects of the category $\mathscr{A}$ are the commutative rings with unity element.

(ii) the morphisms of the category $\mathscr{A}$ are ring homomorphisms preserving the unity element.

Let A and B be two objects of this category. Consider the tensor product $A \otimes B = A \otimes_Z B$. Obviously, $A \otimes B$ has a structure of ring with unity element. The multiplication is defined by setting $(a \otimes b) \cdot (a_1 \otimes b_1) = aa_1 \otimes bb_1$.

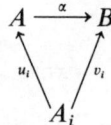

Consider the morphisms $i_A : A \longrightarrow A \otimes B$, $i_B : B \longrightarrow A \otimes B$ defined by $i_A(a) = a \otimes 1$, $i_B(b) = 1 \otimes b$. The object $A \otimes B$ together with the morphisms i_A, i_B is a direct sum of the objects A and B in the category $\mathscr{A}$. If we take $A = Z$, $B = Z_2$, we see immediately, that i_Z is not an injection.

Definition. The category $\mathscr{C}$ is said to be a *category with direct sums* if any finite family of objects of $\mathscr{C}$ possesses at least one direct sum. If any family of objects of the category $\mathscr{C}$ possesses at least one direct sum, then $\mathscr{C}$ is said to be a *category with infinite direct sums.*

Examples. The categories $\mathscr{E}ns$, $\mathscr{A}b$, and $\mathscr{T}op$ possess infinite direct sums. They coincide respectively with the following classical notions: disjoint union, direct sum of Abelian groups, topological sum (in the sense of Bourbaki).

The category of topological spaces with base-point has infinite direct sums. If $(X_i, x_i)_{i \in I}$ is an arbitrary family of topological spaces with base-point, then its direct sum is $\bigvee_{i \in I} X_i$, i.e. the topological sum in which all the base-points have been identified to a single point, the base-point of $\bigvee_{i \in I} X_i$.

The category $\mathscr{G}$ of all groups possesses infinite direct sums. The direct sum of an arbitrary family of groups is the free product of this family.

There exist categories in which there are finite families which possess no direct sum. For example the category Δ, whose objects are the ordered sets $\{0, 1, 2, \cdots, n\}$, n being a non-negative integer, and whose morphisms are non-decreasing maps between such sets. It can be shown that this category has not in general fibered sums, so that in the category Δ/S where S is an object of Δ, there exist pairs of objects which have no direct sum.

Let $\mathscr{C}$ be a category and S an object of $\mathscr{C}$. We consider the category $S/\mathscr{C}$ defined as follows:

(i) the objects of the category $S/\mathscr{C}$ are pairs of the form (A, α) where $\alpha : S \longrightarrow A$ is a morphism in $\mathscr{C}$.

(ii) the morphisms from the object (A, α) to the object (B, β) are the morphisms $\gamma : A \longrightarrow B$ in the category $\mathscr{C}$ such that the diagram

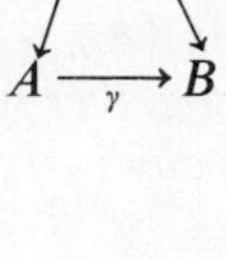

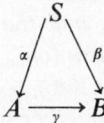

is commutative.

Let (A, α), (B, β) be two objects in the category $S/\mathscr{C}$. A direct sum of these two objects is called *fibered sum* of the objects A, B associated to the scheme

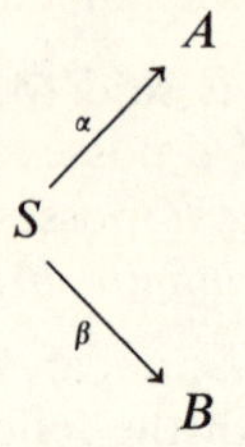

and is denoted by $A \amalg_S B$.

PROPOSITION 2.4. *Let the object A together with the morphisms $(u_i)_{i \in I}$ be a direct product of the family $(A_i)_{i \in I}$. Assume that for each $i \in I$, A_i together with the morphisms $(v_{ij})_{j \in J_i}$ is a direct product of the family $(B_{ij})_{j \in J_i}$. Then A together with the morphisms $(v_{ij}u_i)_{j \in J_i, i \in I}$ is a direct product of the family $(B_{ij})_{j \in J_i, i \in I}$.*

PROOF. Suppose that the object C and the morphisms $w_{ij}: C \longrightarrow B_{ij} (j \in J_i; i \in I)$ are given. Then, for each $i \in I$, there exists a unique morphism $t_i: C \longrightarrow A_i$ such that

$$w_{ij} = v_{ij} t_i \qquad \text{for each } j \in J_i.$$

It follows that there exists a unique morphism $s: C \longrightarrow A$ such that

$$t_i = u_i s \qquad \text{for each } i \in I.$$

Hence

$$w_{ij} = (v_{ij} u_i) s \qquad \text{for each } j \in J_i,\ i \in I.$$

The uniqueness of s follows immediately, whence the proposition.

There is a converse to proposition 2.4, namely

PROPOSITION 2.5. *Let the object A together with the morphisms $(u_i)_{i \in I}$ be a direct product of the family $(A_i)_{i \in I}$. Assume that the set of indices I is the disjoint union of the sets $(I_\mu)_{\mu \in M}$ and that for each $\mu \in M$ there exists an object B_μ and a family of morphism $(v_i)_{i \in I_\mu}$ such that B_μ together with (v_i) is a direct product of the family $(A_i)_{i \in I_\mu}$. Then there exists a family of morphisms $(w_\mu)_{\mu \in M}$ such that A together with (w_μ) is a direct product of the family $(B_\mu)_{\mu \in M}$ and $u_i = v_i w_\mu$ for any $i \in I_\mu$, $\mu \in M$.*

PROOF. Clearly, for each fixed $\mu \in M$, there exists a unique morphism $w_\mu : A \longrightarrow B_\mu$ such that $u_i = v_i w_\mu$ for any $i \in I_\mu$. Assume the object C and the morphisms $t_\mu : C \longrightarrow B_\mu$ are given; then for each $i \in I$ we have the morphisms $v_i t_\mu : C \longrightarrow A_i$ where $i \in I_\mu$, hence there exists a unique morphism $s : C \longrightarrow A$ such that

$$v_i t_\mu = u_i s = v_i w_\mu s, \qquad \text{for } i \in I_\mu, \mu \in M.$$

Now, since B_μ is the direct product of the family $(A_i)_{i \in I_\mu}$ we infer that $t_\mu = w_\mu s$ for every $\mu \in M$. The uniqueness of s is immediate.

Consider now the following condition imposed on a category $\mathscr{C}$:

CAd1. For any pair (A, B) of objects of $\mathscr{C}$ there exists in $\mathrm{Hom}_{\mathscr{C}}(A, B)$ a morphism 0_{BA} such that for any $v : B \longrightarrow C$ and any $u : D \longrightarrow A$ we have

$$v 0_{BA} = 0_{CA}, \qquad 0_{BA} u = 0_{BD}.$$

PROPOSITION 2.6. *Assume that the category $\mathscr{C}$ satisfies the condition CAd1. Then, for any pair A, B of objects of the category $\mathscr{C}$ such $A \amalg B$ and $A \Pi B$ exist, there exists a unique morphism $h(A, B) : A \amalg B \longrightarrow A \Pi B$ such that the following relations are true:*

$$p_B h(A, B) i_A = 0_{BA} \qquad p_A h(A, B) i_A = 1_A$$

$$p_A h(A, B) i_B = 0_{AB} \qquad p_B h(A, B) i_B = 1_B.$$

PROOF. Let $\alpha : A \longrightarrow A \Pi B$, $\beta : B \longrightarrow A \Pi B$ be such that the diagrams

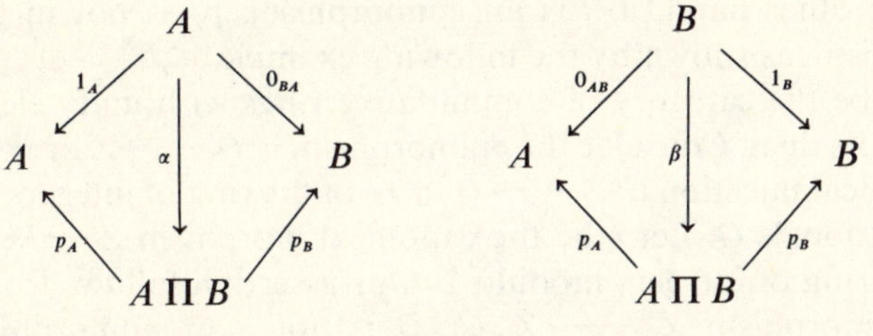

are commutative. $h(A, B)$ is the unique morphism for which the diagram

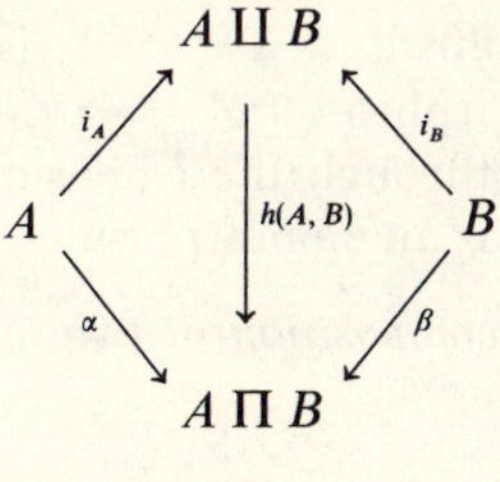

is commutative. We have:

$$p_B h(A, B) i_A = p_B \alpha = 0_{BA} \qquad p_A h(A, B) i_A = p_A \alpha = 1_A.$$

$$p_A h(A, B) i_B = p_A \beta = 0_{AB} \qquad p_B h(A, B) i_B = p_B \beta = 1_B.$$

Definition. Let $\mathscr{C}$ be a category with fibered products, and let $(X \Pi_Y Z, p_X, p_Z)$ be the fibered product associated to the scheme

$$\begin{array}{ccc} X & & \\ & \searrow^{u} & \\ & & Y. \\ & \nearrow_{v} & \\ Z & & \end{array}$$

We shall say that the morphism $p_Z : X \Pi_Y Z \longrightarrow Z$ has been obtained from u through the extension of the base Y to Z by means of the morphism v.

PROPOSITION 2.7. *If u is a monomorphism, then so is p_Z.*

PROOF. Let $\xi_1, \xi_2 : A \longrightarrow X \Pi_Y Z$ be such that $p_Z \xi_1 = p_Z \xi_2$. We have

$$u p_X \xi_1 = v p_Z \xi_1 = v p_Z \xi_2 = u p_X \xi_2,$$

which implies, by hypothesis, that $p_X \xi_1 = p_X \xi_2$. By the definition of fibered products we then have $\xi_1 = \xi_2$.

On the other hand, if u is an epimorphism, p_Z is not in general an epimorphism, as shown by the following example:

Let $\mathscr{A}$ be the category of commutative rings with unity element, and let $\mathscr{A}^0$ be its dual. Consider the epimorphism $u : Q \longrightarrow Z$ in $\mathscr{A}^0$, which is the canonical injection $u : Z \longrightarrow Q$ in $\mathscr{A}$ of the ring of integers Z into the field of rationals Q. Let v be the canonical morphism $Z \longrightarrow Z_2$, where Z_2 is the ring of integers modulo 2. Our assertion follows from the fact that the morphism $Z_2 \longrightarrow Z_2 \otimes_Z Q$ is not a monomorphism in $\mathscr{A}$, since $Z_2 \otimes_Z Q = 0$.

Definition. An epimorphism $u : X \longrightarrow Y$ is said to be a *universal epimorphism*, if for any morphism $v : Z \longrightarrow Y$, the morphism $p_Z : X \Pi_Y Z \longrightarrow Z$ obtained from u through the extension of the base Y to Z by means of the morphism v is an epimorphism.

PROPOSITION 2.8. *The composition of two universal epimorphisms is a universal epimorphism.*

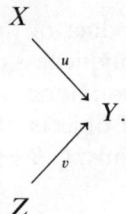

Proof. Let $u_1 : X_1 \longrightarrow X_2$, $u_2 : X_2 \longrightarrow Y$ be universal epimorphisms and let $v : Z \longrightarrow Y$ be an arbitrary morphism. We have to prove that $p_Z : X_1 \Pi_Y Z \longrightarrow Z$ is an epimorphism, where $(X_1 \Pi_Y Z, p_Z, p_{X_1})$ is the fibered product associated to the scheme

$$\begin{array}{ccc} Z & & \\ & \searrow^{v} & \\ & & Y \\ & \nearrow_{u_2u_1} & \\ X_1 & & \end{array}$$

If we denote by $(X_2 \Pi_Y Z, p'_Z, p_{X_2})$ the fibered product associated to the scheme defined by the morphisms u_2, v, then from the fact that u_2 is a universal epimorphism it results that p'_Z is an epimorphism. But by proposition 2.3 $X_1 \Pi_Y Z$ coincides with the fibered product associated to the scheme defined by the morphisms u_1, p_{X_2}, and the following diagram is commutative:

$$\begin{array}{ccc} X_1 \Pi_Y Z & \xrightarrow{\pi} & X_2 \Pi_Y Z \\ \downarrow & & \downarrow{\scriptstyle p_{X_2}} \\ X_1 & \xrightarrow[u_1]{} & X_2 \end{array}$$

Moreover, $p_Z = p'_Z \pi$. But u_1 is a universal epimorphism, so that π is an epimorphism, hence p_Z is an epimorphism.

Proposition 2.9. *If $u_1 : X_1 \longrightarrow X_2$, $u_2 : X_2 \longrightarrow Y$ are such that $u_2 u_1 : X_1 \longrightarrow Y$ is a universal epimorphism, then u_2 is a universal epimorphism.*

The proof, which is analogous to that of proposition 2.3 is left to the reader.

Corollary 2.10. *Any isomorphism is a universal epimorphism.*

2. Kernel and Cokernel

Definition. Let $u_1, u_2 : A \longrightarrow B$ be two morphisms in the category $\mathscr{C}$. The object N together with the morphism $v : N \longrightarrow A$ is said to be a *kernel* for the pair of morphisms u_1, u_2 if the following two conditions are fulfilled:

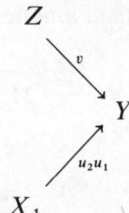

(a) If $u : X \longrightarrow A$ is a morphism such that $u_1 u = u_2 u$, then there exists a unique morphism $v : X \longrightarrow N$ such that $\nu v = u$.

(b) If $u : X \longrightarrow A$ can be factorized through ν, then $u_1 u = u_2 u$.

PROPOSITION 2.11. *If (N, ν), (M, μ) are kernels for the pair of morphisms $u_1, u_2 : A \longrightarrow B$, then there exists a unique isomorphism $\alpha : N \longrightarrow M$ such that the diagram*

$$\begin{array}{ccc} N & \xrightarrow{\alpha} & M \\ & {\scriptstyle\nu}\searrow \quad \swarrow{\scriptstyle\mu} & \\ & A & \end{array} \qquad (1)$$

is commutative.

PROOF. Since (N, ν) is a kernel for the pair (u_1, u_2), it follows from condition (b) in the definition that $u_1 \nu = u_2 \nu$. Hence and from the fact that (M, μ) is also a kernel it follows (condition (a)) that there exists $\alpha : N \longrightarrow M$ such that diagram (1) is commutative. We have to show that α is an isomorphism. By reasoning in the same way we infer that there exists $\beta : M \longrightarrow N$ such that the diagram

$$\begin{array}{ccc} M & \xrightarrow{\beta} & N \\ & {\scriptstyle\mu}\searrow \quad \swarrow{\scriptstyle\nu} & \\ & A & \end{array}$$

is commutative. It will be sufficient to show that $\beta\alpha = 1_N$. This follows from the fact that we have simultaneously $\nu(\beta\alpha) = \nu$, $\nu 1_N = \nu$, and from condition (a) in the definition.

PROPOSITION 2.12. *Let $u_1, u_2 : A \longrightarrow B$ be two morphisms in the category $\mathscr{C}$. Consider the contravariant functor $F : \mathscr{C} \longrightarrow \mathscr{Ens}$ defined as follows:*

$$F(X) = \{f : X \longrightarrow A \mid u_1 f = u_2 f\}$$

If the pair of morphisms u_1, u_2 possesses a kernel (N, ν) then the functor F is represented by the pair (N, ν). Conversely, if the functor F is representable, then the pair of morphisms u_1, u_2 possesses a kernel. Moreover, if (N, ν) is a representation of this functor, then the pair (N, ν) is a kernel for the pair of morphisms u_1, u_2.

PROOF. If (N, ν) is a kernel of the pair of morphisms u_1, u_2, then the functorial morphism

$$\mathrm{Hom}_{\mathscr{C}}(X, N) \longrightarrow F(X)$$

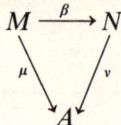

defined by

$$g \longrightarrow vg$$

is evidently a functorial isomorphism.

Conversely, assume that the pair (N, v) represents the functor F. According to the definition, $v \in F(N)$. Hence, $u_1 v = u_2 v$. Furthermore, from the fact that F is represented by (N, v) it follows that the map

$$\mathrm{Hom}_{\mathscr{C}}(X, N) \longrightarrow F(X)$$

defined by

$$g \longrightarrow vg$$

is one-to-one and onto, which is equivalent to the fact that (N, v) is a kernel for the pair of morphisms u_1, u_2.

PROPOSITION 2.13. *If (N, v) is a kernel for the pair of morphisms $u_1, u_2 : A \longrightarrow B$, then $v : N \longrightarrow A$ is a monomorphism.*

PROOF. Let $v_1, v_2 : X \longrightarrow N$ be such that $vv_1 = vv_2$. It follows that the morphism $v = vv_1 = vv_2$ can be factorized uniquely through v, whence $v_1 = v_2$.

The notion dual to that of a kernel of a pair of morphisms is called *cokernel* of a pair of morphisms.

Examples. In the categories $\mathscr{E}ns$, $\mathscr{T}op$, $\mathscr{G}$, $\mathscr{A}b$, each pair of morphisms has a kernel and a cokernel.

If $u_1, u_2 : A \longrightarrow B$ are morphisms in $\mathscr{E}ns$, then the subset of A consisting of the elements x with $u_1(x) = u_2(x)$ together with the inclusion map is a kernel of the pair (u_1, u_2). On the other hand, define a relation R on the set B as follows: two elements $y, z \in B$ are declared to be in the relation R, if there exists an element $x \in A$ such that $u_1(x) = y$, $u_2(x) = z$. Let C be the quotient set of B modulo the equivalence relation generated by R. Then C together with the natural epimorphism $B \longrightarrow C$ is a cokernel of the pair (u_1, u_2).

In the category $\mathscr{T}op$, the kernels and cokernels are those described for $\mathscr{E}ns$, except that the subset topology (respectively, the identification topology) is to be taken into account.

If $f_1, f_2 : G \longrightarrow H$ are morphisms in the category $\mathscr{G}$ of groups, then the subgroup of G consisting of the elements x with $f_1(x) = f_2(x)$ together with the inclusion homomorphism is a kernel of the pair (f_1, f_2). On the other hand, the reduction of H modulo the normal subgroup generated by the elements $f_1(x)f_2(x)^{-1}$ for all $x \in G$ together with the natural epimorphism is a cokernel of the pair (f_1, f_2).

In the category $\mathscr{Ab}$, kernels and cokernels coincide with those described for $\mathscr{G}$.

Definition. The diagram

$$X \xrightarrow{u} Y \underset{w}{\overset{v}{\rightrightarrows}} Z$$

in a category $\mathscr{C}$ is said to be *exact*, if (X, u) is a kernel of the pair of morphisms (v, w).

Dually, the diagram

$$X \underset{v}{\overset{u}{\rightrightarrows}} Y \xrightarrow{w} Z$$

is said to be exact, if (Z, w) is a cokernel of the pair of morphisms (u, v).

3. Grothendieck Topologies and the General Notion of a Sheaf

Definition. A *Grothendieck topology* T consists of a category $\mathscr{Cat}\ T$ and a set $\operatorname{Cov} T$ of families $\{U_i \xrightarrow{\varphi_i} U\}_{i \in I}$ of morphisms in $\mathscr{Cat}\ T$ called *coverings* (where in each covering the range U of the morphisms φ_i is fixed) satisfying the following conditions.

1. If φ is an isomorphism, then $\{\varphi\} \in \operatorname{Cov} T$.

2. If $\{U_i \longrightarrow U\}_{i \in I} \in \operatorname{Cov} T$ and $\{V_{ij} \xrightarrow{\psi_{ij}} U_i\}_{j \in J_i} \in \operatorname{Cov} T$ for each $i \in I$, then $\{V_{ij} \xrightarrow{\varphi_i \psi_{ij}} U\}_{i \in I, j \in J_i} \in \operatorname{Cov} T$.

3. If $\{U_i \xrightarrow{\varphi_i} U\}_{i \in I} \in \operatorname{Cov} T$ and $V \longrightarrow U$ is an arbitrary morphism in $\mathscr{Cat}\ T$, then $U_i \times_U V$ exists for each $i \in I$ and $\{U_i \times_U V \xrightarrow{v_i} V\}_{i \in I} \in \operatorname{Cov} T$, where v_i is the canonical projection.

Definition. Let T be a topology and $\mathscr{C}$ a category with infinite products. A *presheaf* on T with values in $\mathscr{C}$ is a functor $F : (\mathscr{Cat}\ T)^0 \longrightarrow \mathscr{C}$. *A sheaf* F is a presheaf satisfying the following condition:

(*S*) If $\{U_i \xrightarrow{\varphi_i} U\}_{i \in I} \in \operatorname{Cov} T$, then the diagram

$$F(U) \xrightarrow{u} \prod_{i \in I} F(U_i) \underset{w}{\overset{v}{\rightrightarrows}} \prod_{j, k \in I} F(U_j \times_U U_k)$$

is exact, where the morphisms v and w are determined by the condition that the diagram

$$\begin{array}{ccc} \prod_{i \in I} F(U_i) & \xrightarrow{v} & \prod_{(i,j) \in I \times I} F(U_i \times_U U_j) \\ \downarrow{\scriptstyle p_\alpha} & & \downarrow{\scriptstyle p_{(\alpha,\beta)}} \\ F(U_\alpha) & \xrightarrow{F(\pi_\alpha)} & F(U_\alpha \times_U U_\beta) \end{array}$$

is commutative for any $(\alpha, \beta) \in I \times I$, where $p_\alpha, p_\beta, p_{(\alpha,\beta)}, \pi_\alpha, \pi_\beta$, are canonical projections.

Remark. It follows from the definition that the necessary and sufficient condition that the presheaf F on T with values in the category $\mathscr{C}$ be a sheaf, is that for any object A of the category $\mathscr{C}$ the functor G defined by

$$G(U) = \mathrm{Hom}_{\mathscr{C}}(A, F(U))$$

be a sheaf on T with values in $\mathscr{E}ns$.

Examples. 1—Let X be a topological space. We define a Grothendieck topology T as follows:

$\mathscr{C}at\ T$ is the category whose objects are the open sets of X; if U and V are open sets of X, then $\mathrm{Hom}_{\mathscr{C}at\,T}(U, V)$ consists of the inclusion map if $U \subset V$, and is empty otherwise. $\mathrm{Cov}\, T$ is the set of all families $\{U_i \xrightarrow{\varphi_i} U\}_{i \in I}$, where φ_i are inclusion maps and $\bigcup_{i \in I} U_i = U$. Condition (3) is fulfilled since $U_i \times_U V \approx U_i \cap V$. A presheaf (respectively sheaf) on T coincides with the current notion of a presheaf (respectively sheaf) on the space X (see for instance Godement).

To see this, we shall prove the following:

PROPOSITION 2.14. *A presheaf F on T with values in $\mathscr{E}ns$ is a sheaf if and only if it satisfies the following two conditions* (F_1) If $\{U_i \xrightarrow{\varphi_i} U\}_{i \in I} \in \mathrm{Cov}\, T$, *and if $s', s'' \in F(U)$ are such that $[F(\varphi_i)](s') = [F(\varphi_i)](s'')$ for every $i \in I$, then $s' = s''$.* (F_2) If $\{U_i \xrightarrow{\varphi_i} U\}_{i \in I} \in \mathrm{Cov}\ T$ *and if for every $i \in I$ an element $s_i \in F(U_i)$ is given such that for any $i, j \in I$ with $U_i \cap U_j \neq \varnothing$ we have $[F(\psi^i_{ij})](s_i) = [F(\psi^j_{ij})](s_j)$ (where ψ^i_{ij} and ψ^j_{ij} are the inclusion maps of $U_i \cap U_j$ into U_i and U_j respectively), then there exists $s \in F(U)$ such that $[F(\varphi_i)](s) = s_i$ for any $i \in I$.*

PROOF. Necessity. Consider a set $M = \{m\}$ consisting of a single element m, and consider the diagram in condition (S) above

$$F(U) \xrightarrow{u} \prod_{i \in I} F(U_i) \overset{v}{\underset{w}{\rightrightarrows}} \prod_{j,k \in I} F(U_j \cap U_k).$$

Let $s', s'' \in F(U)$ be two elements which satisfy the conditions in (F_1). We have a map $\alpha: M \longrightarrow \prod_{i \in I} F(U_i)$ defined as follows: $\alpha(m) = ([F(\varphi_i)](s'))_{i \in I} = ([F(\varphi_i)](s''))_{i \in I}$. Clearly, this map can be factorized through u by setting $\sigma'(m) = s'$ or $\sigma''(m) = s''$. But this implies $v\alpha = w\alpha$, so that the factorization must be unique: $\sigma'(m) = \sigma''(m)$, i.e. $s' = s''$, so that condition (F_1) is fulfilled.

Consider now elements $s_i \in F(U_i)(i \in I)$ satisfying the conditions in (F_2). Define a map $\alpha : M \longrightarrow \prod_{i \in I} F(U_i)$ as follows: $\alpha(m) = (s_i)_{i \in I}$. Clearly $v\alpha = w\alpha$, so that there exists a (unique) map $\sigma : M \longrightarrow F(U)$ such that $[F(\varphi_i)]\sigma(m) = s_i$ for any $i \in I$. The element $s = \sigma(m)$ satisfies condition (F_2).

Sufficiency. Let M be an arbitrary set and $f_i : M \longrightarrow F(U_i)$, $i \in I$, maps with the property that the maps $g_1, g_2 : M \longrightarrow F(U_i \cap U_j)$ defined to be the compositions

$$M \xrightarrow{f_i} F(U_i) \xrightarrow{F(\psi^i_{ij})} F(U_i \cap U_j)$$

$$M \xrightarrow{f_j} F(U_j) \xrightarrow{F(\psi^j_{ij})} F(U_i \cap U_j)$$

are equal for any pair of indices $i, j \in I$. We have to show that under these conditions the condition (S) is satisfied, i.e. there exists a unique map $f : M \longrightarrow F(U)$ such that the composition

$$M \xrightarrow{f} F(U) \xrightarrow{F(\varphi_i)} F(U_i)$$

coincides with f_i for any $i \in I$. To this end, let x be an arbitrary element of M. The elements $f_i(x) \in F(U_i)$ plainly satisfy the conditions in (F_2), so that there exists an element $s \in F(U)$ such that $[F(\varphi_i)](s) = f_i(x)$ for any $i \in I$. According to (F_1), s is unique. By setting $f(x) = s$ we get the required map.

PROPOSITION 2.15. *Let $\mathscr{C}_1, \mathscr{C}_2$ be two categories and let $H : \mathscr{C}_1 \longrightarrow \mathscr{C}_2$, $K : \mathscr{C}_2 \longrightarrow \mathscr{C}_1$ be two covariant functors such that K is an adjoint functor of the functor H. Assume F is a presheaf on T with values in $\mathscr{C}_2$. Define a presheaf G on T with values in $\mathscr{C}_1$ by setting $G(U) = K(F(U))$ for each object U of $\mathscr{C}_2$. If F is a sheaf, then G is also a sheaf.*

PROOF. According to the remark on p. 47 it will be sufficient to show that for each object A of the category $\mathscr{C}_1$, the presheaf which associates to each object U the set $\mathrm{Hom}_{\mathscr{C}_1}(A, K(F(U)))$ is a sheaf. But, K being an adjoint functor of the functor H, we have the functorial isomorphism

$$\mathrm{Hom}_{\mathscr{C}_1}(A, K(F(U))) \approx \mathrm{Hom}_{\mathscr{C}_2}(H(A), F(U)).$$

Since F is a sheaf, again by the remark on p. 47, the presheaf which associates to each object U the set $\mathrm{Hom}_{\mathscr{C}_2}(H(A), F(U))$ is also a sheaf whence the proposition.

COROLLARY 2.16. *Let F be a presheaf on T with values in the category ${}_\Lambda\mathscr{C}$ of left unitary Λ-modules. Let $K : {}_\Lambda\mathscr{C} \longrightarrow \mathscr{E}ns$ be the covariant functor which associates to each Λ-module its subjacent set. Then F is a sheaf if* *and only if the presheaf G on T with values in $\mathscr{E}ns$ defined by* $G(U) = K(F(U))$, *for each object U, is a sheaf.*

PROOF. It is easy to see that K is an adjoint functor of the functor $H:\mathscr{E}ns \longrightarrow {}_{\Lambda}\mathscr{C}$ which associates to each set M the free Λ-module $H(M)$ generated by the elements of M. By proposition 2.16, if F is a sheaf, then G is also a sheaf.

Conversely, assume G is a sheaf. From the functorial isomorphisms

$$\mathrm{Hom}_{\Lambda}(H(M), N) \approx \mathrm{Hom}_{\mathscr{E}ns}(M, K(N))$$

and from corollary 2.15 it follows that the presheaf on T with values in $\mathscr{E}ns$ which associates to each object U the set $\mathrm{Hom}_{\Lambda}(L, F(U))$ is a sheaf for any free Λ-module L. If we take $L = \Lambda$, then we obtain precisely F, which completes the proof.

2. We obtain a Grothendieck topology T if we set $\mathscr{C}at\ T = \mathscr{E}ns$ and $\mathrm{Cov}\ T =$ set of all families $\{U_i \longrightarrow U\}_{i\in I}$ such that $\bigcup_{i\in I} \varphi_i(U_i) = U$. Condition 3 is satisfied since in $\mathscr{E}ns$ the fibered product of the objects U_i and V relatively to the scheme

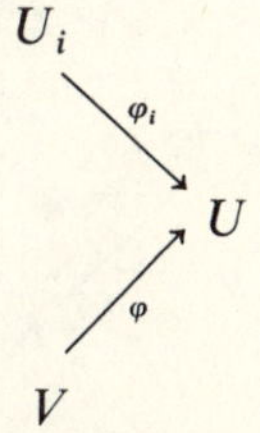

is isomorphic to the subset of the cartesian product $U_i \times V$ consisting of all pairs (x, y) such that $\varphi_i(x) = \varphi(y)$. Every sheaf F on T with values in $\mathscr{E}ns$ is a representable functor, namely $F(U) \approx \mathrm{Hom}(U, F(e))$, where e is a set consisting of a single element.

3. More generally than in (2), let G be a group; we define a Grothendieck topology T_G by setting $\mathscr{C}at\ T =$ category of left G-sets (sets with G operating), $\mathrm{Cov}\ T =$ set of all families $\{U_i \longrightarrow U\}_{i\in I}$, where φ_i are G-maps and $\bigcup_{i\in I} \varphi_i(U_i) = U$. Every sheaf F on T_G with values in $\mathscr{E}ns$ is a representable functor, namely $F(U) \approx \mathrm{Hom}_G(U, F(G))$, where $F(G)$ is G considered as a left G-set which has as operation of $g \in G$ the one induced by right multiplication on G by g^{-1}.

4. Let $\mathscr{C}$ be a category with fibered products. We define a new category $\mathscr{C}'$ as follows:

(i) the objects of the category $\mathscr{C}'$ are families $(U_i)_{i \in I}$ of objects of the category $\mathscr{C}$.

(ii) a morphism from $(U_i)_{i \in I}$ to $(V_j)_{j \in J}$ consists of a map $\varphi : I \longrightarrow J$ and of a family $(\varphi_i)_{i \in I}$ of morphisms $U_i \xrightarrow{\varphi_i} V_{\varphi(i)}$ in the category $\mathscr{C}$. The composition of morphisms is defined in an obvious manner.

PROPOSITION 2.18. *There exists a natural functor* $\mathscr{C} \longrightarrow \mathscr{C}'$.

PROPOSITION 2.19. *The category* $\mathscr{C}'$ *is with fibered products.*

PROOF. Consider the scheme.

$$\begin{array}{ccc} (U_i)_{i \in I} & & \\ & \searrow^{(\varphi, \varphi_i)} & \\ & & (W_k)_{k \in K} \\ & \nearrow_{(\psi, \psi_j)} & \\ (V_j)_{j \in J} & & \end{array}$$

in the category $\mathscr{C}'$. The family $(U_i \prod_{W_k} V_j)_{(i,j) \in I \prod_k J}$, where $\varphi(i) = \psi(j) = k$, and $I \prod_k J$ is the fibered product in the category $\mathscr{E}ns$ associated to the scheme

$$\begin{array}{ccc} I & & \\ & \searrow^{\varphi} & \\ & & K, \\ & \nearrow_{\psi} & \\ J & & \end{array}$$

together with the morphisms $U_i \prod_{W_k} V_j \xrightarrow{\pi_{ij}} U_i$, $U_i \prod_{W_k} V_j \xrightarrow{\sigma_{ij}} V_j$ defined in an obvious manner, constitute a fibered product associated to the given scheme.

Now, if we consider for any object X in $\mathscr{C}$ the universal epimorphisms $(U_i)_{i \in I} \longrightarrow (X)$ in the category $\mathscr{C}'$, we obtain a Grothendieck topology. This follows from the preceding proposition and from the following remark:

If $V_{ij} \xrightarrow{\varphi_{ji}} U_i$, ($i$ fixed) define a universal epimorphism, then $(V_{ij}) \longrightarrow (U_i)$ define a universal epimorphism.

5. If $\mathscr{C}$ is a category, we can define a Grothendieck topology T by taking $\mathscr{C}at\ T = \mathscr{C}$, Cov T = set of universal epimorphisms $\{V \longrightarrow U\}$ in $\mathscr{C}$.

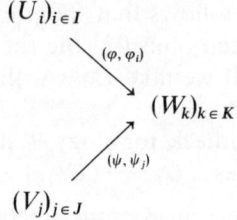

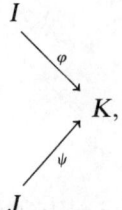

CHAPTER 3

Inductive and Projective Limits

1. The General Notion of a Projective or Inductive Limit

Definition. Let $\mathscr{D}, \mathscr{C}$ be two categories. The *category of functors from $\mathscr{D}$ to $\mathscr{C}$*, denoted by Fonct($\mathscr{D}$, $\mathscr{C}$) is the category whose objects are the covariant functors from $\mathscr{D}$ to $\mathscr{C}$ and whose morphisms are the functorial morphisms between such functors.

If the category $\mathscr{D}$ is small, it is also called a *scheme*, and a functor from $\mathscr{D}$ to $\mathscr{C}$ is called a *diagram* in $\mathscr{C}$ of scheme $\mathscr{D}$. Then Fonct($\mathscr{D}, \mathscr{C}$) is also said to be a category of diagrams.

Example. Let I be a partially ordered set (not necessarily directed). We denote by Ord I the category defined as follows: the objects of Ord I are the elements of I, and, for any pair $x, y \in I$, $\mathrm{Hom}_{\mathrm{Ord}\, I}(x, y)$ is the set consisting of a single element (x, y) if $x \leq y$ in the sense of the order relation in I, and is the void set otherwise.

An object of the category Fonct(Ord I, $\mathscr{C}$) is called *inductive system* in $\mathscr{C}$ over I. An object of the category Fonct((Ord $I)^0$, $\mathscr{C}$) is called *projective system* in $\mathscr{C}$ over I.

Definition. Let $\mathscr{D}$, $\mathscr{C}$ be two categories and $F:\mathscr{D} \longrightarrow \mathscr{C}$ a covariant functor. A *projective limit* of F is an object A of the category $\mathscr{C}$ together with morphisms $u_X : A \longrightarrow F(X)$ defined for each object X of the category $\mathscr{D}$ and such that the following two conditions are satisfied:

1. For any morphisms $\alpha : X \longrightarrow Y$ in the category $\mathscr{D}$, the diagram

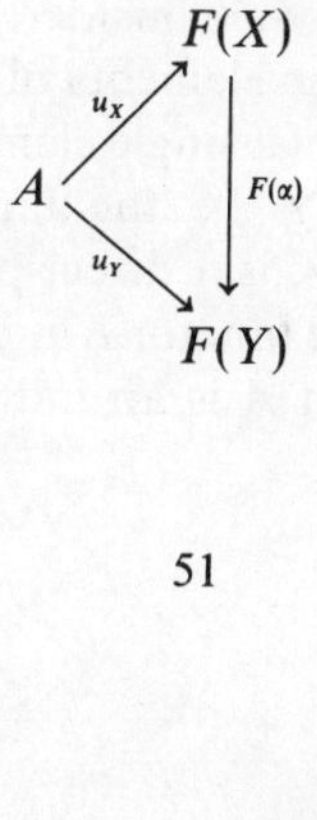

is commutative.

2. If B is an object of the category $\mathscr{C}$ and $v_X: B \longrightarrow F(X)$ are morphisms defined for each object X of the category $\mathscr{D}$ and such that the diagram

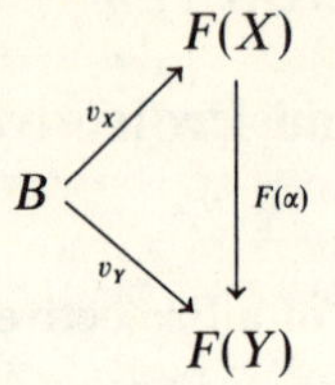

is commutative for any morphism $\alpha: X \longrightarrow Y$ in the category $\mathscr{D}$, then there exists a unique morphism $v: B \longrightarrow A$ such that the diagram

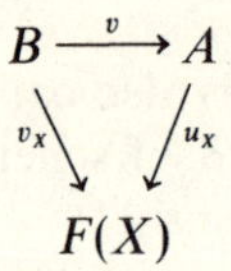

is commutative for any object X of the category $\mathscr{D}$.

The projective limit is sometimes also called inverse limit.

Dually, an *inductive limit* of F is an object A of the category $\mathscr{C}$ together with the morphisms $u_X: F(X) \longrightarrow A$ defined for each object X of the category $\mathscr{D}$ and such that

1′. For any morphism $\alpha: X \longrightarrow Y$ in $\mathscr{D}$ we have $u_X = u_Y F(\alpha)$.

2′. If B is an object in $\mathscr{C}$ and $v_X: F(X) \longrightarrow B$ are morphisms defined for each object X of $\mathscr{D}$, and such that $v_X = v_Y F(\alpha)$ for any morphism $\alpha: X \longrightarrow Y$ in $\mathscr{D}$, then there exists a unique morphism $v: A \longrightarrow B$ such that $v_X = v u_X$ for any object X of $\mathscr{D}$.

The inductive limit is sometimes also called direct limit.

The notions of projective and inductive limit for contravariant functors are defined likewise.

Examples. 1. Let $\mathscr{C}$ be an arbitrary category, M an arbitrary set and $(A_i)_{i \in M}$ a family of objects of $\mathscr{C}$. Consider the category $\mathscr{M}$ defined as follows: the objects of $\mathscr{M}$ are the elements of M, and for any pair $i, j \in M$, $\mathrm{Hom}_{\mathscr{M}}(i, j)$ is the set consisting of a single element 1_i if $i = j$, and is the void set otherwise. Let $F: \mathscr{M} \longrightarrow \mathscr{C}$ be the functor defined by $F(i) = A_i$, $i \in M$. Then if the object A of $\mathscr{C}$ is a direct product of the family $(A_i)_{i \in M}$ then A is a projective limit of the functor F and conversely. If A is the direct sum of the family $(A_i)_{i \in M}$, then A is an inductive limit of the functor F and conversely.

2. Let $\mathscr{C}$ be an arbitrary category and $u_1, u_2 : A \longrightarrow B$ two morphisms in $\mathscr{C}$. Consider the category $\mathscr{D}$ which has two objects X, Y and two morphisms $v_1, v_2 : X \longrightarrow Y$.

Let $F : \mathscr{D} \longrightarrow \mathscr{C}$ be the functor defined by $F(X) = A$, $F(Y) = B$, $F(v_i) = u_i (i = 1, 2)$. Then, if the object K of $\mathscr{C}$ is a kernel of the pair of morphisms u_1, u_2, then K is a projective limit of the functor F and conversely. If K is a cokernel of the pair of morphisms u_1, u_2, then K is an inductive limit of the functor F and conversely.

3. Let $\mathscr{C}$ be an arbitrary category and $u_1 : A \longrightarrow B$, $u_2 : C \longrightarrow B$ two morphisms in $\mathscr{C}$. Consider the category $\mathscr{D}$ which has three objects X, Y, Z and two morphisms: $v_1 : X \longrightarrow Y$, $v_2 : Z \longrightarrow Y$. Let $F : \mathscr{D} \longrightarrow \mathscr{C}$ be the functor defined by: $F(X) = A$, $F(Y) = B$, $F(Z) = C$, $F(v_i) = u_i (i = 1, 2)$. If the object D is a fibered product of A and C with respect to B, then D is a projective limit of the functor F and conversely.

There exists, of course, a similar description of the notion of a fibered sum.

4. Let $\mathscr{C}$ be one of the categories $\mathscr{E}ns$, $\mathscr{A}b$, ${}_{\Lambda}\mathscr{C}$ and let I be an arbitrary partially ordered set. If $F : \operatorname{Ord} I \longrightarrow \mathscr{C}$ is a functor, then the notion of a projective (or inductive) limit of the functor F coincides with the current notion of a projective or inverse (or inductive (or direct)) limit of the family $(F(i))_{i \in I}$.

PROPOSITION 3.1. *Let $\mathscr{C}$, $\mathscr{D}_1$, $\mathscr{D}_2$ be three categories and let $F_i : \mathscr{D}_i \longrightarrow \mathscr{C}$ $(i = 1, 2)$, $G : \mathscr{D}_2 \longrightarrow \mathscr{D}_1$ be functors. Assume that A_i together with the morphisms u^i_X defined for each object X of $\mathscr{D}_i$ is a projective limit of $F_i (i = 1, 2)$. Finally, let f be a functorial morphism from F_1G to F_2. Under these conditions, there exists a unique morphism $u : A_1 \longrightarrow A_2$ such that the diagram*

$$\begin{array}{ccc} A_1 & \xrightarrow{u} & A_2 \\ {\scriptstyle u^1_{G(X)}}\downarrow & & \downarrow{\scriptstyle u^2_X} \\ F_1(G(X)) & \xrightarrow[f(X)]{} & F_2(X) \end{array} \qquad (1)$$

is commutative for any object X of the category $\mathscr{D}_2$.

Furthermore, if $\mathscr{D}_1 = \mathscr{D}_2 = \mathscr{D}$, $G = 1_{\mathscr{D}}$ and $\operatorname{Fonct}(\mathscr{D}, \mathscr{C})$ *is a category and if each object F of* $\operatorname{Fonct}(\mathscr{D}, \mathscr{C})$ *possesses a projective limit A, then the assignments*

$$F \longrightarrow A$$

$$f \longrightarrow u$$

define a covariant functor denoted by $\varprojlim : \operatorname{Fonct}(\mathscr{D}, \mathscr{C}) \longrightarrow \mathscr{C}$.

Proof. For any morphism $v: X \longrightarrow Y$ in $\mathscr{D}_2$, the following relations hold:

$$f(Y)u^1_{G(Y)} = f(Y)F_1(G(v))u^1_{G(X)} = F_2(v)f(X)u^1_{G(X)}.$$

Hence, if we consider the morphisms $f(X)u^1_{G(X)}: A_1 \longrightarrow F_2(X)$ and apply condition (2) in the definition of projective limits, we get a unique morphism $u: A_1 \longrightarrow A_2$ such that diagram (1) is commutative. The second part of the proposition is immediate.

COROLLARY 3.2. *If (A_1, u^1_X) and (A_2, u^2_X) are two projective limits of the functor $F: \mathscr{D} \longrightarrow \mathscr{C}$, then there exists a unique isomorphism $u: A_1 \longrightarrow A_2$ such that $u^1_X = u^2_X u$ for any object X of the category $\mathscr{D}$.*

Using this corollary, for each functor F which possesses at least one projective limit we choose once for all a projective limit which we denote by $\varprojlim F$.

Plainly, the duals of proposition 3.1 and of corollary 3.2 are also true. Thus, for each functor F which possesses at least one inductive limit we choose once for all an inductive limit which we denote by $\varinjlim F$.

Remark. If the category $\mathscr{D}$ is of the form Ord I, where I is a partially ordered set, we sometimes use the notations $\varprojlim_{i \in I} F(i)$ and $\varinjlim_{i \in I} F(i)$ for the projective, respectively inductive limit of the functor F.

2. Existence of Inductive or Projective Limits

PROPOSITION 3.3. *Let $\mathscr{C}$ be a category. In order that every functor from a small category to $\mathscr{C}$ have a projective limit it is necessary and sufficient that $\mathscr{C}$ be a category with infinite products and that any pair of morphisms in $\mathscr{C}$ have a kernel.*

PROOF. The necessity is immediate in view of examples 1 and 2 above.

To prove the sufficiency, let $F: \mathscr{D} \longrightarrow \mathscr{C}$ be a functor, where $\mathscr{D}$ is a small category. Let P be the direct product of the objects $F(X)$, where X runs through the set of objects of $\mathscr{D}$, and let $p_X: P \longrightarrow F(X)$ be the canonical projections. Finally, let Q be the direct product of the objects $F(Y)$, where we take a copy of $F(Y)$ for each morphism $\alpha: X \longrightarrow Y$ of the category $\mathscr{D}$ and let $q^\alpha_Y: Q \longrightarrow F(Y)$ be the canonical projections.

Consider the morphisms $F(\alpha)p_X: P \longrightarrow F(Y)$ defined for each Y and each $\alpha: X \longrightarrow Y$. There exists a unique morphism w such that all the diagrams

$$\begin{array}{ccc} Q & \xleftarrow{w} & P \\ {\scriptstyle q_Y^\alpha}\downarrow & & \downarrow{\scriptstyle p_X} \\ F(Y) & \xleftarrow[F(\alpha)]{} & F(X) \end{array} \tag{2}$$

are commutative. On the other hand, consider the morphisms $p_Y:P \longrightarrow F(Y)$ defined also for each Y and each $\alpha:X \longrightarrow Y$. There exists a unique morphism $\overline{w}$ such that all the diagrams

$$\begin{array}{ccc} Q & \xleftarrow{\overline{w}} & P \\ {\scriptstyle q_Y^\alpha}\downarrow & \swarrow{\scriptstyle p_Y} & \\ F(Y) & & \end{array} \tag{3}$$

are commutative. Let A together with the morphism $s:A \longrightarrow P$ be the kernel of the pairs of morphisms $(w, \overline{w})$. We assert that A together with the morphisms $u_X = p_X s:A \longrightarrow F(X)$ defined for each object X of $\mathscr{D}$ is a projective limit of the functor F.

For, given any $\alpha:X \longrightarrow Y$, we may write, in view of the commutativity of diagrams (2) and (3):

$$F(\alpha)u_X = F(\alpha)p_X s = q_Y^\alpha ws = q_Y^\alpha \overline{w}s = p_Y s = u_Y.$$

Furthermore, assume there is given an object B of $\mathscr{C}$ together with morphisms $v_X:B \longrightarrow F(X)$ defined for each object X of $\mathscr{D}$, such that the diagram

$$\begin{array}{ccc} & & F(X) \\ & {\scriptstyle v_X}\nearrow & \downarrow{\scriptstyle F(\alpha)} \\ B & & \\ & {\scriptstyle v_Y}\searrow & \\ & & F(Y) \end{array}$$

is commutative for any $\alpha:X \longrightarrow Y$. Then there exists a unique morphism $t:B \longrightarrow P$ such that $p_X t = v_X$ for any X. We may write for any $\alpha:X \longrightarrow Y$:

$$q_Y^\alpha wt = F(\alpha)p_X t = F(\alpha)v_X = v_Y,$$

$$q_Y^\alpha \overline{w}t = p_Y t = v_Y.$$

By the definition of the direct product, we infer that $wt = \overline{w}t$; according to the definition of the kernel of a pair of morphisms, there exists then a unique morphism $v:B \longrightarrow A$ such that $sv = t$.

Hence $u_X v = p_X s v = p_X t = v_X$ for any X in $\mathscr{D}$, and the proof is complete.

Dually we have the following:

PROPOSITION 3.4. *Let $\mathscr{C}$ be a category. In order that every functor from a small category to $\mathscr{C}$ have an inductive limit it is necessary and sufficient that $\mathscr{C}$ be a category with infinite direct sums, and that any pair of morphisms in $\mathscr{C}$ have a cokernel.*

COROLLARY 3.5. *Every functor from a small category into each of the categories $\mathscr{E}ns$, $\mathscr{T}op$, $\mathscr{G}$, $\mathscr{A}b$, ${}_{\Lambda}\mathscr{C}$ has a projective limit and an inductive limit.*

PROOF. It follows immediately from the examples pp. 35, 39, and 45.

We recall the current constructions of projective and inductive limits in the category of sets for functors from categories of the form Ord I, where I is an arbitrary partially ordered set.

If (A_i, u_{ij}) is an arbitrary projective system in $\mathscr{E}ns$ over I, let A be subset of the cartesian product $\prod_{i \in I} A_i$ consisting of all families $(x_i)_{i \in I}$ with $x_i \in A_i$ $(i \in I)$ and $u_{ij}(x_j) = x_i$ for any pair $i, j \in I$ with $i \leq j$. Let also $u_i : A \longrightarrow A_i (i \in I)$ be defined by $u_i((x_j)_{j \in I}) = x_i$. Then A together with the maps u_i is a projective limit of (A_i, u_{ij}).

If (A_i, u_{ij}) is an arbitrary inductive system in $\mathscr{E}ns$ over I, let A be the quotient set of the disjoint sum $\coprod_{i \in I} A_i$ with respect to the equivalence relation generated by the following relation R: $x_i \in A_i$ and $x_j \in A_j$ are in relation R if there exists $k \in I$ with: $k \geq i$, $k \geq j$ and $u_{ki}(x_i) = u_{kj}(x_j)$. Let also $u_i : A_i \longrightarrow A (i \in I)$ be the composition of the injection into the disjoint sum and the canonical map of this sum onto A. Then A together with the maps u_i is an inductive limit of (A_i, u_{ij}).

3. Commutation of Functors with Projective and Inductive Limits

Definition. Let $\mathscr{B}$, $\mathscr{C}$, $\mathscr{D}$ be three categories and let $F : \mathscr{B} \longrightarrow \mathscr{C}$ be a covariant functor. We say that the functor F *preserves projective limits of functors from* $\mathscr{D}$ if the following condition is satisfied: If $G : \mathscr{D} \longrightarrow \mathscr{B}$ is a functor and the object B of $\mathscr{B}$ together with the morphisms $u_D : B \longrightarrow G(D)$ defined for each object D of $\mathscr{D}$ is a projective limit of G, then the object $F(B)$ of $\mathscr{C}$ together with the morphisms $F(u_D) : F(B) \longrightarrow F(G(D))$ is a projective limit of FG.

The definition for covariant functors which preserve inductive limits is similar.

Assume now that $F:\mathscr{B} \longrightarrow \mathscr{C}$ is a contravariant functor. We say that the functor F *carries projective limits of functors from* $\mathscr{D}$ *into inductive limits* if the following condition is satisfied: If $G:\mathscr{D} \longrightarrow \mathscr{B}$ is a functor and B together with the morphisms $u_D:B \longrightarrow G(D)$ is a projective limit of G, then $F(B)$ together with the morphisms $F(u_D):F(G(D)) \longrightarrow F(B)$ is an inductive limit of FG.

The definition for contravariant functors which carry inductive limits into projective limits is similar.

Let now $\mathscr{C}$ be an arbitrary category and let X be a fixed object of the category $\mathscr{C}$. We recall (see example 3, p. 8) that the contravariant functor $h_X:\mathscr{C} \longrightarrow \mathscr{E}ns$ is defined by

$$h_X(Y) = \mathrm{Hom}_{\mathscr{C}}(Y, X) \qquad \text{for any object } Y \text{ of } \mathscr{C},$$

and the covariant functor $\bar{h}_X:\mathscr{C} \longrightarrow \mathscr{E}ns$ is defined by

$$\bar{h}_X(Y) = \mathrm{Hom}_{\mathscr{C}}(X, Y) \qquad \text{for any object } Y \text{ of } \mathscr{C}.$$

Note that if $\mathrm{Fonct}(\mathscr{C}, \mathscr{E}ns)$ is a category, then the correspondence $\bar{h}:\mathscr{C} \longrightarrow \mathrm{Fonct}(\mathscr{C}, \mathscr{E}ns)$ defined by $\bar{h}(X) = \bar{h}_X$ for any object X of $\mathscr{C}$ is a contravariant functor. Similarly, if $\mathrm{Fonct}(\mathscr{C}^0, \mathscr{E}ns)$ is a category, then the correspondence $h:\mathscr{C} \longrightarrow \mathrm{Fonct}(\mathscr{C}^0, \mathscr{E}ns)$ defined by $h(X) = h_X$ for any X in $\mathscr{C}$ is a covariant functor.

PROPOSITION 3.6. *Let* $\mathscr{C}$ *be an arbitrary category and* A *an object of* $\mathscr{C}$. *Then the functor* $\bar{h}_A:\mathscr{C} \longrightarrow \mathscr{E}ns$ *preserves projective limits of functors from any category.*

PROOF. Let $\mathscr{D}$ be a category and $F:\mathscr{D} \longrightarrow \mathscr{C}$ a functor. Assume that the object B of $\mathscr{C}$ together with the morphisms $u_X:B \longrightarrow F(X)$ defined for each object X of $\mathscr{D}$ is a projective limit of F. We have for each morphism $\alpha:X \longrightarrow Y$ in $\mathscr{D}$ the following commutative diagram

$$\begin{array}{ccc} & \overset{\bar{h}_A(u_X)}{\nearrow} & \bar{h}_A(F(X)) \\ \bar{h}_A(B) & & \big\downarrow \bar{h}_A(F(\alpha)) \\ & \underset{\bar{h}_A(u_Y)}{\searrow} & \bar{h}_A(F(Y)) \end{array}$$

Suppose we are given a set M and maps $f(X):M \longrightarrow \bar{h}_A(F(X))$ such that for each morphism $\alpha:X \longrightarrow Y$ in $\mathscr{D}$, the diagram

$$\begin{array}{ccc} & \overset{f(X)}{\nearrow} & \bar{h}_A(F(X)) \\ M & & \big\downarrow \bar{h}_A(F(\alpha)) \\ & \underset{f(Y)}{\searrow} & \bar{h}_A(F(Y)) \end{array}$$

is commutative. For each $m \in M$, $[f(X)](m)$ is an element of $\bar{h}_A(F(X))$ = $\mathrm{Hom}_{\mathscr{C}}(A, F(X))$, and from the last diagram we infer that the diagram

$$\begin{array}{ccc} & [f(X)](m) & F(X) \\ A & \nearrow & \\ & \searrow & \downarrow F(\alpha) \\ & [f(Y)](m) & F(Y) \end{array}$$

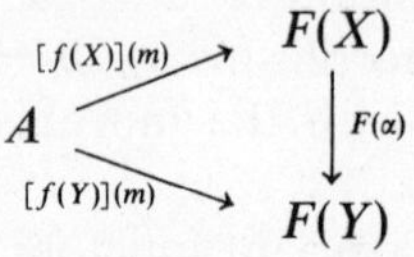

is commutative. Hence there exists a unique morphism $v_m: A \longrightarrow B$ such that $u_X \cdot v_m = [f(X)](m)$ for any X. We define a map $g: M \longrightarrow \bar{h}_A(B)$ by $g(m) = v_m$, $m \in M$. One can see that for any object X of $\mathscr{D}$, $\bar{h}_A(u_X)g = f(X)$, which shows that $\bar{h}_A(B)$ together with the morphisms $\bar{h}_A(u_X)$ is a projective limit of the functor $\bar{h}_A \cdot F$.

We also have the dual:

PROPOSITION 3.7. *Let $\mathscr{C}$ be an arbitrary category and A an object of $\mathscr{C}$. Then the functor $h_A: \mathscr{C} \longrightarrow \mathscr{E}ns$ carries inductive limits of functors from any category into projective limits.*

COROLLARY 3.8. *Let $\mathscr{C}$ be a category and $F: \mathscr{C} \longrightarrow \mathscr{E}ns$ a contravariant functor. A necessary condition for F to be representable is that it carries inductive limits of functors from any category into projective limits.*

PROOF. If F is representable, there exists an object A of $\mathscr{C}$ and a functorial isomorphism:

$$F \approx h_A.$$

The corollary now follows from proposition 3.7.

COROLLARY 3.9. *Let $F: \mathscr{C} \longrightarrow \mathscr{C}'$ be a covariant functor. A necessary condition for F to possess a left-adjoint (or right-adjoint) functor is that it preserves projective (or inductive) limits of functors from any category.*

PROOF. Let $G: \mathscr{C}' \longrightarrow \mathscr{C}$ be a left-adjoint functor of F. This means that there exists a functorial isomorphism

$$\bar{h}_{G(B)}(A) \approx \bar{h}_B(F(A))$$

for any A in $\mathscr{C}$ and B in $\mathscr{C}'$. Let now $\mathscr{D}$ be a category, $L: \mathscr{D} \longrightarrow \mathscr{C}$ a functor and assume that $\varprojlim L$ exists. We have the following sequence of isomorphisms, in view of proposition 3.6:

$$\bar{h}_B(F(\varprojlim L)) \approx \bar{h}_{G(B)}(\varprojlim L) \approx \varprojlim[\bar{h}_{G(B)} \cdot L]$$

$$\approx \varprojlim[\bar{h}_B \cdot FL) \approx \bar{h}_B(\varprojlim FL).$$

Since this holds for any object B of the category $\mathscr{C}'$, we infer that

$$F(\varprojlim L) \approx \varprojlim(FL)$$

PROPOSITION 3.10. *Let $\mathscr{B}, \mathscr{C}, \mathscr{D}$ be three categories such that* Fonct$(\mathscr{C}, \mathscr{B})$ *is a category and let $F: \mathscr{D} \longrightarrow$* Fonct$(\mathscr{C}, B)$ *be a functor. For each object A of the category $\mathscr{C}$ let $H_A: \mathscr{D} \longrightarrow \mathscr{B}$ be the functor defined by*

$$H_A(X) = [F(X)](A), \qquad \text{for any object } X \text{ of } \mathscr{D}.$$

Let $G: \mathscr{C} \longrightarrow \mathscr{B}$ be a functor such that, for any object A of $\mathscr{C}$, $G(A)$ together with the morphisms $u_X^A: G(A) \longrightarrow H_A(X)$ defined for each object X of $\mathscr{D}$, is a projective limit of H_A and assume, moreover, that for any morphism $\alpha: A \longrightarrow B$ in $\mathscr{C}$, the following diagram is commutative:

$$\begin{array}{ccc} G(A) & \xrightarrow{u_X^A} & H_A(X) \\ {\scriptstyle G(\alpha)}\downarrow & & \downarrow{\scriptstyle [F(X)](\alpha)} \\ G(B) & \xrightarrow[u_X^B]{} & H_B(X) \end{array}$$

Under these conditions, G together with the functorial morphisms $v_X: G \longrightarrow F(X)$ defined by $v_X(A) = u_X^A$ for each X in $\mathscr{D}$ and each A in $\mathscr{C}$, is a projective limit of F.

PROOF. Let $L: \mathscr{C} \longrightarrow \mathscr{B}$ be a functor and let $w_X: L \longrightarrow F(X)$ be functorial morphisms defined for each X in $\mathscr{D}$ and such that for any morphism $\mu: X \longrightarrow Y$ in $\mathscr{D}$, $F(\mu)w_Y = w_X$. We then have for each X in $\mathscr{D}$ and each A in $\mathscr{C}$ a commutative diagram

$$\begin{array}{ccc} & \nearrow^{w_X(A)} & [F(X)](A) = H_A(X) \\ L(A) & & \downarrow{\scriptstyle [F(\mu)](A)} \\ & \searrow_{w_Y(A)} & [F(Y)](A) = H_A(Y) \end{array}$$

Hence, by the definition of a projective limit, we obtain for each object A of the category $\mathscr{C}$ a unique morphism $s(A): L(A) \longrightarrow G(A)$ such that $v_X(A) \cdot s(A) = u_X^A \cdot s(A) = w_X(A)$ for each X in $\mathscr{D}$. It is readily seen that s is a functorial morphism from L to G and that $v_X s = w_X$ for each X in $\mathscr{D}$, whence the conclusion of the proposition.

COROLLARY 3.11. *For every category $\mathscr{C}$, the functor*

$$h: \mathscr{C} \longrightarrow \text{Fonct}(\mathscr{C}^0, \mathscr{E}ns)$$

preserves projective limits from any category.

PROOF. Let $\mathscr{D}$ be an arbitrary category and $L:\mathscr{D} \longrightarrow \mathscr{C}$ a functor. Assume that the object C of $\mathscr{C}$ is a projective limit of L. Now apply proposition 3.10, taking $\mathscr{B} = \mathscr{E}ns$, $F = h \cdot L$ and $G = h_C$. The condition that $G(A)$ be a projective limit of H_A for any A in $\mathscr{C}$ is verified according to proposition 3.6.

4. Characterization of Adjoint Functors

Definition. A category $\mathscr{C}$ is said to be *well-powered* if the family of subobjects of any object of $\mathscr{C}$ is a *set*.

LEMMA 3.12. *Let $\mathscr{C}$ be a well-powered category such that every functor from a small category to $\mathscr{C}$ has a projective limit and let $T: \mathscr{C} \longrightarrow \mathscr{D}$ be a covariant functor which preserves projective limits. Let A be an object of $\mathscr{C}$, B an object of $\mathscr{D}$ and $u: B \longrightarrow T(A)$ a morphism in $\mathscr{D}$. Then there exists a sub-object of A, $i: A' \longrightarrow A$ such that u may be factored as $u = T(i)v$ where $v: B \longrightarrow T(A')$, and such that there exists no proper subobject of A', $j: A'' \longrightarrow A'$ with the property that v may be factored through $T(j)$.*

PROOF. Consider the category $\mathscr{A}$ whose objects are all the subobjects of A, $i: C \longrightarrow A$ such that u may be factored through $T(i)$. If $j: D \longrightarrow A$ is another such subobject, a morphism in $\mathscr{A}$ is a morphism $s: C \longrightarrow D$ such that $js = i$. Since the category $\mathscr{C}$ is well-powered, $\mathscr{A}$ is a small category. Note that $1_A: A \longrightarrow A$ is an object of the category $\mathscr{A}$.

Next, consider the functor $F: \mathscr{A} \longrightarrow \mathscr{C}$ which associates to each $i: C \longrightarrow A$ the object C of $\mathscr{C}$. Let A' together with the morphisms $w_C: A' \longrightarrow C$ be a projective limit of the functor F, and let $i = w_A$. It is easy to see that i is a monomorphism. Since T preserves projective limits, $T(A')$ together with the morphisms $T(w_C): T(A') \longrightarrow T(C)$ is a projective limit of the functor TF. But u factors for each C as $B \xrightarrow{z_C} T(C) \xrightarrow{T(i)} T(A)$ so that there exists a unique morphism $v: B \longrightarrow T(A')$ such that the diagram

$$\begin{array}{c} B \xrightarrow{v} T(A') \\ \searrow_{z_C} \downarrow_{T(w_C)} \\ T(C) \end{array}$$

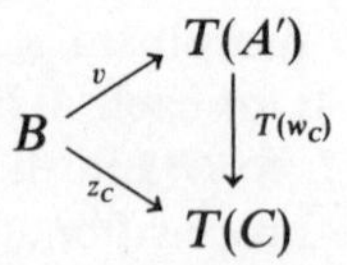

is commutative for any C. In particular, since $z_A = u$, we get $u = T(i)v$.

Suppose that $j: A'' \longrightarrow A'$ is a subobject of A' such that v may be factored as $B \longrightarrow T(A'') \xrightarrow{T(j)} T(A')$. This implies that u factors through $T(ij)$, i.e. $ij: A'' \longrightarrow A$ is an object of the category $\mathscr{A}$. We have therefore the morphism $w_{A''}:A' \longrightarrow A''$; it is easy to check that the morphisms j and $w_{A''}$ establish an equivalence between the following subobjects of $A: j:A'' \longrightarrow A'$ and $1_{A'}:A' \longrightarrow A'$, i.e. A'' is not a proper subobject of A'. This completes the proof.

THEOREM 3.13 (P. Freyd). *Let $\mathscr{C}$ be a well-powered category such that every functor from a small category to $\mathscr{C}$ has a projective limit. Let $T:\mathscr{C} \longrightarrow \mathscr{D}$ be a covariant functor. Then T has an adjoint functor if and only if the following conditions are fulfilled:*

1. *For every object B of $\mathscr{D}$ there exists an object A of $\mathscr{C}$ and a morphism $B \longrightarrow T(A)$ in $\mathscr{D}$.*
2. *T preserves projective limits.*
3. *For every object B in $\mathscr{D}$ there exists a set $\mathscr{S}_B$ of objects of $\mathscr{C}$ such that for any object A in $\mathscr{C}$ and any morphism $v:B \longrightarrow T(A)$ in $\mathscr{D}$ there exists an object $A' \in \mathscr{S}_B$ and morphisms $u:A' \longrightarrow A$ in $\mathscr{C}$ and $w:B \longrightarrow T(A')$ in $\mathscr{D}$ such that the diagram*

$$\begin{array}{ccc} & & T(A') \\ & \overset{w}{\nearrow} & \\ B & & \downarrow T(u) \\ & \underset{v}{\searrow} & \\ & & T(A) \end{array}$$

is commutative.

PROOF. Necessity. Let $S:\mathscr{D} \longrightarrow \mathscr{C}$ be an adjoint functor of T. According to corollary 3.9, condition 2 is satisfied.

To check condition 3, take $\mathscr{S}_B = \{S(B)\}$. Let A be an object of $\mathscr{C}$ and $v:B \longrightarrow T(A)$ a morphism in $\mathscr{D}$. By the definition of adjoint functors, we have a commutative diagram

$$\begin{array}{ccc} \operatorname{Hom}_{\mathscr{D}}(S(B), S(B)) & \xrightarrow{\psi(B, S(B))} & \operatorname{Hom}_{\mathscr{C}}(B, T(S(B))) \\ \downarrow {\scriptstyle \operatorname{Hom}_{\mathscr{D}}(S(B), u)} & & \downarrow {\scriptstyle \operatorname{Hom}_{\mathscr{C}}(B, T(u))} \\ \operatorname{Hom}_{\mathscr{D}}(S(B), A) & \xrightarrow{\psi(B, A)} & \operatorname{Hom}_{\mathscr{C}}(B, T(A)) \end{array} \qquad (4)$$

where $\psi(B, S(B))$ and $\psi(B, A)$ are isomorphisms and the morphism $u:S(B) \longrightarrow A$ is defined by $[\psi(B, A)](u) = v$. Let $w:B \longrightarrow T(S(B))$ be defined by $w = [\psi(B, S(B))](1_{S(B)})$. We claim that the diagram

$$\begin{array}{ccc} & & T(S(B)) \\ & \overset{w}{\nearrow} & \\ B & & \downarrow T(u) \\ & \underset{v}{\searrow} & \\ & & T(A) \end{array}$$

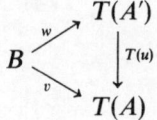

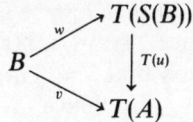

is commutative. Indeed, by the commutativity of diagram (4), $T(u)w = [\psi(B, A)](u) = v$.

Sufficiency. Let B be an arbitrary object of $\mathscr{D}$ and $\mathscr{S}_B$ be the set described in condition 3. Set

$$\bar{A} = \prod_{A' \in \mathscr{S}_B} \prod_{v \in \operatorname{Hom}_{\mathscr{D}}(B, T(A'))} A'_v$$

where $A'_v = A'$ for each $v \in \operatorname{Hom}_{\mathscr{D}}(B, T(A'))$. Since T preserves projective limits, we have $T(\bar{A}) = \Pi\Pi T(A')$. Let

$$t = \prod_{A' \in \mathscr{S}_B} \prod_{v \in \operatorname{Hom}_{\mathscr{D}}(B, T(A'))} v : B \longrightarrow T(\bar{A}).$$

Then, by using the canonical projections from $\bar{A}$ to its factors and condition 3, it is readily seen that for any object A of $\mathscr{C}$ and any morphism $v : B \longrightarrow T(A)$ in $\mathscr{D}$, there exists a morphism $u : \bar{A} \longrightarrow A$ in $\mathscr{C}$ such that the diagram

$$\begin{array}{ccc} & & T(\bar{A}) \\ & \overset{t}{\nearrow} & \\ B & & \downarrow T(u) \\ & \underset{v}{\searrow} & \\ & & T(A) \end{array} \tag{5}$$

is commutative. We emphasize that u is in general not unique.

According to lemma 3.12, there exists a subobject $i : \bar{\bar{A}} \longrightarrow \bar{A}$ such that t may be factored as $B \xrightarrow{s} T(\bar{\bar{A}}) \xrightarrow{T(i)} T(\bar{A})$ and such that $\bar{\bar{A}}$ is minimal with respect to this property. We assert that for any object A of $\mathscr{C}$ and any morphism $v : B \longrightarrow T(A)$ in $\mathscr{D}$, there exists a *unique* morphism $z : \bar{\bar{A}} \longrightarrow A$ such that the diagram

$$\begin{array}{ccc} & & T(\bar{\bar{A}}) \\ & \overset{s}{\nearrow} & \\ B & & \downarrow T(z) \\ & \underset{v}{\searrow} & \\ & & T(A) \end{array}$$

is commutative. Indeed, we know already that there exists a morphism $u : \bar{A} \longrightarrow A$ such that equation (5) is commutative. It is sufficient to take $z = ui$.

To prove the uniqueness, let $z' : \bar{\bar{A}} \longrightarrow A$ be such that $T(z')s = v$. Let $k : K \longrightarrow \bar{\bar{A}}$ be the kernel of the pair of morphisms (z, z'). Since T preserves projective limits, $T(k) : T(K) \longrightarrow T(\bar{\bar{A}})$ is a kernel of the pair of morphisms $(T(z), T(z'))$. Since $T(z)s = T(z')s$, there exists a unique morphism $y : B \longrightarrow T(N)$ such that $s = T(k)y$. But by the minimality of $\bar{\bar{A}}$, this implies $N = \bar{\bar{A}}$, $k = 1_A$ and hence, $z' = z$.

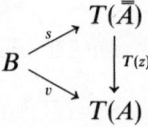

We now define a functor $S:\mathscr{D} \longrightarrow \mathscr{C}$ by setting $S(B) = \bar{\bar{A}}$, and, if $r:B_1 \longrightarrow B_2$ is a morphism in $\mathscr{D}$, by setting $S(r) = z$, where z is the unique morphism from $S(B_1)$ to $S(B_2)$ such that the diagram

$$\begin{array}{ccc} B_1 & \longrightarrow & T(S(B_1)) \\ {\scriptstyle r}\big\downarrow & & \big\downarrow{\scriptstyle T(z)} \\ B_2 & \longrightarrow & T(S(B_2)) \end{array}$$

is commutative.

We define for any object A of $\mathscr{C}$ and any object B of $\mathscr{D}$ two maps

$$\varphi(A, B):\mathrm{Hom}_{\mathscr{D}}(B, T(A)) \longrightarrow \mathrm{Hom}_{\mathscr{C}}(S(B), A)$$

$$\psi(A, B):\mathrm{Hom}_{\mathscr{C}}(S(B), A) \longrightarrow \mathrm{Hom}_{\mathscr{D}}(B, T(A))$$

as follows:

For every morphism $v:B \longrightarrow T(A)$, we set $[\varphi(A, B)](v) = z$ where z is the unique morphism from $S(B)$ to A with the property that the diagram

$$\begin{array}{ccc} & & T(S(B)) \\ & \overset{s}{\nearrow} & \\ B & & \big\downarrow{\scriptstyle T(z)} \\ & \underset{v}{\searrow} & \\ & & T(A) \end{array}$$

is commutative.

For every morphism $w:S(B) \longrightarrow A$, we set $[\psi(A, B)](w) = T(w)s$, where $s:B \longrightarrow T(S(B))$ is the morphism occurring in the above construction of $\bar{\bar{A}} = S(B)$.

It is not difficult to see that $\varphi(A, B)$ and $\psi(A, B)$ are functorial, and that $\varphi(A, B) \circ \psi(A, B) = 1_{\mathrm{Hom}(S(B), A)}$, $\psi(A, B) \circ \varphi(A, B) = 1_{\mathrm{Hom}(B, T(A))}$. But this means that S is an adjoint functor of T, and the proof of the theorem is complete.

We shall now show that every contravariant functor is an inductive limit of representable functors. More precisely, we have the following:

Proposition 3.14. *Let $\mathscr{C}$ be an arbitrary category and $F:\mathscr{C} \longrightarrow \mathscr{E}ns$ a contravariant functor. Then there exists a category $\mathscr{D}$ and a functor $G:\mathscr{D} \longrightarrow \mathscr{C}$ such that F is an inductive limit of the functor $h \cdot G$.*

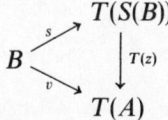

Proof. Let $\mathscr{D}$ be the category whose objects are all the morphisms $h_A \longrightarrow F$ in $\mathrm{Fonct}(\mathscr{C}^0, \mathscr{E}ns)$, for all the objects A of the category $\mathscr{C}$. A morphism in $\mathscr{D}$ is a morphism $v: A \longrightarrow B$ in $\mathscr{C}$ such that the diagram

$$\begin{array}{ccc} h_A & & \\ {\scriptstyle h(v)}\downarrow & \searrow & \\ & & F \\ & \nearrow & \\ h_B & & \end{array}$$

is commutative. Let $G: \mathscr{D} \longrightarrow \mathscr{C}$ be the functor which associates to each $h_A \longrightarrow F$ the object A of $\mathscr{C}$.

If $D = (h_A \longrightarrow F)$ is an arbitrary object of the category $\mathscr{D}$, define a morphism $u_D: h_{G(D)} \longrightarrow F$ in Fonct($\mathscr{C}^0, \mathscr{E}ns$) by $u_D = \alpha$. We now prove that F, together with the morphisms u_D is an inductive limit of the functor hG.

To this end, assume a functor $H: \mathscr{C}^0 \longrightarrow \mathscr{E}ns$ is given, together with functorial morphisms $v_D: h_{G(D)} \longrightarrow H$ defined for each object D in $\mathscr{D}$ and such that for each morphism $\mu: D \longrightarrow E$ in $\mathscr{D}$, the relation $v_D = v_E hG(\mu)$ holds. We must define a functorial morphism $v: F \longrightarrow H$ such that the diagram

$$\begin{array}{ccc} & & F \\ & \overset{u_D}{\nearrow} & \\ h_{G(D)} & & \downarrow{\scriptstyle v} \\ & \underset{v_D}{\searrow} & \\ & & H \end{array}$$

is commutative for any D in $\mathscr{D}$.

Let X be an arbitrary object of $\mathscr{C}$ and x an element of $F(X)$. According to theorem 1.6, we have an isomorphism $F(X) \approx \mathrm{Hom}_{\mathrm{Fonct}(\mathscr{C}^0, \mathscr{E}ns)}(h_X, F)$. The element x corresponds under this isomorphism to the functorial morphism $w_x: h_X \longrightarrow F$ characterized by

$$[w_x(X)](1_X) = x. \tag{6}$$

Let D_x be the following object of $\mathscr{D}: h_X \longrightarrow F$. We define $v(X)$ by the following relation:

$$[v(X)](x) = [v_{D_x}(X)](1_X), \qquad (x \in F(X)). \tag{7}$$

We now show that for any object $D = (h_A \overset{\alpha}{\longrightarrow} F)$ of $\mathscr{D}$, the following diagram is commutative:

$$\begin{array}{ccc} & & F(X) \\ & \overset{u_D(X)}{\nearrow} & \\ h_{G(D)}(X) & & \downarrow{\scriptstyle v(X)} \\ & \underset{v_D(X)}{\searrow} & \\ & & H(X). \end{array} \tag{8}$$

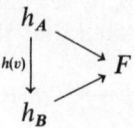

Let $z \in h_{G(D)}(X) = h_A(X) = \mathrm{Hom}_{\mathscr{C}}(X, A)$, and let $y = [u_D(X)](z) \in F(X)$. Then we have:

$$u_D h(z) = w_y. \tag{9}$$

For, let $s: Y \longrightarrow X$ be an arbitrary element of $\mathrm{Hom}_{\mathscr{C}}(Y, X)$. We have

$$\begin{aligned} u_D(Y)[(h(z))(Y)](s) &= [u_D(Y)](zs) = u_D(Y)[h_A(s)](z) \\ &= F(s)[u_D(X)](z) = F(s)y = F(s)[w_y(X)](1_X) \\ &= w_y(Y)[h_X(s)](1_X) = [w_y(Y)](s). \end{aligned}$$

Equation (9) implies that z is a morphism of the category $\mathscr{D}$; using this fact as well as equations (6), (7), (9) we may write:

$$\begin{aligned} [v(X)u_D(X)](z) &= v(X)u_D(X)[(h(z))(X)](1_X) \\ &= v(X)w_y(X)(1_X) = [v_{D_y}(X)](1_X) \\ &= v_D(X)[(h(z))(X)](1_X) = [v_D(X)](z), \end{aligned}$$

so that diagram (8) is commutative.

The uniqueness of v follows immediately from relations (6) and (7). We leave it to the reader to check that v is a functorial morphism.

PROPOSITION 3.15. *Let $\mathscr{C}$ be an arbitrary category and $\mathscr{B}$ a category with inductive limits. Let $F: \mathrm{Fonct}(\mathscr{C}^0, \mathscr{E}ns) \longrightarrow \mathscr{B}$ be a covariant functor. Then F preserves inductive limits if and only if it has a right-adjoint functor.*

Moreover, the functor

$$H: \mathrm{Fonct}(\mathrm{Fonct}(\mathscr{C}^0, \mathscr{E}ns), \mathscr{B}) \longrightarrow \mathrm{Fonct}(\mathscr{C}, \mathscr{B})$$

defined by $H(F) = Fh$ establishes an equivalence between the full subcategory of $\mathrm{Fonct}(\mathrm{Fonct}(\mathscr{C}^0, \mathscr{E}ns), \mathscr{B})$ *consisting of the functors which preserve inductive limits, and the category,* $\mathrm{Fonct}(\mathscr{C}, \mathscr{B})$.

PROOF. If F has a right-adjoint functor, then by corollary 3.9, F preserves inductive limits.

Conversely, suppose that $F: \mathrm{Fonct}(\mathscr{C}^0, \mathscr{E}ns) \longrightarrow \mathscr{B}$ preserves inductive limits. Define a functor $L: \mathscr{B} \longrightarrow \mathrm{Fonct}(\mathscr{C}^0, \mathscr{E}ns)$ by

$$L(Y) = h_Y^{\mathscr{B}} \cdot F \cdot h^{\mathscr{C}} \qquad \text{for each object } Y \text{ of } \mathscr{B}.$$

In other words, $L(Y)$ associates to each object X of $\mathscr{C}$ the set

$$\mathrm{Hom}_{\mathscr{B}}(F(h_X^{\mathscr{C}}), Y).$$

Furthermore, define for each object E of $\mathrm{Fonct}(\mathscr{C}^0, \mathscr{E}ns)$ and each object Y

of $\mathscr{B}$ a map

$$\varphi(E, Y): \mathrm{Hom}_{\mathscr{B}}(F(E), Y) \longrightarrow \mathrm{Hom}_{\mathrm{Fonct}(\mathscr{C}^0, \mathscr{E}ns)}(E, L(Y))$$

as follows: Let $u: F(E) \longrightarrow Y$ be an arbitrary morphism in $\mathscr{B}$, X an object of $\mathscr{C}$ and x an arbitrary element of the set $E(X)$. Let $v: h_X \longrightarrow E$ be the functorial morphism which corresponds to x according to theorem 1.6. Then $[\varphi(E, Y)](u)$ is the functorial morphism $w: E \longrightarrow L(Y)$ defined by:

$$[w(X)](x) = uF(v).$$

If the functor E is representable, i.e. of the form $E = h_Z$ for a suitable object Z of $\mathscr{C}$, then

$$\mathrm{Hom}_{\mathscr{B}}(F(E), Y) = \mathrm{Hom}_{\mathscr{B}}(F(h_Z), Y) = [L(Y)](Z)$$

and it is easy to see that in this case $\varphi(E, Y)$ is the bijection of $[L(Y)](Z)$ onto $\mathrm{Hom}_{\mathrm{Fonct}(\mathscr{C}^0, \mathscr{E}ns)}(h_Z, L(Y))$ described in theorem 1.6.

Let now E be an arbitrary object of $\mathrm{Fonct}(\mathscr{C}^0, \mathscr{E}ns)$. According to proposition 3.14, there exists a category $\mathscr{D}$ and a functor $G: \mathscr{D} \longrightarrow \mathscr{C}$ such that the functor E is an inductive limit of $h^{\mathscr{C}}G$. By the preceding remark, we have for each object D of $\mathscr{D}$ a functorial bijection

$$h_Y[F(h_{G(D)})] \xrightarrow{\approx} h_{L(Y)}(h_{G(D)}).$$

According to proposition 3.1, these bijections induce a morphism

$$\varprojlim[h_Y \cdot F \cdot h^{\mathscr{C}} \cdot G] \longrightarrow \varprojlim[h_{L(Y)} \cdot h^{\mathscr{C}} \cdot G]$$

which is easily seen to be a bijection. But by proposition 3.7 the functor h_A carries inductive limits into projective limits, so that we have a bijection

$$h_Y(\varinjlim(F \cdot h^{\mathscr{C}} \cdot G)) \xrightarrow{\approx} h_{L(Y)}(\varinjlim(h^{\mathscr{C}} \cdot G)).$$

Since by assumption F preserves inductive limits, this yields a bijection

$$h_Y(F(\varinjlim h^{\mathscr{C}} \cdot G)) \xrightarrow{\approx} h_{L(Y)}(\varinjlim h^{\mathscr{C}} \cdot G)$$

which coincides with $\varphi(\varinjlim h^{\mathscr{C}} . G, Y) = \varphi(E, Y)$ since $\varphi(E, Y)$ is functorial in E. Thus L is a right-adjoint functor of F.

To prove the second part of the proposition, define a functor

$$K: \mathrm{Fonct}(\mathscr{C}, \mathscr{B}) \longrightarrow \mathrm{Fonct}(\mathrm{Fonct}(\mathscr{C}^0, \mathscr{E}ns), \mathscr{B})$$

as follows: Let $M: \mathscr{C} \longrightarrow \mathscr{B}$ and $E: \mathscr{C}^0 \longrightarrow \mathscr{E}ns$ be arbitrary covariant functors. Let $G: \mathscr{D} \longrightarrow \mathscr{C}$ be the functor supplied by proposition 3.14

such that $E = \varinjlim h^{\mathscr{C}} \cdot G$. We set

$$[K(M)](E) = \varinjlim(MG).$$

Moreover, if $u:E \longrightarrow E'$ is a morphism in Fonct($\mathscr{C}^0, \mathscr{E}ns$) and if $G':\mathscr{D}' \longrightarrow \mathscr{C}$ is the functor supplied by proposition 3.14 for the functor E', there obviously exists a functor $P:\mathscr{D} \longrightarrow \mathscr{D}'$ such that $G = G'P$ (namely, P is the functor which associates to the object α of $\mathscr{D}$ the object $u\alpha$ of $\mathscr{D}'$). By applying the dual of proposition 3.1 to the identity morphism $MG \longrightarrow MG'P$ we obtain a morphism $v:\varinjlim (MG) \longrightarrow \varinjlim (MG')$. We set

$$[K(M)](u) = v.$$

Now, for any F in Fonct(Fonct($\mathscr{C}^0, \mathscr{E}ns$), $\mathscr{B}$) which preserves inductive limits and for any E in Fonct($\mathscr{C}^0, \mathscr{E}ns$) we have

$$[K(H(F))](E) = \varinjlim (H(F)G) = \varinjlim (Fh^{\mathscr{C}}G) = F(\varinjlim h^{\mathscr{C}}G) = F(E).$$

Conversely, for any M in Fonct($\mathscr{C}, \mathscr{B}$) and for any X in $\mathscr{C}$ we may write:

$$[H(K(M))](X) = K(M)(h_X) = \varinjlim (MG),$$

where $G:\mathscr{D} \longrightarrow \mathscr{C}$ is the functor supplied by proposition 3.14 such that $h_X = \varinjlim h^{\mathscr{C}}G$. But it is easy to see that $\varinjlim (MG) = M(X)$, which completes the proof of the proposition.

5. Prorepresentable Functors

Let I be a directed partially ordered set, and let $\mathbf{X} = (X_i, \varphi_{kl})$ be an inductive system over I with values in a category $\mathscr{C}$. We associate to this system a functor $h_{\mathbf{X}}:\mathscr{C} \longrightarrow \mathscr{E}ns$ defined as follows:

(i) $h_{\mathbf{X}}(Y) = \varinjlim_{i \in I} \mathrm{Hom}_{\mathscr{C}}(Y, X_i)$, for any object Y of $\mathscr{C}$.

(ii) for any morphism $u:Y \longrightarrow Z$ in $\mathscr{C}$, $h_{\mathbf{X}}(u):h_{\mathbf{X}}(Z) \longrightarrow h_{\mathbf{X}}(Y)$ is defined as follows: For each $i \in I$ we have the morphisms $h_{X_i}(u):h_{X_i}(Z) \longrightarrow h_{X_i}(Y)$, and if $i \leq j$ the diagram

$$\begin{array}{ccc} h_{X_i}(Z) & \longrightarrow & h_{X_j}(Z) \\ \downarrow{\scriptstyle h_{X_i}(u)} & & \downarrow{\scriptstyle h_{X_j}(u)} \\ h_{X_i}(Y) & \longrightarrow & h_{X_j}(Y) \end{array}$$

is commutative, where the horizontal morphisms are induced by the morphism $\varphi_{ji}: X_i \longrightarrow X_j$. This yields a unique morphism $h_{\mathbf{X}}(u): h_{\mathbf{X}}(Z) \longrightarrow h_{\mathbf{X}}(Y)$ such that the diagram

$$\begin{array}{ccc} h_{\mathbf{X}}(Z) & \xrightarrow{h_{\mathbf{X}}(u)} & h_{\mathbf{X}}(Y) \\ \uparrow & & \uparrow \\ h_{X_i}(Z) & \xrightarrow[h_{X_i}(u)]{} & h_{X_i}(Z) \end{array}$$

is commutative for any $i \in I$, where the vertical morphisms are the canonical injections.

Let now $F: \mathscr{C} \longrightarrow \mathscr{E}ns$ be a contravariant functor. We wish to determine the class of all functorial morphisms from $h_{\mathbf{X}}$ to F. The result generalizes theorem 1.6.

Proposition 3.16. *There exists a canonical map* $\alpha: \mathrm{Hom}(h_{\mathbf{X}}, F) \longrightarrow \varprojlim_{i \in I} F(X_i)$ *which is one-to-one and onto, where* $\varprojlim_{i \in I} F(X_i)$ *is the projective limit of the projective system* $(F(X_i), F(\varphi_{kl}))$ *over* I *with values in* $\mathscr{E}ns$.

Proof. Let $\varphi \in \mathrm{Hom}(h_{\mathbf{X}}, F)$. We define $\alpha(\varphi) \in \varprojlim F(X_i)$ as follows: For each $i \in I$, denote by ε_i the canonical image of the element $1_{X_i} \in \mathrm{Hom}_{\mathscr{C}}(X_i, X_i)$ in $h_{\mathbf{X}}(X_i)$ and let $\eta_i = [\varphi(X_i)](\varepsilon_i)$. The family $(\eta_i)_{i \in I}$ determines an element of $\varprojlim_{i \in I} F(X_i)$. For, let $i \leq j$; we have to show that $F(\varphi_{ji})\eta_j = \eta_i$. To this end, consider the commutative diagram

$$\begin{array}{ccc} h_{\mathbf{X}}(X_j) & \xrightarrow{\varphi(X_j)} & F(X_j) \\ {\scriptstyle h_{\mathbf{X}}(\varphi_{ji})}\downarrow & & \downarrow{\scriptstyle F(\varphi_{ji})} \\ h_{\mathbf{X}}(X_i) & \xrightarrow[\varphi(X_i)]{} & F(X_i). \end{array}$$

We have:

$$\begin{aligned} \eta_i = [\varphi(X_i)](\varepsilon_i) &= [\varphi(X_i) h_{\mathbf{X}}(\varphi_{ji})](\varepsilon_j) \\ &= [F(\varphi_{ji})\varphi(X_j)](\varepsilon_j) = F(\varphi_{ji})(\eta_j). \end{aligned}$$

Conversely, let $\xi = (\xi_i)_{i \in I} \in \varprojlim_{i \in I} F(X_i)$. We define a functorial morphism $\beta(\xi): h_{\mathbf{X}} \longrightarrow F$ as follows: For each $i \in I$, there exists according to theorem 1.6. a functorial morphism $\beta(\xi_i): h_{X_i} \longrightarrow F$ such that $\{[\beta(\xi_i)](X_i)\}(1_{X_i}) = \xi_i$. If Y is an arbitrary object in $\mathscr{C}$ and $i \leq j$, then the following diagram is commutative:

$$\begin{array}{ccc} h_{X_i}(Y) & & \\ \Big\downarrow & \searrow^{[\beta(\xi_i)](Y)} & \\ & & F(Y), \\ & \nearrow_{[\beta(\xi_j)](Y)} & \\ h_{X_j}(Y) & & \end{array}$$

whence there obviously results a functorial morphism $\beta(\xi): h_{\mathbf{X}} \longrightarrow F$.

One shows straightforwardly that $\beta[\alpha(\varphi)] = \varphi$ and $\alpha[\beta(\xi)] = \xi$, which completes the proof.

Definition. The inductive system $\mathbf{X} = (X_i, \varphi_{kl})$ together with the element $\xi = (\xi_i) \in \varprojlim_{i \in I} F(X_i)$ is said to form a prorepresentation system for the functor F if the functorial morphism $\beta(\xi): h_{\mathbf{X}} \longrightarrow F$ is a functorial isomorphism. If a functor F possesses a prorepresentation system, then F is said to be *prorepresentable.* If the morphisms φ_{kl} are monomorphisms for any k, l then F is said to be *strictly prorepresentable.*

Definition. A morphism $u: X \longrightarrow Y$ in a category $\mathscr{C}$ is said to be a *strict epimorphism* if it is an epimorphism, if the fibered product $X \Pi_Y X$ exists and if the sequence

$$X \Pi_Y X \overset{p_1}{\underset{p_2}{\rightrightarrows}} X \overset{u}{\longrightarrow} Y$$

is exact, where p_1 and p_2 are the canonical projections.

Definition. If $F: \mathscr{C} \longrightarrow \mathscr{E}ns$ is a contravariant functor, then the couple (X, ξ) where X is an object of $\mathscr{C}$ and $\xi \in F(X)$ is called a *minimal couple*, if any strict epimorphism $X \overset{u}{\longrightarrow} X_1$ with the property that there exists $\xi_1 \in F(X_1)$ such that $F(u)\xi_1 = \xi$ is an isomorphism.

LEMMA 3.17. *If a morphism $u: X \longrightarrow Y$ is both a strict epimorphism and a monomorphism, then it is an isomorphism.*

PROOF. Consider the diagram

$$\begin{array}{ccccc} X \Pi_Y X & \overset{p_1}{\underset{p_2}{\rightrightarrows}} & X & \overset{u}{\longrightarrow} & Y. \\ & & \Big\downarrow{\scriptstyle 1_X} & & \\ & & X & & \end{array}$$

Since by definition $up_1 = up_2$ and u is a monomorphism, we have $p_1 = p_2$. Then, again by definition, there exists a unique morphism

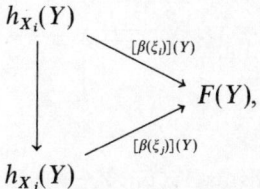

$v: Y \longrightarrow X$ such that $vu = 1_X$. Now consider the diagram

$$\begin{array}{ccccc} X \Pi_Y X & \overset{p_1}{\underset{p_2}{\rightrightarrows}} & X & \xrightarrow{u} & Y. \\ & & {\scriptstyle u}\downarrow & & \\ & & Y & & \end{array}$$

There are two morphisms $1_Y, uv: Y \longrightarrow Y$ which make this diagram commutative. Since the top line is exact, these two morphisms must coincide: $uv = 1_Y$.

LEMMA 3.18. *If the sequence*

$$A \overset{u_1}{\underset{u_2}{\rightrightarrows}} B \xrightarrow{\gamma} C$$

is exact, then the morphism γ is a strict epimorphism.

PROOF. γ is obviously an epimorphism. We have to show that the sequence

$$B \Pi_C B \overset{p_1}{\underset{p_2}{\rightrightarrows}} B \xrightarrow{\gamma} C$$

is exact, where p_1 and p_2 are the canonical projections from the fibered product. Plainly $\gamma p_1 = \gamma p_2$. Let $v: B \longrightarrow D$ be such that $vp_1 = vp_2$. There exists a unique morphism $\alpha: A \longrightarrow B \Pi_C B$ such that the diagram

$$\begin{array}{ccc} & & B \\ & \nearrow^{u_1} & \uparrow{\scriptstyle p_1} \\ A & \xrightarrow{\alpha} & B \Pi_C B \\ & \searrow_{u_2} & \downarrow{\scriptstyle p_2} \\ & & B \end{array}$$

is commutative. Hence $vu_1 = vu_2$, and there exists therefore a unique morphism $\delta: C \longrightarrow D$ such that $\delta\gamma = v$.

Definition. We say that the contravariant functor $F: \mathscr{C} \longrightarrow \mathscr{E}ns$ is left-exact if:

1. F transforms direct sums into direct products.
2. If the sequence $A \overset{u}{\underset{v}{\rightrightarrows}} B \xrightarrow{w} C$ is exact, then the sequence $F(C) \xrightarrow{F(w)} F(B) \overset{F(u)}{\underset{F(v)}{\rightrightarrows}} F(A)$ is exact.

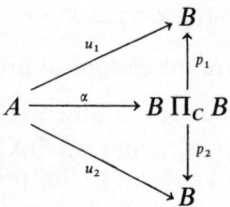

LEMMA 3.19. *Let $\mathscr{C}$ be an arbitrary category and I a directed partially ordered set. Let (F_1, φ_{kl}) be an inductive system over I with values in the class* Fonct$(\mathscr{C}^0, \mathscr{E}ns)$. *Finally, let $F: \mathscr{C} \longrightarrow \mathscr{E}ns$ be the contravariant*

functor defined by

$$F(X) = \varinjlim_{i \in I} F_1(X) \qquad \textit{for any object } X \textit{ of } \mathscr{C}.$$

Under these conditions, if F_i is left-exact for any $i \in I$, then F is also left-exact.

PROOF. We verify condition 2 in the definition leaving to the reader the verification of condition 1.

Let

$$A \overset{u}{\underset{v}{\rightrightarrows}} B \xrightarrow{w} C$$

be an exact sequence in the category $\mathscr{C}$. We must show that the sequence

$$F(A) \overset{F(u)}{\underset{F(v)}{\leftleftarrows}} F(B) \xleftarrow{F(w)} F(C)$$

is exact. This is equivalent (see the example on p. 45) to the following statement: For any element $x \in F(B)$ such that $F(u)x = F(v)x$, there exists a unique element $y \in F(C)$ such that $F(w)y = x$.

To prove this, let

$$\varphi_i^X : F_i(X) \longrightarrow F(X)$$

be the canonical map into the inductive limit, defined for each object X of $\mathscr{C}$ and each $i \in I$.

Let $x \in F(B)$ be such that $F(u)x = F(v)x$. There plainly exists an index $i_0 \in I$ and an element $x_0 \in F_{i_0}(B)$ such that $\varphi_{i_0}^{\beta}(x_0) = x$ and $F_{i_0}(u)x_0 = F_{i_0}(v)x_0$. Since the sequence

$$F_{i_0}(A) \overset{F_{i_0}(u)}{\underset{F_{i_0}(v)}{\leftleftarrows}} F_{i_0}(B) \xleftarrow{F_{i_0}(w)} F_{i_0}(C)$$

is exact by hypothesis, there exists $y_0 \in F_{i_0}(C)$ such that $F_{i_0}(w)y_0 = x_0$. Set $y = \varphi_{i_0}^C(y_0)$. Then $F(w)y = x$.

To prove the uniqueness of y, assume that $y, z \in F(C)$ are such that $F(w)y = F(w)z$. Since I is directed, it is easy to see that we can select an index $i_1 \in I$ and elements $y_1, z_1 \in F_{i_1}(C)$ such that $\varphi_{i_1}^C(y_1) = y$, $\varphi_{i_1}^C(z_1) = z$ and, moreover, $F_{i_1}(w)y_1 = F_{i_1}(w)z_1$. But since the sequence

$$F_{i_1}(A) \overset{F_{i_1}(u)}{\underset{F_{i_1}(v)}{\leftleftarrows}} F_{i_1}(B) \xleftarrow{F_{i_1}(w)} F_{i_1}(C)$$

is exact by hypothesis, we infer that $y_1 = z_1$, whence $y = z$.

LEMMA 3.20. *Assume $\mathscr{C}$ is a category such that any pair of morphisms has a cokernel and $F: \mathscr{C} \longrightarrow \mathscr{E}ns$ is a left-exact contravariant functor. Let (X, ξ) and (X', ξ') be two couples such that (X', ξ') is minimal. If $u, v: X \longrightarrow X'$ are such that $F(u)\xi' = F(v)\xi' = \xi$, then $u = v$.*

PROOF. Let $X' \xrightarrow{w} X''$ be a cokernel of the pair of morphisms (u, v). By lemma 3.18, w is a strict epimorphism. It is sufficient to prove that w is an isomorphism, in other words it is sufficient to show that there exists $\xi'' \in F(X'')$ such that $F(w)\xi'' = \xi'$. But F is left-exact, so that the sequence

$$F(X'') \xrightarrow{F(w)} F(X') \overset{F(u)}{\underset{F(v)}{\rightrightarrows}} F(X)$$

is exact. Taking into account the relation $F(u)\xi' = F(v)\xi'$, and the form of the kernel in the category $\mathscr{E}ns$ (see the example on p. 45), we infer that there exists $\xi'' \in F(X'')$ such that $F(w)\xi'' = \xi'$. This completes the proof.

Definition. We say that the couple (X, ξ) is *dominated* by the couple (X', ξ') if there exists $u: X \longrightarrow X'$ such that $F(u)\xi' = \xi$.

THEOREM 3.21. (A. Grothendieck). *Let $\mathscr{C}$ be a category with direct sums and such that any pair of morphisms has a cokernel, and let $F: \mathscr{C} \longrightarrow \mathscr{E}ns$ be a contravariant functor. Then the necessary and sufficient condition for F to be strictly prorepresentable is that*

1. *F be left-exact,*
2. *Any couple (Y, η) where $\eta \in F(Y)$ be dominated by a minimal couple.*

PROOF. To prove the sufficiency, let I be the set of all minimal couples. According to lemma 3.20, the relation of domination defines a partial order structure on the set I. Moreover, the partially ordered set so obtained is directed. To see this, let (X, ξ), (X', ξ') be two minimal couples. The functor F being left-exact, we have $F(X \amalg X') = F(X) \Pi F(X')$ and if $i_1: X \longrightarrow X \amalg X'$, $i_2: X' \longrightarrow X \amalg X'$ are the canonical injections, then $F(i_1): F(X) \Pi F(X') \longrightarrow F(X)$, $F(i_2): F(X) \Pi F(X') \longrightarrow F(X')$ are the canonical projections. Thus $(\xi, \xi') \in F(X \amalg X')$ and the couple $(X \amalg X', (\xi, \xi'))$ dominates both (X, ξ) and (X', ξ'). Since by condition 2 there exists a minimal couple which dominates $(X \amalg X', (\xi, \xi'))$, it follows that there exists a minimal couple which dominates both (X, ξ) and (X', ξ').

Note that by lemma 3.20, if (X, ξ) and (X', ξ') are minimal couples and if (X', ξ') dominates (X, ξ) then there exists a unique morphism $u: X \longrightarrow X'$ such that $F(u)\xi' = \xi$. This gives rise to an inductive system over I with values in $\mathscr{C}$. The family $(\xi)_{(X, \xi) \in I}$ defines an element of $\varinjlim_{(X, \xi) \in I} F(X)$, hence by proposition 3.16, a functorial morphism

$$h_{\mathbf{X}} \xrightarrow{\varphi} F,$$

where $\mathbf{X} = (X, u)_{(X, \xi) \in I}$.

We have to check that φ is a functorial isomorphism. To do this, let Y be an arbitrary object of $\mathscr{C}$, and let η be an arbitrary element of $F(Y)$.

By condition 2, there exists a minimal couple (X, ξ) and a morphism $v: Y \longrightarrow X$ such that $F(v)\xi = \eta$. If $\dot{v}$ is the canonical image of v in $h_{\mathbf{X}}(Y)$, we have $[\varphi(Y)](\dot{v}) = F(v)\xi = \eta$, which means that $\varphi(Y)$ is a surjection. Now suppose that there are two morphisms $u_1, u_2 : Y \longrightarrow X$, where (X, ξ) is minimal, such that $[\varphi(Y)](\dot{u}_1) = [\varphi(Y)](\dot{u}_2)$. This implies $F(u_1)\xi = F(u_2)\xi$, which in turn implies by lemma 3.20 that $u_1 = u_2$. This means that $\varphi(Y)$ is an injection.

We must also show that for each pair of minimal couples $(X, \xi), (X', \xi')$, the morphism $u: X \longrightarrow X'$ is a monomorphism. Let $v_1, v_2 : Z \longrightarrow X$ be such that $uv_1 = uv_2$, and let $w: X \longrightarrow U$ be the cokernel of the pair of morphisms (v_1, v_2). Then there exists a unique morphism $t: U \longrightarrow X'$ such that $tw = u$. This implies $F(w)[F(t)\xi'] = \xi$, i.e. the couple (X, ξ) is dominated by the couple $(U, F(t)\xi')$. Since by lemma 3.18 w is a strict epimorphism, we infer that w is an isomorphism, whence $v_1 = v_2$ and u is a monomorphism.

To prove the necessity, we may suppose F to be of the form $F = h_{\mathbf{X}}$, where $\mathbf{X} = (X_i, \varphi_{kl})$ is an inductive system over I with values in $\mathscr{C}$ such that φ_{kl} is a monomorphism for any k, l. According to proposition 3.7, the functor h_{X_i} is left-exact for each $i \in I$, and therefore, by lemma 3.19, F is also left-exact.

To check condition 2, let $(Y, \eta), \eta \in F(Y) = \varinjlim_{i \in I} \mathrm{Hom}_{\mathscr{C}}(Y, X_i)$ be an arbitrary couple, and let $v: Y \longrightarrow X_j$ represent η. Now consider the couple (X_j, ξ_j), where ξ_j is the canonical image of $1_{X_j}: X_j \longrightarrow X_j$ in $F(X_j) = \varinjlim_{i \in I} \mathrm{Hom}_{\mathscr{C}}(X_j, X_i)$. Since $F(v)\xi_j = \eta$, the couple (Y, η) is dominated by the couple (X_j, ξ_j). We have still to show that (X_j, ξ_j) is a minimal couple. To this end, let $u: X_j \longrightarrow Z$ be a strict epimorphism with the property that there exists $\zeta \in F(Z) = \varinjlim_{i \in I} \mathrm{Hom}_{\mathscr{C}}(Z, X_i)$ such that $F(u)\zeta = \xi_j$. Let $w: Z \longrightarrow X_k$ represent ζ; then wu represents $F(u)\zeta$, and the relation $F(u)\zeta = \xi_j$ means that there exists an index $l \in I$, $l \geq j, k$ such that $\varphi_{kl} wu = \varphi_{jl} 1_{X_j} = \varphi_{jl}$. Since φ_{jl} is by assumption a monomorphism, it follows by proposition 1.2 that u is a monomorphism. This implies by lemma 3.17 that u is an isomorphism, i.e. (X_j, ξ_j) is a minimal couple. This concludes the proof of the theorem.

CHAPTER 4

Structures on the Objects of a Category

1. Algebraic Operations on the Objects of a Category. Homomorphisms

Let $\mathscr{C}$ be an arbitrary category and X an object of $\mathscr{C}$. Recall that the contravariant functor $h_X : \mathscr{C} \longrightarrow \mathscr{E}ns$ is defined by:

$$h_X(Y) = \mathrm{Hom}_{\mathscr{C}}(Y, X), \qquad h_X(u) = vu \qquad \text{for any } v : Z \longrightarrow X,$$

where $u : Y \longrightarrow Z$.

Let $\mathscr{G}$ be the category of groups and let $c : \mathscr{G} \longrightarrow \mathscr{E}ns$ be the 'forgetful' functor, which assigns to each group its underlying set.

Definition. We say that a group structure has been defined on the object X in the category $\mathscr{C}$ if a functor $\sigma : \mathscr{C} \longrightarrow \mathscr{G}$ has been given such that the diagram

$$\begin{array}{ccc} \mathscr{C} & \xrightarrow{h_X} & \mathscr{E}ns \\ & \searrow^{\sigma} \quad {}^{c}\nearrow & \\ & \mathscr{G} & \end{array} \qquad \text{(D)}$$

is commutative. X is then said to be a *group in* $\mathscr{C}$.

This is clearly equivalent to the fact that $\mathrm{Hom}_{\mathscr{C}}(Y, X)$ is a group for any object Y of $\mathscr{C}$ and $\mathrm{Hom}_{\mathscr{C}}(u, X)$ is a homomorphism for any morphism u in $\mathscr{C}$.

We emphasize that σ is in general not unique.

Examples. If $\mathscr{C} = \mathscr{E}ns$ then X has a group structure in $\mathscr{C}$ if and only if X is a group in the current sense of the word; this is easily seen by taking for Y a set consisting of a single element.

If $\mathscr{C} = \mathscr{T}op$, then X has a group structure in $\mathscr{C}$ if and only if it is a topological group.

If $\mathscr{C}$ is the category of analytic manifolds, then X has a group structure in $\mathscr{C}$ if and only if it is a Lie group.

If $\mathscr{C}$ is the category of based topological spaces and of based homotopy classes of continuous maps, then X has a group structure in $\mathscr{C}$ if and only if it is a homotopy-associative H-space with homotopy-inverse.

Definition. An object A of a category $\mathscr{C}$ is said to be *initial*, if the set $\mathrm{Hom}_{\mathscr{C}}(A, X)$ consists of a single element for any object X of $\mathscr{C}$.

Dually, an object B of a category $\mathscr{C}$ is said to be *final* if the set $\mathrm{Hom}_{\mathscr{C}}(X, B)$ consists of a single element for any object X of $\mathscr{C}$.

An object which is both initial and final is called a *zero-object*.

Examples. In the category $\mathscr{E}ns$, any set consisting of a single element is a final object. In the category $\mathscr{G}$ of groups, the group which reduces to the neutral element is a zero-object.

THEOREM 4.1. *Let $\mathscr{C}$ be a category with direct products and having a final object e, and let X be an object of $\mathscr{C}$. Let η be the unique element of* $\mathrm{Hom}_{\mathscr{C}}(X, e)$. *Under these conditions, a group structure can be defined on X in the category $\mathscr{C}$, if and only if there exist three morphisms $m: X \times X \longrightarrow X$, $a: X \longrightarrow X$, and $\varepsilon: e \longrightarrow X$ such that the diagrams*

$$\begin{array}{ccc} X \times X \times X & \xrightarrow{m \times 1_X} & X \times X \\ {\scriptstyle 1_X \times m}\downarrow & & \downarrow{\scriptstyle m} \\ X \times X & \xrightarrow[m]{} & X \end{array} \qquad (1)$$

$$\begin{array}{ccc} X \times X & \xrightarrow{(\varepsilon\eta) \times 1_X} & X \times X \\ {\scriptstyle \Delta}\uparrow & & \downarrow{\scriptstyle m} \\ X & \xrightarrow[1_X]{} & X \end{array} \qquad (2)$$

$$\begin{array}{ccc} X \times X & \xrightarrow{a \times 1_X} & X \times X \\ {\scriptstyle \Delta}\uparrow & & \downarrow{\scriptstyle m} \\ X & \xrightarrow[\varepsilon\eta]{} & X \end{array} \qquad (3)$$

be commutative, where Δ is the 'diagonal' morphism (see p. 35).

PROOF. First, recall that by proposition 3.6 we have for any objects A, B, C of $\mathscr{C}$ a functorial isomorphism

$$\mathrm{Hom}_{\mathscr{C}}(A, B \times C) \approx \mathrm{Hom}_{\mathscr{C}}(A, B) \times \mathrm{Hom}_{\mathscr{C}}(A, C).$$

Assume that a group structure can be defined on X in the category $\mathscr{C}$. Then we define for each object Y of $\mathscr{C}$ maps

$$\bar{m}(Y): \mathrm{Hom}_{\mathscr{C}}(Y, X \times X) \longrightarrow \mathrm{Hom}_{\mathscr{C}}(Y, X)$$

$$\bar{a}(Y): \mathrm{Hom}_{\mathscr{C}}(Y, X) \longrightarrow \mathrm{Hom}_{\mathscr{C}}(Y, X)$$

$$\bar{\varepsilon}(Y): \mathrm{Hom}_{\mathscr{C}}(Y, e) \longrightarrow \mathrm{Hom}_{\mathscr{C}}(Y, X)$$

as follows:

$$[\overline{m}(Y)](u_1, u_2) = u_1 u_2, \qquad [\overline{a}(Y)](u) = u^{-1},$$

$[\overline{\varepsilon}(Y)](\xi) =$ neutral element of the group $\mathrm{Hom}_{\mathscr{C}}(Y, X)$, where $u_1, u_2, u \in \mathrm{Hom}_{\mathscr{C}}(Y, X)$, $u_1 u_2$ denotes the product and u^{-1} denotes the inverse in the group $\mathrm{Hom}_{\mathscr{C}}(Y, X)$, and ξ is the unique element of $\mathrm{Hom}_{\mathscr{C}}(Y, e)$.

It is easy to verify that $\overline{m}(Y)$, $\overline{a}(Y)$, $\overline{\varepsilon}(Y)$ define three functorial morphisms

$$\overline{m}: h_{X \times X} \longrightarrow h_X, \qquad \overline{a}: h_X \longrightarrow h_X, \qquad \overline{\varepsilon}: h_e \longrightarrow h_X.$$

According to corollary 1.7 there exist three uniquely determined morphisms

$$m: X \times X \longrightarrow X, \qquad a: X \longrightarrow X, \qquad \varepsilon: e \longrightarrow X$$

which induce respectively $\overline{m}$, $\overline{a}$, $\overline{\varepsilon}$.

To prove the commutativity of (1) observe that (D) induces for each object Y of $\mathscr{C}$ a diagram

$$\begin{array}{ccc}
\begin{pmatrix}\mathrm{Hom}_{\mathscr{C}}(Y, X) \times \mathrm{Hom}_{\mathscr{C}}(Y, X) \times \\ \times \mathrm{Hom}_{\mathscr{C}}(Y, X)\end{pmatrix} & \xrightarrow{\overline{m \times 1_X}(Y)} & \begin{pmatrix}\mathrm{Hom}_{\mathscr{C}}(Y, X) \times \\ \times \mathrm{Hom}_{\mathscr{C}}(Y, X)\end{pmatrix} \\
\Big\downarrow{\scriptstyle \overline{1_X \times m}(Y)} & & \Big\downarrow{\scriptstyle \overline{m}(Y)} \\
\mathrm{Hom}_{\mathscr{C}}(Y, X) \times \mathrm{Hom}_{\mathscr{C}}(Y, X) & \xrightarrow[\overline{m}(Y)]{} & \mathrm{Hom}_{\mathscr{C}}(Y, X)
\end{array}$$

which is commutative, since we have for any $u_1, u_2, u_3 \in \mathrm{Hom}_{\mathscr{C}}(Y, X)$, in view of the associativity in the group $\mathrm{Hom}_{\mathscr{C}}(Y, X)$:

$$\begin{aligned}
[\overline{m}(Y)\overline{m \times 1_X}(Y)](u_1, u_2, u_3) &= \overline{m}(Y)(u_1 u_2, u_3) = (u_1 u_2) u_3 \\
&= u_1(u_2 u_3) = \overline{m}(Y)(u_1, u_2 u_3) \\
&= [\overline{m}(Y)\overline{1_X \times m}(Y)](u_1, u_2, u_3).
\end{aligned}$$

Applying corollary 1.7 we obtain $m(m \times 1_X) = m(1_X \times m)$, i.e. the commutativity of (1) is established.

The morphism $\varepsilon\eta: X \longrightarrow X$ induces a functorial morphism $\overline{\varepsilon\eta}: h_X \longrightarrow h_X$. We claim that for any Y in $\mathscr{C}$ the homomorphism $\overline{\varepsilon\eta}(Y)$ maps $\mathrm{Hom}_{\mathscr{C}}(Y, X)$ into the neutral element ω of $\mathrm{Hom}_{\mathscr{C}}(Y, X)$. For, if $u \in \mathrm{Hom}_{\mathscr{C}}(Y, X)$

$$\overline{\varepsilon\eta}(Y)(u) = \varepsilon\eta u = \overline{\varepsilon}(Y)(\eta u) = \omega.$$

Diagram (2) induces for each object Y of $\mathscr{C}$ a diagram

$$\begin{array}{ccc} \mathrm{Hom}_{\mathscr{C}}(Y, X) \times \mathrm{Hom}_{\mathscr{C}}(Y, X) & \xrightarrow{\overline{(\varepsilon\eta) \times 1_X}(Y)} & \mathrm{Hom}_{\mathscr{C}}(Y, X) \times \mathrm{Hom}_{\mathscr{C}}(Y, X) \\ \uparrow{\scriptstyle \bar{\Delta}(Y)} & & \downarrow{\scriptstyle \overline{m}(Y)} \\ \mathrm{Hom}_{\mathscr{C}}(Y, X) & \xrightarrow[\overline{1_X}(Y)]{} & \mathrm{Hom}_{\mathscr{C}}(Y, X) \end{array}$$

This diagram is commutative:

$$\begin{aligned} [\overline{m}(Y)\overline{(\varepsilon\eta) \times 1_X}(Y)\bar{\Delta}(Y)](u) &= [\overline{m}(Y)\overline{(\varepsilon\eta) \times 1_X}(Y)](u, u) \\ &= \overline{m}(Y)(\omega, u) = \omega u = u \end{aligned}$$

for any $u \in \mathrm{Hom}_{\mathscr{C}}(Y, X)$.

Again by corollary 1.7 we deduce that (2) is commutative.

Similarly, diagram (3) induces a diagram

$$\begin{array}{ccc} \mathrm{Hom}_{\mathscr{C}}(Y, X) \times \mathrm{Hom}_{\mathscr{C}}(Y, X) & \xrightarrow{\overline{a \times 1_X}(Y)} & \mathrm{Hom}_{\mathscr{C}}(Y, X) \times \mathrm{Hom}_{\mathscr{C}}(Y, X) \\ \uparrow{\scriptstyle \bar{\Delta}(Y)} & & \downarrow{\scriptstyle \overline{m}(Y)} \\ \mathrm{Hom}_{\mathscr{C}}(Y, X) & \xrightarrow[\overline{\varepsilon\eta}(Y)]{} & \mathrm{Hom}_{\mathscr{C}}(Y, X) \end{array}$$

This diagram is commutative, because

$$[\overline{m}(Y)\overline{a \times 1_X}(Y)\bar{\Delta}(Y)](u) = [\overline{m}(Y)\overline{a \times 1_X}(Y)](u, u) = \overline{m}(Y)(u^{-1}, u) = \omega$$

for any $u \in \mathrm{Hom}_{\mathscr{C}}(Y, X)$, which, by corollary 1.7, implies that (3) is commutative.

Conversely, assume we are given three morphisms m, a, ε such that diagrams (1), (2), (3) are commutative. For each object Y of $\mathscr{C}$ we define a multiplication on $\mathrm{Hom}_{\mathscr{C}}(Y, X)$, denoted by '$\perp$' as follows:

The morphism $m: X \times X \longrightarrow X$ induces a functorial morphism $\overline{m}: h_{X \times X} \longrightarrow h_X$. Thus we have a map

$$\mathrm{Hom}_{\mathscr{C}}(Y, X) \times \mathrm{Hom}_{\mathscr{C}}(Y, X) \approx \mathrm{Hom}_{\mathscr{C}}(Y, X \times X) \xrightarrow{\overline{m}(Y)} \mathrm{Hom}_{\mathscr{C}}(Y, X).$$

We set by definition $u_1 \perp u_2 = [\overline{m}(Y)](u_1, u_2) (u_1, u_2 \in \mathrm{Hom}_{\mathscr{C}}(Y, X))$. Clearly $u_1 \perp u_2 = mv$, where $v: Y \longrightarrow X \times X$ is the unique morphism such that $p_1 v = u_1, p_2 v = u_2, p_1, p_2: X \times X \longrightarrow X$ being the canonical projections. It is easy to verify that for any morphism u in $\mathscr{C}$, $h_X(u)$ preserves the multiplication just defined.

We now prove the associativity of the multiplication '$\perp$'. Let $u_1, u_2, u_3 \in \mathrm{Hom}_{\mathscr{C}}(Y, X)$. Then, by the above observation,

$$(u_1 \perp u_2) \perp u_3 = mw,$$

where $w: Y \longrightarrow X \times X$ is the unique morphism such that $p_1 w = mv$, $p_2 w = u_3$, where in turn $v: Y \longrightarrow X \times X$ is the unique morphism such that $p_1 v = u_1$, $p_2 v = u_2$.

Let $s: Y \longrightarrow X \times X \times X$ be the unique morphism such that $q_i s = u_i (i = 1, 2, 3)$, where $q_1, q_2, q_3 : X \times X \times X \longrightarrow X$ are the canonical projections.

We assert that $w = (m \times 1_X)s$. For, consider the commutative diagram

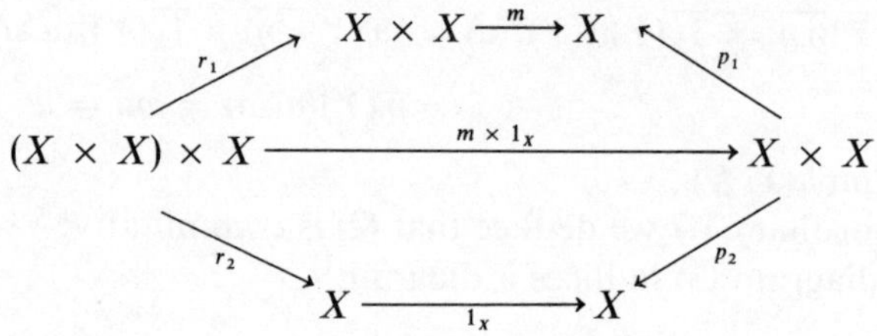

where r_1, r_2 are canonical projections. We may write:

$$p_1(m \times 1_X)s = mr_1 s, \qquad p_2(m \times 1_X)s = r_2 s = q_3 s = u_3,$$

and

$$p_1(r_1 s) = q_1 s = u_1, \qquad p_2(r_1 s) = q_2 s = u_2.$$

Quite analogously it can be shown that

$$u_1 \perp (u_2 \perp u_3) = m(1_X \times m)s,$$

so that, in view of the commutativity of diagram (1)

$$(u_1 \perp u_2) \perp u_3 = u_1 \perp (u_2 \perp u_3).$$

To verify the existence of the neutral element for the multiplication '$\perp$', let ζ be the unique element of the set $\mathrm{Hom}_{\mathscr{C}}(Y, e)$. We wish to show that $\varepsilon\zeta$ is the neutral element of $\mathrm{Hom}_{\mathscr{C}}(Y, X)$. Thus, we must show that

$$(\varepsilon\zeta) \perp u = u \perp (\varepsilon\zeta) = u \quad \text{for any} \quad u \in \mathrm{Hom}_{\mathscr{C}}(Y, X). \tag{4}$$

By definition, $(\varepsilon\zeta) \perp u = mv$, where $v: Y \longrightarrow X \times X$ is the unique morphism such that $p_1 v = \varepsilon\zeta$, $p_2 v = u$. We claim that $v = ((\varepsilon\eta) \times 1_X)\Delta u$. Indeed, by using the commutativity of the diagrams

$$\begin{array}{ccc} X \times X & \xrightarrow{(\varepsilon\eta) \times 1_X} & X \times X \\ {\scriptstyle p_1}\downarrow & & \downarrow{\scriptstyle p_1} \\ X & \xrightarrow{\varepsilon\eta} & X \end{array} \qquad \begin{array}{ccc} X \times X & \xrightarrow{(\varepsilon\eta) \times 1_X} & X \times X \\ {\scriptstyle p_2}\downarrow & & \downarrow{\scriptstyle p_2} \\ X & \xrightarrow{1_X} & X \end{array}$$

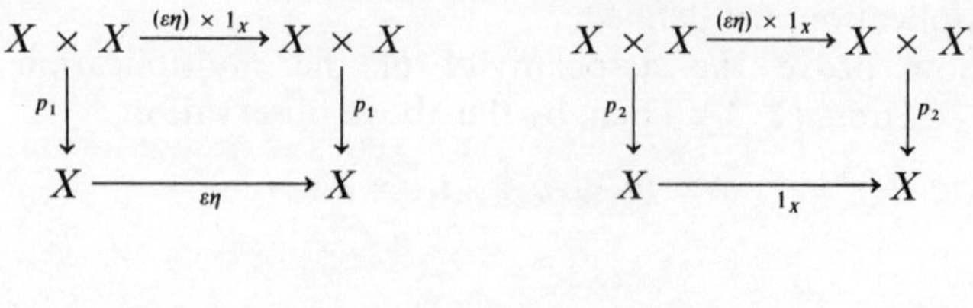

we obtain:

$$p_1((\varepsilon\eta) \times 1_X)\Delta u = \varepsilon\eta p_1 \Delta u = \varepsilon\eta u = \varepsilon\zeta$$

$$p_2((\varepsilon\eta) \times 1_X)\Delta u = 1_X p_2 \Delta u = u.$$

Hence, (4) follows from the commutativity of diagram (2).

Finally, we prove the existence of the inverse for the multiplication '$\perp$' by showing that

$$(au) \perp u = u \perp (au) = \varepsilon\zeta \qquad \text{for any } u \in \mathrm{Hom}_{\mathscr{C}}(Y, X). \qquad (5)$$

By definition, $(au) \perp u = mw$, where $w: Y \longrightarrow X \times X$ is the unique morphism such that $p_1 w = au$, $p_2 w = u$. We claim that $w = (a \times 1_X)\Delta u$. Indeed, by using the commutativity of the diagrams

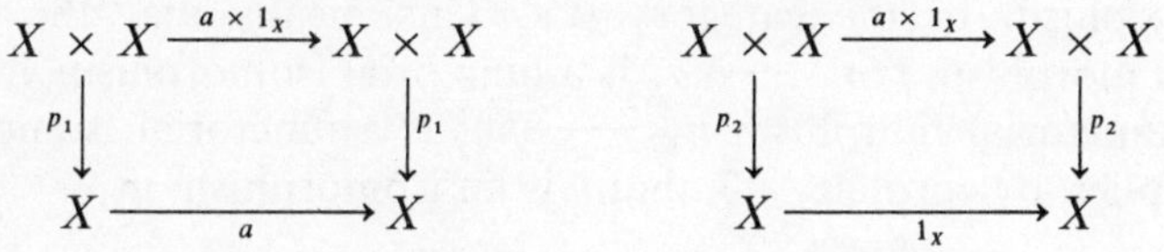

we obtain

$$p_1(a \times 1_X)\Delta u = a p_1 \Delta u = au$$

$$p_2(a \times 1_X)\Delta u = 1_X p_2 \Delta u = u.$$

Hence, (5) follows from the commutativity of diagram (3). This completes the proof of the theorem.

Remark. It is not hard to see that the group structures which can be defined on X in $\mathscr{C}$ and the triples (m, a, ε) are in a one-to-one correspondence via the assignments we have defined in the proof.

Let now $\mathscr{C}$ be an arbitrary category. We define the category of the groups of $\mathscr{C}$, denoted by $\mathscr{G}\mathscr{C}$ as follows:

(i) The objects of $\mathscr{G}\mathscr{C}$ are the couples (X, σ), where X is an object of $\mathscr{C}$ and σ is a functorial morphism such that diagram (D) above is commutative.

(ii) The morphisms from (X_1, σ_1) to (X_2, σ_2) are simply the functorial morphisms from σ_1 to σ_2.

A morphism from (X_1, σ_1) to (X_2, σ_2) in $\mathscr{G}\mathscr{C}$ is also called a *homomorphism* from X_1 to X_2.

By using corollary 1.7, it is easy to see that the homomorphisms from X_1 to X_2 are in one-to-one correspondence with the morphisms

$u: X_1 \longrightarrow X_2$ in $\mathscr{C}$ such that the diagram

$$\begin{array}{ccc} X_1 \times X_1 & \xrightarrow{m_1} & X_1 \\ {\scriptstyle u \times u}\downarrow & & \downarrow{\scriptstyle u} \\ X_2 \times X_2 & \xrightarrow[m_2]{} & X_2 \end{array}$$

is commutative, where m_1, m_2 are the morphisms attached to the group structures of X_1, X_2 in $\mathscr{C}$ as in theorem 4.1.

PROPOSITION 4.2. *Let $u: X_1 \longrightarrow X_2$ be a homomorphism. In order that u be an isomorphism in the category $\mathscr{G}\mathscr{C}$ it is necessary and sufficient that u be an isomorphism in the category $\mathscr{C}$.*

PROOF. The sufficiency is obvious. To prove the necessity suppose u is an isomorphism in the category $\mathscr{G}\mathscr{C}$. This means that the induced functorial morphism $f: \sigma_1 \longrightarrow \sigma_2$ is a functorial isomorphism, hence the induced functorial morphism $h_{X_1} \longrightarrow h_{X_2}$ is a functorial isomorphism, which implies, by corollary 1.7, that u is an isomorphism in $\mathscr{C}$.

PROPOSITION 4.3. *If the category $\mathscr{C}$ has direct products then so does the category $\mathscr{G}\mathscr{C}$.*

PROOF. Let X_1 and X_2 be two groups in the category $\mathscr{C}$. Now, $X_1 \times X_2$ is also a group, since

$$\mathrm{Hom}_{\mathscr{C}}(Y, X_1 \times X_2) = \mathrm{Hom}_{\mathscr{C}}(Y, X_1) \times \mathrm{Hom}_{\mathscr{C}}(Y, X_2)$$

for any object Y of $\mathscr{C}$, and the multiplication is defined componentwise. Let $p_i: X_1 \times X_2 \longrightarrow X_i (i = 1, 2)$ be the canonical projections in $\mathscr{C}$. We assert that $X_1 \times X_2$ together with the morphisms p_1, p_2 is a direct product of X_1 and X_2 in $\mathscr{G}\mathscr{C}$. Indeed, first of all, p_1 and p_2 are homomorphisms, since the diagram

$$\begin{array}{ccc} \begin{array}{c}[\mathrm{Hom}_{\mathscr{C}}(Y, X_1) \times \mathrm{Hom}_{\mathscr{C}}(Y, X_2)] \\ \times\, [\mathrm{Hom}_{\mathscr{C}}(Y, X_1) \times \mathrm{Hom}_{\mathscr{C}}(Y, X_2)]\end{array} & \xrightarrow{\alpha} & \begin{array}{c}\mathrm{Hom}_{\mathscr{C}}(Y, X_1) \\ \times\, \mathrm{Hom}_{\mathscr{C}}(Y, X_2)\end{array} \\ {\scriptstyle \bar{p}_i(Y) \times \bar{p}_i(Y)}\downarrow & & \downarrow{\scriptstyle \bar{p}_i(Y)} \\ \mathrm{Hom}_{\mathscr{C}}(Y, X_i) \times \mathrm{Hom}_{\mathscr{C}}(Y, X_i) & \xrightarrow[\mu_i]{} & \mathrm{Hom}_{\mathscr{C}}(Y, X_i) \end{array}$$

is commutative for $i = 1, 2$, where α is the multiplication of the group structure on $h_{X_1 \times X_2}(Y)$ and is given by $\alpha((u_1, u_2), (v_1, v_2)) = (u_1 v_1, u_2 v_2)$, μ_i is the multiplication on $h_{X_i}(Y)$, and $\bar{p}_i: h_{X_1 \times X_2} \longrightarrow h_{X_i}$ is the functorial morphism induced by $p_i (i = 1, 2)$.

Now, let $s_i : Z \longrightarrow X_i$ $(i = 1, 2)$ be two homomorphisms where Z is a group in $\mathscr{C}$. Then there exists a unique morphism $\gamma : Z \longrightarrow X_1 \times X_2$ in $\mathscr{C}$ such that the diagram

$$\begin{array}{ccccc} & & X_1 & & \\ & \overset{s_1}{\nearrow} & & \overset{p_1}{\nwarrow} & \\ Z & & \xrightarrow{\quad\gamma\quad} & & X_1 \times X_2 \\ & \underset{s_2}{\searrow} & & \underset{p_2}{\swarrow} & \\ & & X_2 & & \end{array}$$

is commutative. We have still to show that γ is a homomorphism. This follows from the fact that there exists a unique map τ in the category $\mathscr{Ens}$ such that the diagram

$$\begin{array}{ccccc} & & \mathrm{Hom}_{\mathscr{C}}(Y, X_1) & & \\ & \overset{\bar{s}_1(Y)}{\nearrow} & & \overset{\bar{p}_1(Y)}{\nwarrow} & \\ \mathrm{Hom}_{\mathscr{C}}(Y, Z) & & \xrightarrow{\quad\tau\quad} & & \mathrm{Hom}_{\mathscr{C}}(Y, X_1) \times \mathrm{Hom}_{\mathscr{C}}(Y, X_2) \\ & \underset{\bar{s}_2(Y)}{\searrow} & & \underset{\bar{p}_2(Y)}{\swarrow} & \\ & & \mathrm{Hom}_{\mathscr{C}}(Y, X_2) & & \end{array}$$

is commutative, where $\bar{s}_i : h_Z \longrightarrow h_{X_i}$ is the functorial morphism induced by s_i $(i = 1, 2)$. Then clearly we must have $\bar{\gamma}(Y) = \tau$, where $\bar{\gamma} : h_Z \longrightarrow h_{X_1 \times X_2}$ is the functorial morphism induced by γ. But considering the above diagram in the category $\mathscr{G}$, we see that τ is a homomorphism. This completes the proof of the proposition.

2. The Existence of Kernels for Homomorphisms

Let $\mathscr{C}$ be an arbitrary category and let $f : X \longrightarrow Y$ be a homomorphism, where X and Y are groups in $\mathscr{C}$. The morphism f induces a functorial morphism $\bar{f} : h_X \longrightarrow h_Y$, and we have for each object Z of $\mathscr{C}$ a group homomorphism $\bar{f}(Z) : \mathrm{Hom}_{\mathscr{C}}(Z, X) \longrightarrow \mathrm{Hom}_{\mathscr{C}}(Z, Y)$.

We define a functor $k : \mathscr{C} \longrightarrow \mathscr{G}$ as follows: for each object Z of $\mathscr{C}$, let $k(Z)$ be the set of all morphisms $u : Z \longrightarrow X$ with the property that $[\bar{f}(Z)](u) = \omega$, where ω is the neutral element of $\mathrm{Hom}_{\mathscr{C}}(Z, Y)$, i.e. $k(Z)$ is the kernel of $\bar{f}(Z)$ in the group-theoretical sense.

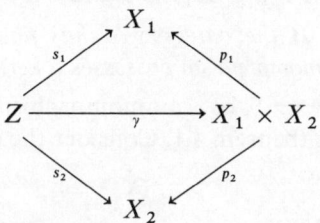

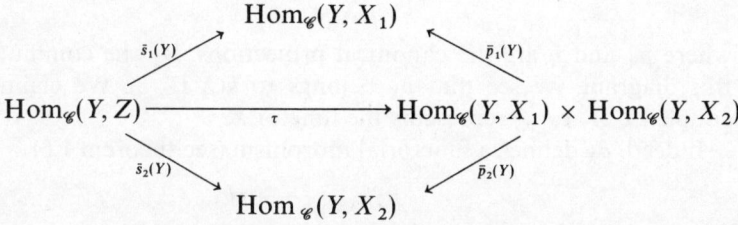

Definition. We say that the homomorphism f *possesses a kernel*, if the functor k is representable. If (K, ξ) represents the functor k, then we say that K together with the morphism $\xi : K \longrightarrow X$ is a *kernel* of f.

It is obvious that K is a group in $\mathscr{C}$ and ξ is a homomorphism.

PROPOSITION 4.4. *If the category $\mathscr{C}$ has fibered products and a final object, then every homomorphism possesses a kernel.*

PROOF. Let $f : X \longrightarrow Y$ be a homomorphism, e the final object of $\mathscr{C}$ and $\varepsilon : e \longrightarrow Y$ as in theorem 4.1. Consider the diagram

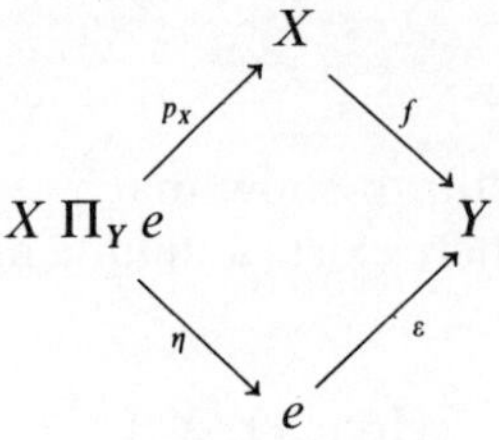

where p_X and η are the canonical projections. By the commutativity of this diagram, we see that p_X belongs to $k(X \Pi_Y e)$. We claim that the couple $(X \Pi_Y e, p_X)$ represents the functor k.

Indeed, p_X defines a functorial morphism (see theorem 1.6)

$$\beta : h_{X \Pi_Y e} \longrightarrow k$$

given by $[\beta(Z)](\zeta) = p_X \zeta$, for any $\zeta : Z \longrightarrow X \Pi_Y e$. Let the morphisms $\zeta_1, \zeta_2 : Z \longrightarrow X \Pi_Y e$ be such that $p_X \zeta_1 = p_X \zeta_2$. We then have the commutative diagram

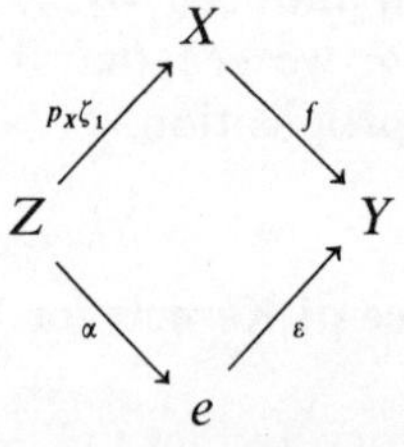

where α is the unique element of $\mathrm{Hom}_{\mathscr{C}}(Z, e)$. This implies the existence of a unique morphism $\zeta : Z \longrightarrow X \Pi_Y e$ such that $p_X \zeta = p_X \zeta_1$, whence $\zeta_1 = \zeta_2$ and $\beta(Z)$ is a monomorphism. The fact that $\beta(Z)$ is an epimorphism is immediate. Thus β is a functorial isomorphism, i.e. f possesses a kernel.

We wish now to relate the notion of a kernel of a homomorphism just introduced to that of a kernel of a pair of morphisms in the sense of chapter 2.

Let $f: X \longrightarrow Y$ be a homomorphism and assume $K \xrightarrow{\xi} X$ to be the kernel of f. Let η be the unique element of $\mathrm{Hom}_{\mathscr{C}}(X, e)$ and let ε be the neutral element of the group $\mathrm{Hom}_{\mathscr{C}}(e, Y)$. Then

PROPOSITION 4.5. *The object K together with the morphism ξ is a kernel of the pair of morphisms $(f, \varepsilon\eta)$ both in the category $\mathscr{C}$ and in the category $\mathscr{G}\mathscr{C}$.*

PROOF. Let $\zeta: Z \longrightarrow X$ be a morphism in the category $\mathscr{C}$ such that $f\zeta = \varepsilon\eta\zeta$. Then clearly ζ belongs actually to $k(Z)$ so that, k being represented by the couple (K, ξ), there exists a unique morphism $\alpha: Z \longrightarrow K$ in $\mathscr{C}$ such that $\zeta = \xi\alpha$. Since on the other hand it is straightforward to prove that $f\xi = \varepsilon\eta\xi$, this means that K together with ξ is a kernel of the pair of morphisms $(f, \varepsilon\eta)$ in $\mathscr{C}$.

Assume now that $\zeta: Z \longrightarrow X$ is a morphism in $\mathscr{G}\mathscr{C}$, i.e. a homomorphism. By the previous argument, there exists a unique morphism $\alpha: Z \longrightarrow K$ in $\mathscr{C}$, such that $\zeta = \xi\alpha$. It remains to show that α is a homomorphism. For each object T of $\mathscr{C}$, we have the commutative diagram of groups

$$\begin{array}{ccc} \mathrm{Hom}_{\mathscr{C}}(T, K) & \xrightarrow{\bar{\xi}(T)} & \mathrm{Hom}_{\mathscr{C}}(T, X) \\ {\scriptstyle \bar{\alpha}(T)} \nwarrow & & \nearrow {\scriptstyle \bar{\xi}(T)} \\ & \mathrm{Hom}_{\mathscr{C}}(T, Z) & \end{array}$$

where $\bar{\xi}(T)$ is a monomorphism. Since $\bar{\zeta}(T)$ is a homomorphism, it is immediate that $\bar{\alpha}(T)$ is also a homomorphism.

3. Equivalence Relations

As known, an equivalence relation on a set M is a subset R of the cartesian product $M \times M$ such that:

1. $(x, x) \in R$ for any $x \in M$.
2. If $(x, y) \in R$, then $(y, x) \in R$.
3. If $(x, y) \in R$ and $(y, z) \in R$, then $(x, z) \in R$.

We denote by (M, R) the set M together with the structure given by the equivalence relation R.

Let $\mathscr{R}$ be the category described as follows:

(i) the objects of $\mathscr{R}$ are all couples (M, R), where M is a set and R is an equivalence relation on M.

(ii) a morphism from (M, R) to (N, S) is a map $u: M \longrightarrow N$ such that $(x, y) \in R$ implies $(u(x), u(y)) \in S$.

Now, we have two functors:

$$C:\mathscr{R} \longrightarrow \mathscr{E}ns \quad \text{defined by} \quad C(M, R) = M,$$

$$Q:\mathscr{R} \longrightarrow \mathscr{E}ns \quad \text{defined by} \quad Q(M, R) = M/R,$$

where M/R is the set of equivalence classes with respect to the equivalence relation R.

We also have a functorial morphism $\varphi: C \longrightarrow Q$ defined as follows: for each object (M, R) of $\mathscr{R}$, $\varphi(M, R): M \longrightarrow M/R$ is the canonical map which assigns to each element $x \in M$ its equivalence class.

Definition. Let $\mathscr{C}$ be an arbitrary category and X an object of $\mathscr{C}$. We say that an equivalence relation has been defined on the object X if a functor $R:\mathscr{C} \longrightarrow \mathscr{R}$ has been given such that the diagram

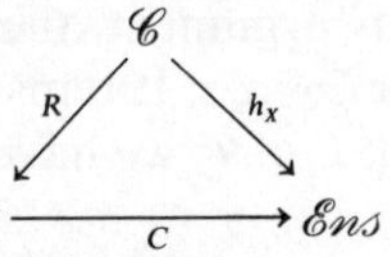

is commutative.

This is clearly equivalent to the fact that an equivalence relation $\rho(Y)$ on the set $\mathrm{Hom}_{\mathscr{C}}(Y, X)$ has been given for each object Y of $\mathscr{C}$, such that, for any morphism $u: Y \longrightarrow Z$ in $\mathscr{C}$, the relation $(v, w) \in \rho(Y)$ implies $(uv, uw) \in \rho(Z)$.

4. The General Notion of a Structure on the Objects of a Category

The notions introduced above can be generalized as follows, according to an idea due to Ehresmann:

Consider the category Bij $\mathscr{E}ns$ defined as follows:

(i) The objects of Bij $\mathscr{E}ns$ are all the sets.

(ii) If M and N are two arbitrary sets, then the set of morphisms from M to N in the category Bij $\mathscr{E}ns$ is identical by definition with the set of all bijections (in the category $\mathscr{E}ns$) from M to N. We take as composition of morphisms the usual composition of maps.

It follows in particular that every morphism is invertible in the category Bij $\mathscr{E}ns$.

Definition. A covariant functor

$$F: \text{Bij } \mathscr{E}ns \longrightarrow \text{Bij } \mathscr{E}ns$$

is said to be a *species of structure.*

Examples. 1. Let M be a set and let $G(M)$ be the set of maps $M \times M \longrightarrow M$ which define on M integral algebraic operations which are associative, with a unity element, and where each element is invertible. This clearly gives rise to a functor $G: \text{Bij } \mathcal{E}ns \longrightarrow \text{Bij } \mathcal{E}ns$, the species of structure of group.

2. Let $\mathscr{P}(M)$ denote the set of all subsets of the set M. If we denote by $G(M)$ the set of all subsets $\mathscr{D}$ of $\mathscr{P}(M)$ which satisfy the conditions

(a) $\varnothing \in \mathscr{D}$, $M \in \mathscr{D}$;

(b) $D_1, D_2 \in \mathscr{D}$ implies $D_1 \cap D_2 \in \mathscr{D}$;

(c) $D_i \in \mathscr{D}$ for any $i \in I$ implies $\bigcup_{i \in I} D_i \in \mathscr{D}$, we obtain a functor $G: \text{Bij } \mathcal{E}ns \longrightarrow \text{Bij } \mathcal{E}ns$, the species of structure of topological space.

If F is a species of structure and $s \in F(M)$, then s is said to be a structure of species F on the set M. If $f: M \longrightarrow N$ is a morphism in Bij $\mathcal{E}ns$, i.e. a bijection, then $t = (F(f))(s)$ is a structure of species F on N which is called the structure of species F obtained by transporting to N the structure s by means of f. We also say in this case that f realizes an isomorphism between the set M provided with the structure s and the set N provided with the structure t.

We say that a species of structure with morphisms (F, σ) has been given if the following are given:

(a) A species of structure $F: \text{Bij } \mathcal{E}ns \longrightarrow \text{Bij } \mathcal{E}ns$.

(b) For every quadruple (M, s, N, t) where $s \in F(M)$, $t \in F(N)$, a set $\sigma(M, s, N, t)$ of maps from the set M to the set N such that the following conditions are satisfied:

1. $f \in \sigma(M, s, N, t)$, $g \in \sigma(N, t, R, u)$ imply $gf \in \sigma(M, s, R, u)$.

2. if $f: M \longrightarrow N$ realizes an isomorphism between the set M provided with the structure s and the set N provided with the structure t then $f \in \sigma(M, s, N, t)$ and $f^{-1} \in \sigma(N, t, M, s)$.

To any species of structure with morphisms we can obviously associate a category $\mathscr{K}$ as follows:

(i) An object of the category $\mathscr{K}$ is a pair (M, s) where M is a set and s a structure of the considered species on M.

(ii) $\text{Hom}_{\mathscr{K}}((M, s), (N, t)) = \sigma(M, s, N, t)$.

(iii) The composition of morphisms is the usual composition of maps.

There obviously exists a canonical functor

$$C: \mathscr{K} \longrightarrow \mathcal{E}ns$$

which ignores the considered structure.

Assume a species of structure with morphisms (F, σ) is given and let $\mathscr{C}$ be a category and X an object of $\mathscr{C}$. We denote as above by $\mathscr{K}$ the category associated to (F, σ) and $C : \mathscr{K} \longrightarrow \mathscr{E}ns$ the canonical functor.

We say that a structure of species (F, σ) has been defined on the object X if a contravariant functor $S : \mathscr{C} \longrightarrow \mathscr{K}$ has been given such that the diagram of categories and functors

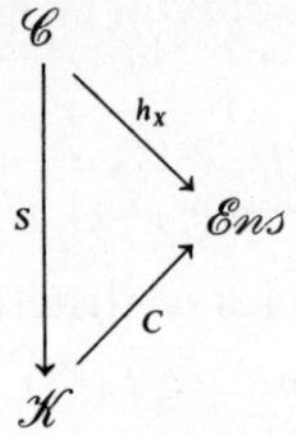

is commutative.

Let X and Y be two objects of the category $\mathscr{C}$, each provided with a structure of the same species. Assume the respective structures are defined by the functors $S, T : \mathscr{C} \longrightarrow \mathscr{K}$. Under these conditions a functorial morphism $\varphi : S \longrightarrow T$ is simply said to be a homomorphism from the object X provided with the structure S to the object Y provided with the structure T.

CHAPTER 5

General Theory of Abelian Categories

1. Additive Categories

A category $\mathscr{C}$ is said to be *additive* if the following conditions are satisfied:

1. For any pair of objects (A, B) in $\mathscr{C}$, there exist the direct product $A \amalg B$ and the direct sum $A \coprod B$.

2. For any pair of objects (A, B) in $\mathscr{C}$, $\mathrm{Hom}_{\mathscr{C}}(A, B)$ has an Abelian group structure such that the composition map

$$\mathrm{Hom}_{\mathscr{C}}(A, B) \times \mathrm{Hom}_{\mathscr{C}}(B, C) \longrightarrow \mathrm{Hom}_{\mathscr{C}}(A, C)$$

is bilinear (when there will be no possibility of confusion we shall denote simply by 0 the neutral element of the group $\mathrm{Hom}_{\mathscr{C}}(A, B)$).

3. There exists in the category $\mathscr{C}$ an object A such that $1_A = 0$.

Plainly every additive category satisfies condition CAd 1 on p. 41.

It is also evident that if A and B are such that $1_A = 0, 1_B = 0$, then A and B are canonically isomorphic. We shall denote by 0 one of the objects of $\mathscr{C}$ which satisfy condition 3.

PROPOSITION 5.1. *Let A, B, C be three objects of an additive category $\mathscr{C}$ and $i_A : A \longrightarrow C$, $i_B : B \longrightarrow C$ morphisms in $\mathscr{C}$. The necessary and sufficient condition that C together with the morphisms i_A, i_B be the direct sum of the family of objects (A, B) is that there exist two morphisms $p_A : C \longrightarrow A$, $p_B : C \longrightarrow B$ such that the following conditions are satisfied:*

(i) $p_A i_A = 1_A$, $\quad p_B i_A = 0_{BA}$

(ii) $p_B i_B = 1_B$, $\quad p_A i_B = 0_{AB}$

(iii) $i_A p_A + i_B p_B = 1_C$.

PROOF. Necessity. The definition of direct sums yields the morphisms p_A, p_B such that the diagrams

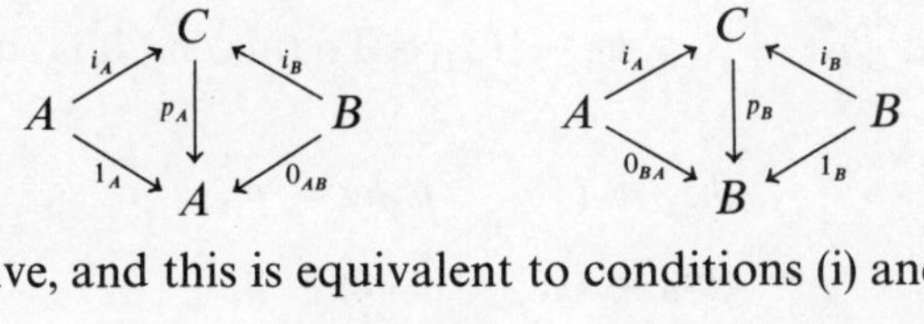

are commutative, and this is equivalent to conditions (i) and (ii).

Also from the definition of direct sums it follows that the map

$$\mathrm{Hom}_{\mathscr{C}}(C, X) \longrightarrow \mathrm{Hom}_{\mathscr{C}}(A, X) \times \mathrm{Hom}_{\mathscr{C}}(B, X)$$

which associates to each morphism $u: C \longrightarrow X$ the pair (ui_A, ui_B) is a bijection. If we take $X = C$ and $u = 1_C$, then it results that the pair associated by this map to the morphism 1_C is (i_A, i_B). It is therefore sufficient to show that the same pair is associated to the morphism $i_A p_A + i_B p_B$.

We have, by virtue of relations (i) and (ii) already proved:

$$(i_A p_A + i_B p_B) i_A = i_A$$

$$(i_A p_A + i_B p_B) i_B = i_B.$$

Sufficiency. Let $u: A \longrightarrow X$, $v: B \longrightarrow X$. If we set $\xi: C \longrightarrow X$, where $\xi = v p_B + u p_A$, then the diagram

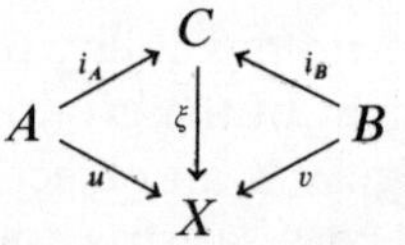

is commutative, by conditions (i) and (ii). If $\xi': C \longrightarrow X$ is also such that this diagram is commutative, then we have $\xi' i_A = u$, $\xi' i_B = v$ and hence

$$\xi' i_A p_A + \xi' i_B p_B = u p_A + v p_B,$$

which, according to (iii), implies $\xi' = \xi$.

PROPOSITION 5.2. *Let $\mathscr{C}$ be an additive category and A, B two objects of $\mathscr{C}$. Then the morphism $h(A, B): A \amalg B \longrightarrow A \Pi B$ constructed in proposition 2.6 is an isomorphism.*

PROOF. To avoid confusions, we denote by p'_A, p'_B the canonical projections of the product $A \Pi B$ and by p_A, p_B the morphisms in proposition 5.1. We shall also abbreviate $h(A, B)$ to h.

Let $g: A \Pi B \longrightarrow A \amalg B$ be defined by

$$g = i_A p'_A + i_B p'_B.$$

We show that $gh = 1_{A \amalg B}$, $hg = 1_{A \Pi B}$. To this end it is enough to show that

$$ghi_A = i_A, \qquad p_A hg = p_A,$$

$$ghi_B = i_B, \qquad p_B hg = p_B.$$

But we have, according to proposition 2.6

$$ghi_A = (i_A p'_A + i_B p'_B)hi_A = i_A.$$

The other three relations are proved analogously.

Definition. Let $\mathscr{C}, \mathscr{C}'$ be two additive categories and $F: \mathscr{C} \longrightarrow \mathscr{C}'$ a functor. F is said to be *additive* if for any pair of morphisms $u, v \in \operatorname{Hom}_{\mathscr{C}}(A, B)$ we have

$$F(u + v) = F(u) + F(v).$$

Unless otherwise stated, whenever we will consider functors defined on additive categories and taking their values in additive categories, we will assume them to be additive.

PROPOSITION 5.3. *If the covariant functor $F: \mathscr{C} \longrightarrow \mathscr{C}'$ is additive and if the object C together with the morphisms $i_A: A \longrightarrow C$, $i_B: B \longrightarrow C$ is a direct sum for the family (A, B), then the object $F(C)$ together with the morphisms $F(i_A): F(A) \longrightarrow F(C)$, $F(i_B): F(B) \longrightarrow F(C)$ is a direct sum for the family $(F(A), F(B))$.*

PROOF. Let $p_A: C \longrightarrow A$, $p_B: C \longrightarrow B$ be the canonical projections (see proposition 5.1). Then $F(i_A)$, $F(i_B)$, $F(p_A)$, $F(p_B)$ satisfy the conditions in proposition 5.1, in view of the additivity of F, whence our proposition.

2. Kernel and Cokernel

Let $\mathscr{C}$ be an additive category and $u_1, u_2: A \longrightarrow B$ a pair of morphisms in $\mathscr{C}$. If (N, v) is a kernel of this pair of morphisms, then (N, v) is also a kernel for the pair of morphisms $(u_1 - u_2, 0_{BA})$. Conversely, if (N, v) is a kernel for the pair of morphisms $(u_1 - u_2, 0_{BA})$, then (N, v) is also a kernel for the pair of morphisms (u_1, u_2).

Obviously, a similar statement holds for cokernels. We conclude that in the case of additive categories, we may confine ourselves in the study of the notions of kernel and cokernel to the pairs of the form $(u, 0)$, which justifies the following.

Definition. Let $K \xrightarrow{i} A \xrightarrow{u} B$ be two morphisms in the category $\mathscr{C}$. We say that the pair (K, i) is a *kernel for the morphism* u if it is a kernel for the pair of morphisms $(u, 0_{BA})$ or, equivalently, if the following sequence of abelian groups and homomorphisms of Abelian groups:

$$0 \longrightarrow \operatorname{Hom}_{\mathscr{C}}(X, K) \xrightarrow{\alpha} \operatorname{Hom}_{\mathscr{C}}(X, A) \xrightarrow{\beta} \operatorname{Hom}_{\mathscr{C}}(X, B) \qquad (1)$$

is exact for any object X of $\mathscr{C}$. In the sequence (1) the homomorphisms α, β are defined by

$$\alpha(\xi) = i\xi, \qquad \beta(\eta) = u\eta.$$

Clearly, if they exist, the various kernels of the morphism u define the same subobject of A, which we denote by ker u and call the kernel of the morphism u.

Dually, if $A \xrightarrow{u} B \xrightarrow{j} L$ are morphisms in the category $\mathscr{C}$, the pair (L, j) is said to be a cokernel for the morphism u if (L, j) is a cokernel for the pair $(u, 0)$ or, equivalently, if the sequence of abelian groups and homomorphisms

$$0 \longrightarrow \mathrm{Hom}_{\mathscr{C}}(L, X) \xrightarrow{\varphi} \mathrm{Hom}_{\mathscr{C}}(B, X) \xrightarrow{\psi} \mathrm{Hom}_{\mathscr{C}}(A, X) \qquad (2)$$

is exact. In the sequence (2) the homomorphisms φ, ψ are defined by

$$\varphi(\xi) = \xi j, \qquad \psi(\eta) = \eta u.$$

Clearly, if they exist, the various cokernels of u define the same quotient object of B which we denote by coker u and call the cokernel of the morphism u.

Let now $u : A \longrightarrow B$ be a morphism in the additive category $\mathscr{C}$. Suppose the morphism u possesses at least one cokernel and let coker $u = (L, j)$. If it exists, ker j is called the *image* of the morphism u and is denoted by im u. Dually, if ker u exists and ker $u = (K, i)$, and if coker i exists, then the quotient object of A coker i is called the *coimage* of the morphism u and is denoted by coim u.

To formulate the next proposition we need the following general definition.

Definition. Let C, B be two objects of an arbitrary category $\mathscr{C}$. We say that B is a *direct summand* of C if there exists an object A of $\mathscr{C}$ and the morphisms $i_A : A \longrightarrow C$, $i_B : B \longrightarrow C$ such that C together with the morphisms i_A, i_B is a direct sum for the family of objects (A, B).

PROPOSITION 5.4. *Let C, B be two objects and $p : C \longrightarrow B$ a morphism in the additive category $\mathscr{C}$, and assume the following conditions are satisfied*:

(a) *the morphism p possesses at least one kernel,*

(b) *there exists a morphism $j : B \longrightarrow C$ such that $pj = 1_B$.*

Then B is a direct summand of C.

PROOF. Let (A, i) be a kernel for p. According to proposition 5.1, it is sufficient to prove that there exists a morphism $q : C \longrightarrow A$ such that $qi = 1_A$, $jp + iq = 1_C$, $qj = 0$.

To do this, consider the diagram

$$\begin{array}{ccccc} A & \xrightarrow{i} & C & \xrightarrow{p} & B. \\ & & \uparrow{\scriptstyle 1_C - jp} & & \\ & & C & & \end{array}$$

Since $p(1_C - jp) = 0$, the definition of a kernel yields a unique morphism $q : C \longrightarrow A$ such that $iq + jp = 1_C$. It remains only to show that $qi = 1_A$ and $qj = 0$. Since i is an injection, it is sufficient to show that $i(qj) = 0$ and $i(qi) = i$.

Indeed, we have

$$i(qj) = (iq)j = (1_C - jp)j = 0, \qquad i(qi) = (iq)i = (1_C - jp)i = i.$$

We state also the dual of proposition 5.4 which is often useful:

Let A, C be two objects and $i : A \longrightarrow C$ a morphism in the additive category $\mathscr{C}$ and assume the following conditions are satisfied:

(a) the morphism i possesses at least a cokernel;

(b) there exists a morphism $p : C \longrightarrow A$ such that $pi = 1_A$.

Then A is a direct summand of C.

PROPOSITION 5.5. *If $u : A \longrightarrow B$ is a morphism in the additive category $\mathscr{C}$, then the following assertions are equivalent:*

(a) *u is injective*

(b) *$0 \longrightarrow A$ is a kernel for u, in other words* $\ker u = 0$.

(c) *$1_A : A \longrightarrow A$ is a coimage for u, in other words* $\operatorname{coim} u = A$.

PROOF. (a) $\Rightarrow$ (b). If $\xi : X \longrightarrow A$ is such that $u\xi = 0$, then from the fact that u is injective it follows that $\xi = 0$, hence ξ factorizes uniquely through 0. Therefore $0 \longrightarrow A$ is a kernel for u.

(b) $\Rightarrow$ (c). For any $\xi : A \longrightarrow X$, we have $\xi 0 = 0$ and $\xi = \xi 1_A$. Hence $\operatorname{coker}(0 \longrightarrow A) = 1_A$.

(b) $\Rightarrow$ (a). If $\xi : X \longrightarrow A$ is such that $u\xi = 0$, it results that $\xi = 0$, hence u is injective.

(c) $\Rightarrow$ (b). Let $i : K \longrightarrow A$ be such that $\operatorname{coker} i = 1_A$. We have $1_A i = 0 = i$.

3. The Canonical Factorization of a Morphism

PROPOSITION 5.6. *Let $\mathscr{C}$ be an additive category and $u : A \longrightarrow B$ a morphism in $\mathscr{C}$. Assume that there exist* $\ker u$, $\operatorname{coker} u$, $\operatorname{im} u$, $\operatorname{coim} u$. *Then there exists a unique morphism $\bar{u} : \operatorname{coim} u \longrightarrow \operatorname{im} u$ such that u is the composition of the following morphisms:*

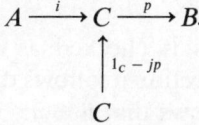

$A \xrightarrow{\lambda} \operatorname{coim} u \xrightarrow{\bar{u}} \operatorname{im} u \xrightarrow{\mu} B$, *where* λ *is the canonical surjection and* μ *the canonical injection.*

PROOF. The uniqueness is checked as follows: if $\mu\bar{u}\lambda = \mu\bar{\bar{u}}\lambda$, then from the fact that μ is an injection it follows that $\bar{u}\lambda = \bar{\bar{u}}\lambda$ and from the fact that λ is a surjection it follows that $\bar{u} = \bar{\bar{u}}$.

To prove the existence of $\bar{u}$, consider the following diagram

$$\begin{array}{ccccc} \ker u & \xrightarrow{i} & A & \xrightarrow{\lambda} & \operatorname{coker} i = \operatorname{coim} u \\ & & \Big\downarrow{\scriptstyle u} & & \\ \operatorname{coker} u & \xleftarrow[j]{} & B & \xleftarrow[\mu]{} & \ker j = \operatorname{im} u. \end{array}$$

Since $ui = 0$, there exists a morphism $u' : \operatorname{coim} u \longrightarrow B$ such that $u'\lambda = u$. But $ju' = 0$. Since λ is an epimorphism, it is enough to prove that $(ju')\lambda = 0$, which is immediate, since $j(u'\lambda) = ju = 0$.

Then there exists $\bar{u} : \operatorname{coim} u \longrightarrow \operatorname{im} u$ such that $\mu\bar{u} = u'$.

4. Pre-Abelian Categories

Definition. An additive category $\mathscr{C}$ is said to be *pre-Abelian* if for any morphism $u \in \operatorname{Hom}_{\mathscr{C}}(A, B)$ in $\mathscr{C}$ there exist $\ker u$ and $\operatorname{coker} u$.

It follows immediately that in a pre-Abelian category any morphism possesses an image and a coimage.

The dual of a pre-Abelian category is pre-Abelian.

In order to study the functorial properties of kernels, cokernels, images and coimages in a pre-Abelian category $\mathscr{C}$ we define a new category $\mathscr{C}_1$ as follows:

(i) the objects of $\mathscr{C}_1$ are ordered pairs (A, B) of objects of the category $\mathscr{C}$ together with a morphism $u : A \longrightarrow B$ in $\mathscr{C}$, in other words a diagram in $\mathscr{C}$ of the form $A \xrightarrow{u} B$.

(ii) a morphism from the diagram $A \xrightarrow{u} B$ to the diagram $A_1 \xrightarrow{u_1} B_1$ is by definition a pair of morphisms $\alpha : A \longrightarrow A_1$, $\beta : B \longrightarrow B_1$ such that the diagram

$$\begin{array}{ccc} A & \xrightarrow{u} & B \\ {\scriptstyle \alpha}\Big\downarrow & & \Big\downarrow{\scriptstyle \beta} \\ A_1 & \xrightarrow[u_1]{} & B_1 \end{array} \qquad (3)$$

is commutative.

We define a functor $\ker : \mathscr{C}_1 \longrightarrow \mathscr{C}_1$ as follows: To the diagram $A \xrightarrow{u} B$ we associate the diagram $\ker u \xrightarrow{i} A$, where i is the canonical injection. If the pair of morphisms (α, β) defines a morphism from the diagram $A \longrightarrow B$ to the diagram $A_1 \longrightarrow B_1$, then there exists a unique morphism $\ker u \longrightarrow \ker u_1$ such that the diagram

$$\begin{array}{ccc} \ker u & \xrightarrow{i} & A \\ \downarrow & & \downarrow{\scriptstyle\alpha} \\ \ker u_1 & \xrightarrow{i_1} & A_1 \end{array}$$

is commutative. In other words, we obtain a uniquely determined morphism from the diagram $\ker u \xrightarrow{i} A$ to the diagram $\ker u_1 \xrightarrow{i_1} A_1$.

One defines similarly coker, im, $\operatorname{coim} : \mathscr{C}_1 \longrightarrow \mathscr{C}_1$.

The canonical factorization of a morphism in a pre-Abelian category also leads us to a functor defined on the category $\mathscr{C}_1$, but taking its values in a more complicated category $\mathscr{C}_2$ which is defined as follows:

(i) the objects of the category $\mathscr{C}_2$ are diagrams in $\mathscr{C}$ of the form

$$A_1 \xrightarrow{u_1} A_2 \xrightarrow{u_2} A_3 \xrightarrow{u_3} A_4.$$

(ii) a morphism from the diagram $A_1 \xrightarrow{u_1} A_2 \xrightarrow{u_2} A_3 \xrightarrow{u_3} A_4$ to the diagram $B_1 \xrightarrow{v_1} B_2 \xrightarrow{v_2} B_3 \xrightarrow{v_3} B_4$ is, by definition, an ordered system $(\alpha_1, \alpha_2, \alpha_3, \alpha_4)$ of morphisms $\alpha_i : A_i \longrightarrow B_i$ $(i = 1, 2, 3, 4)$ such that the diagram

$$\begin{array}{ccccccc} A_1 & \xrightarrow{u_1} & A_2 & \xrightarrow{u_2} & A_3 & \xrightarrow{u_3} & A_4 \\ {\scriptstyle\alpha_1}\downarrow & & {\scriptstyle\alpha_2}\downarrow & & {\scriptstyle\alpha_3}\downarrow & & {\scriptstyle\alpha_4}\downarrow \\ B_1 & \xrightarrow[v_1]{} & B_2 & \xrightarrow[v_2]{} & B_3 & \xrightarrow[v_3]{} & B_4 \end{array}$$

is commutative.

We define a functor $F : \mathscr{C}_1 \longrightarrow \mathscr{C}_2$ as follows: To the diagram $A \xrightarrow{u} B$, we associate the diagram

$$A \xrightarrow{\lambda} \operatorname{coim} u \xrightarrow{\bar{u}} \operatorname{im} u \xrightarrow{\mu} B.$$

If $A_1 \xrightarrow{u_1} B_1$ is another diagram in $\mathscr{C}_1$ and $\alpha : A \longrightarrow A_1$, $\beta : B \longrightarrow B_1$ are such that diagram (3) is commutative, then there exist $\alpha_1 : \operatorname{coim} u \longrightarrow \operatorname{coim} u_1$, $\beta_1 : \operatorname{im} u \longrightarrow \operatorname{im} u_1$ such that the following diagrams are commutative:

$$\begin{array}{ccc} A & \xrightarrow{\lambda} & \operatorname{coim} u \\ {\scriptstyle\alpha}\downarrow & & \downarrow{\scriptstyle\alpha_1} \\ A_1 & \xrightarrow[\lambda_1]{} & \operatorname{coim} u_1 \end{array} \qquad \begin{array}{ccc} \operatorname{im} u & \xrightarrow{\mu} & B \\ {\scriptstyle\beta_1}\downarrow & & \downarrow{\scriptstyle\beta} \\ \operatorname{im} u_1 & \xrightarrow[\mu_1]{} & B_1. \end{array}$$

It remains to show that the diagram

$$\begin{array}{ccc} \operatorname{coim} u & \xrightarrow{\bar{u}} & \operatorname{im} u \\ {\scriptstyle \alpha_1}\downarrow & & \downarrow{\scriptstyle \beta_1} \\ \operatorname{coim} u_1 & \xrightarrow[\bar{u}_1]{} & \operatorname{im} u_1 \end{array}$$

is also commutative. We have

$$\mu_1\bar{u}_1\alpha_1\lambda = \mu_1\bar{u}_1\lambda_1\alpha = u_1\alpha$$

$$\mu_1\beta_1\bar{u}\lambda = \beta\mu\bar{u}\lambda = \beta u.$$

In other words, $\mu_1\bar{u}_1\alpha_1\lambda = \mu_1\beta_1\bar{u}\lambda$, whence $\bar{u}_1\alpha_1 = \beta_1\bar{u}$ because μ_1 is an injection and λ a surjection.

PROPOSITION 5.7. *In a pre-Abelian category there exist fibered products and sums.*

PROOF. Let $\mathscr{C}$ be a pre-Abelian category. Since the dual of a pre-Abelian category is pre-Abelian, we shall confine ourselves to prove the existence of fibered products. Consider the diagram

$$\begin{array}{ccc} B & & \\ & \searrow^{f} & \\ & & A. \\ & \nearrow_{g} & \\ C & & \end{array} \tag{4}$$

Since $\mathscr{C}$ is with direct sums, consider $B \oplus C$ and the canonical projections $p_B : B \oplus C \longrightarrow B$, $p_C : B \oplus C \longrightarrow C$. Let $\mu : B \oplus C \longrightarrow A$ be defined by

$$\mu = fp_B - gp_C,$$

and let (N, ν) be the kernel of μ. We assert that the object N together with the morphisms $p_B\nu$, $p_C\nu$ is a fibered product associated to the diagram (4).

The relation $\mu\nu = 0$ implies $f(p_B\nu) = g(p_C\nu)$. Moreover, if $\xi : X \longrightarrow B$, $\eta : X \longrightarrow C$ are such that $f\xi = g\eta$, then there exists a unique $\zeta : X \longrightarrow N$ such that $p_B\nu\zeta = \xi$, $p_C\nu\zeta = \eta$. For, consider the morphism $i_B\xi + i_C\eta : X \longrightarrow B \oplus C$. We have $\mu(i_B\xi + i_C\eta) = (fp_B - gp_C)(i_B\xi + i_C\eta) = f\xi - g\eta = 0$. It results that there exists a unique ζ such that $\nu\zeta = i_B\xi + i_C\eta$. Hence it follows immediately that $p_B\nu\zeta = \xi$, $p_C\nu\zeta = \eta$. The uniqueness follows from the fact that $p_B\nu\zeta = \xi$ and $p_C\nu\zeta = \eta$ imply $\nu\zeta = i_B\xi + i_C\eta$.

Examples of pre-Abelian categories

1. Consider the category $\mathscr{C}$ defined as follows:

(i) the objects of $\mathscr{C}$ are Hausdorff Abelian topological groups.

(ii) the morphisms of $\mathscr{C}$ are the continuous homomorphisms.

Obviously $\mathscr{C}$ is an additive category. Moreover, if $f: A \longrightarrow B$ is a morphism in the category $\mathscr{C}$, then $\ker f = f^{-1}(\{0\})$ and $\operatorname{coker} f = B/\overline{f(A)}$.

In this category there exist bijections which are not isomorphisms. Such are, for instance, the injective continuous homomorphisms $f: A \longrightarrow B$ such that $f(A) \neq B$ but $\overline{f(A)} = B$.

2. Let A be a ring. We say that a positive filtration has been defined on A, or that A is a filtered ring, if a sequence $A_0, A_1, A_2, \cdots, A_n, \cdots$ of ideals of A has been given such that the following conditions are fulfilled:

(a) $A_0 = A$.

(b) $A_n \supset A_{n+1}$ for every $n \geq 0$.

(c) $A_n \cdot A_m \subset A_{n+m}$.

For example, the powers of an ideal m of A define evidently a positive filtration on A.

Assume now a fixed filtration is given on A, defined by the sequence of ideals $(A_i)_{i \geq 0}$.

An A-module M is said to be a filtered module with respect to the filtered ring A if a sequence $(M_i)_{i \geq 0}$ of submodules of M has been given such that the following conditions are satisfied:

(a) $M_0 = M$.

(b) $M_n \supset M_{n+1}$.

(c) $A_p \cdot M_q \subset M_{p+q}$.

We now define a category $\mathscr{C}$ as follows:

(i) the objects of $\mathscr{C}$ are filtered modules with respect to the filtered ring A.

(ii) let M and N be two filtered modules with respect to A. Suppose the $(M_i)_{i \geq 0}$ and $(N_i)_{i \geq 0}$ define the filtration on M respectively on N. A morphism from M to N in the category $\mathscr{C}$ is a homomorphism from the A-module M to the A-module N with the further property that $f(M_n) \subset N_n$ for any $n \geq 0$.

It is not difficult to verify that $\mathscr{C}$ is a pre-Abelian category. In this category, too, there exist bijections which are not isomorphisms.

3. The category ${}_\Lambda\mathscr{C}$ of unitary left Λ-modules (Λ = ring with a unit element) and of homomorphisms of Λ-modules is a pre-Abelian category. In this category every bijection is an isomorphism.

We now wish to characterize strict epimorphisms and monomorphisms in pre-Abelian categories.

LEMMA 5.8. *Let* $f: M' \longrightarrow M$ *be a morphism in a pre-Abelian category* $\mathscr{C}$, $(\ker f, \lambda)$ *its kernel,* $M' \Pi_M M'$ *the fibered product associated to the diagram*

$$\begin{array}{ccc} M' & & \\ & \searrow^{f} & \\ & & M \\ & \nearrow_{f} & \\ M' & & \end{array}$$

and let p_1, p_2 *be the canonical projections of this fibered product. Then* $\operatorname{coker}(p_1 - p_2) = \operatorname{coim} f$.

PROOF. There exists a monomorphism $\sigma_1 : \ker f \longrightarrow M' \Pi_M M'$ such that $p_1\sigma_1 = \lambda$, $p_2\sigma_1 = 0$. For, we know that there exists a canonical monomorphism $v: M' \Pi_M M' \longrightarrow M' \oplus M'$ such that the pair $(M' \Pi_M M', v)$ is the kernel of the morphism $f\pi_1 - f\pi_2$, where π_1, π_2 are the canonical projections of $M' \oplus M'$. If i_1, i_2 are the canonical injections of $M' \oplus M'$, then $(f\pi_1 - f\pi_2)i_1\lambda = 0$, hence there exists $\sigma_1 : \ker f \longrightarrow M' \Pi_M M'$ such that $v\sigma_1 = i_1\lambda$, whence $p_1\sigma_1 = \lambda$, $p_2\sigma_1 = 0$.

There also exists a morphism $\alpha: M' \Pi_M M' \longrightarrow \ker f$ such that $\lambda\alpha = p_1 - p_2$. For, it is sufficient to notice that $f(p_1 - p_2) = 0$. Now, consider the following diagram

$$\begin{array}{ccccc} M' \Pi_M M' & \xrightarrow{p_1 - p_2} & M' & \xrightarrow{\mu} & \operatorname{coker}(p_1 - p_2) \\ {\scriptstyle \sigma_1}\uparrow\ \downarrow{\scriptstyle \alpha} & & \downarrow{\scriptstyle 1_{M'}} & & \\ \ker f & \xrightarrow[\lambda]{} & M' & \xrightarrow[\rho]{} & \operatorname{coim} f = \operatorname{coker}(\ker f). \end{array}$$

We have $\rho(p_1 - p_2) = \rho\lambda\alpha = 0$; there exists therefore $\tau: \operatorname{coker}(p_1 - p_2) \longrightarrow \operatorname{coim} f$ such that $\tau\mu = \rho$. Furthermore, $\mu\lambda = \mu(p_1 - p_2)\sigma_1 = 0$ and there exists therefore $\tau_1 : \operatorname{coim} f \longrightarrow \operatorname{coker}(p_1 - p_2)$ such that $\tau_1\rho = \mu$. From this we deduce immediately that τ, τ_1 define an isomorphism between $\operatorname{coker}(p_1 - p_2)$ and $\operatorname{coim} f$.

PROPOSITION 5.9. *Let* $f: M' \longrightarrow M$ *be an epimorphism and* $M' \xrightarrow{j} \operatorname{coim} f \xrightarrow{\bar{f}} M$ *its canonical factorization. The necessary and sufficient condition for* f *to be a strict epimorphism is that* $\bar{f}$ *be an isomorphism.*

PROOF. If f is a strict epimorphism, then the relation $\operatorname{coim} f = \operatorname{coker}(p_1 - p_2)$ implies that $\bar{f}$ is an isomorphism. If $\bar{f}$ is an isomorphism, the same relation implies the fact that f is a strict epimorphism.

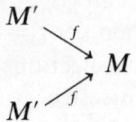

5. Abelian Categories

If $\mathscr{C}$ is a pre-Abelian category and $u: A \longrightarrow B$ is a morphism in $\mathscr{C}$, then we shall denote by $\bar{u}$ the morphism which occurs in the canonical factorization of u:

$$A \longrightarrow \operatorname{coim} u \xrightarrow{\bar{u}} \operatorname{im} u \longrightarrow B.$$

Definition. A pre-Abelian category $\mathscr{C}$ is called *Abelian* if for any morphism $u: A \longrightarrow B$ in $\mathscr{C}$ the morphism $\bar{u}: \operatorname{coim} u \longrightarrow \operatorname{im} u$ is an isomorphism.

PROPOSITION 5.10. *A pre-Abelian category $\mathscr{C}$ is Abelian if and only if the following conditions are fulfilled:*
(a) *for any* $u: A \longrightarrow B, \bar{u}: \operatorname{coim} u \longrightarrow \operatorname{im} u$ *is bijective;*
(b) *any bijection is an isomorphism.*

PROOF. If $\mathscr{C}$ is Abelian, then (a) is obviously satisfied and (b) follows from the fact that, if $u: A \longrightarrow B$ is a bijection, then $\bar{u} = u$ (see proposition 5.5). The converse is clear.

The second duality principle for Abelian categories

Let $\mathscr{C}$ be an Abelian category and A an object of $\mathscr{C}$. We denote by $\mathscr{P}(A)$ the class of subobjects of A and by $\mathscr{Q}(A)$ the class of quotient objects of A. These two classes are provided with natural partial order relations (see pp. 6–7). Consider the function which associates to each subobject (A', i) of A the quotient object of A defined by the pair $(\operatorname{coker} i, j)$, where j is the canonical surjection of A onto $\operatorname{coker} i$. Dually, consider the function which associates to each quotient object (A'', k) of A the subobject of A defined by the pair $(\ker k, l)$, where l is the canonical injection of $\ker k$ into A. The aim of the following propositions is to show that these two functions are inverse to each other. Thus, it will result in particular that there exists a natural one-to-one correspondence between the classes $\mathscr{P}(A)$ and $\mathscr{Q}(A)$; this correspondence reverses the order relations on the two classes.

PROPOSITION 5.11. *Let A be an object of an Abelian category $\mathscr{C}$ and (A', i) a subobject of A. Let (A'', j) be the quotient object of A defined by the cokernel of i. Under these conditions, the kernel of j coincides with (A', i).*

PROPOSITION 5.12. *Let A be an object of an Abelian category $\mathscr{C}$ and (A'', j) a quotient object of A. Let (A', i) be the subobject of A defined by the kernel of j. Under these conditions, the cokernel of i coincides with (A'', j).*

Clearly proposition 5.12 is dual to proposition 5.11, so in view of the fact that the dual of an Abelian category is Abelian it is sufficient to prove proposition 5.11.

PROOF OF PROPOSITION 5.11. By definition $\ker(\operatorname{coker} i) = \operatorname{im} i$. Let $\lambda : \operatorname{im} i \longrightarrow A$. It will be sufficient to exhibit an isomorphism $\mu : A' \longrightarrow \operatorname{im} i$ such that the diagram

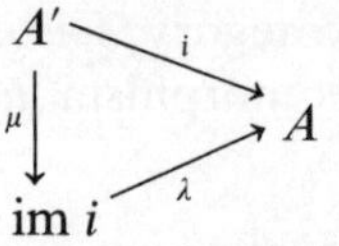

is commutative. To do this, consider the canonical factorization of i:

$$A' \xrightarrow{\nu} \operatorname{coim} i \xrightarrow{\bar{i}} \operatorname{im} i \xrightarrow{\lambda} A.$$

Since i is injective, $\operatorname{coim} i = A'$ and $\nu = 1_{A'}$. $\mathscr{C}$ being Abelian, $\bar{i}$ is an isomorphism. The proof is complete.

If A' is a subobject of the object A in the Abelian category $\mathscr{C}$, we shall denote by A/A' the quotient object of A which corresponds to A' under the bijection described above. If A'' is a quotient object of A we shall denote by $A \backslash A''$ the subobject of A which corresponds to A'' under the mentioned bijection.

6. Exact Functors

Definition. A sequence of morphisms of an Abelian category $\mathscr{C}$

$$A \xrightarrow{u} B \xrightarrow{v} C$$

is said to be *exact*, if $\operatorname{im} u = \ker v$. An arbitrary sequence of consecutive morphisms is said to be exact, if the subsequence formed by any couple of consecutive morphisms is an exact sequence.

The following propositions are immediate:

PROPOSITION 5.13. *The necessary and sufficient condition that the sequence*

$$0 \longrightarrow A \xrightarrow{u} B \xrightarrow{v} C$$

be exact is that the sequence of Abelian groups and homomorphisms of Abelian groups

$$0 \longrightarrow \operatorname{Hom}_{\mathscr{C}}(X, A) \longrightarrow \operatorname{Hom}_{\mathscr{C}}(X, B) \longrightarrow \operatorname{Hom}_{\mathscr{C}}(X, C)$$

be exact for any object X of $\mathscr{C}$.

PROPOSITION 5.14. *The necessary and sufficient condition that the sequence*

$$C \xrightarrow{u} B \xrightarrow{v} A \longrightarrow 0$$

be exact is that the sequence of Abelian groups and homomorphisms of Abelian groups

$$0 \longrightarrow \mathrm{Hom}_{\mathscr{C}}(A, X) \longrightarrow \mathrm{Hom}_{\mathscr{C}}(B, X) \longrightarrow \mathrm{Hom}_{\mathscr{C}}(C, X)$$

be exact for any object X of $\mathscr{C}$.

Propositions 5.13 and 5.14 are dual to each other.

PROPOSITION 5.15. *In order that the sequence*

$$0 \longrightarrow A' \xrightarrow{u} A \xrightarrow{v} A'' \longrightarrow 0$$

be exact it is necessary and sufficient that (A', u) be a kernel of v and (A'', v) be a cokernel of u.

Definition. If $\mathscr{C}_1$, $\mathscr{C}_2$ are Abelian categories and if $F:\mathscr{C}_1 \longrightarrow \mathscr{C}_2$ is an additive convariant functor, we say that F is *left-exact* (or *right-exact*) if for any exact sequence (E)

$$0 \longrightarrow A' \xrightarrow{\alpha'} A \xrightarrow{\alpha''} A'' \longrightarrow 0 \tag{E}$$

in the category $\mathscr{C}_1$, the sequence

$$0 \longrightarrow F(A') \xrightarrow{F(\alpha')} F(A) \xrightarrow{F(\alpha'')} F(A'')$$

(or $F(A') \xrightarrow{F(\alpha')} F(A) \xrightarrow{F(\alpha'')} F(A'') \longrightarrow 0$, for right-exactness) is exact.

If for any exact sequence (E) the sequence,

$$0 \longrightarrow F(A') \xrightarrow{F(\alpha')} F(A) \xrightarrow{F(\alpha'')} F(A'') \longrightarrow 0$$

is exact, then we say that F is *exact*.

We obtain by dualization the notion of a left-exact (or right-exact) contravariant functor and that of an exact contravariant functor.

PROPOSITION 5.16. *The necessary and sufficient condition that the functor $F:\mathscr{C}_1 \longrightarrow \mathscr{C}_2$ be exact is that it transform any exact sequence in the category $\mathscr{C}_1$ into an exact sequence in the category $\mathscr{C}_2$.*

PROOF. The sufficiency is evident. To prove the necessity, let

$$A \xrightarrow{u} B \xrightarrow{v} C$$

be an exact sequence in the category $\mathscr{C}_1$. We have to show that the sequence

$$F(A) \xrightarrow{F(u)} F(B) \xrightarrow{F(v)} F(C)$$

is exact. However, it is clear that the sequence

$$0 \longrightarrow \operatorname{im} u \xrightarrow{\lambda} B \xrightarrow{\mu} \operatorname{coim} v \longrightarrow 0$$

is exact. Furthermore, it follows from the hypothesis that F transforms a monomorphism into a monomorphism and an epimorphism into an epimorphism. Also, it is evident that $\operatorname{im} u = \operatorname{im} \lambda$ and $\ker v = \ker \mu$. The hypothesis implies that the sequence

$$0 \longrightarrow F(\operatorname{im} u) \xrightarrow{F(\lambda)} F(B) \xrightarrow{F(\mu)} F(\operatorname{coim} v) \longrightarrow 0.$$

is exact. On the other hand we have $u = \lambda u'$ where u' is an epimorphism and $v = v'\mu$ where v' is a monomorphism. Hence $F(u) = F(\lambda)F(u')$ where $F(u')$ is an epimorphism and $F(v) = F(v')F(\mu)$ where $F(v')$ is a monomorphism. Hence $\operatorname{im}(F(u)) = \operatorname{im}(F(\lambda))$, $\ker(F(v)) = \ker(F(\mu))$. But by the exactness of the previous sequence we have $\operatorname{im}(F(\lambda)) = \ker(F(\mu))$, whence $\operatorname{im}(F(u)) = \ker(F(v))$.

In the same manner it can be proved that:

If $F:\mathscr{C}_1 \longrightarrow \mathscr{C}_2$ is left-exact then it transforms any exact sequence of the form

$$0 \longrightarrow A' \longrightarrow A \longrightarrow A''$$

into an exact sequence.

If F is right-exact, it transforms any exact sequence of the form

$$A' \longrightarrow A \longrightarrow A'' \longrightarrow 0$$

into an exact sequence.

PROPOSITION 5.17. *If $\mathscr{C}_1, \mathscr{C}_2$ are Abelian categories and if $F:\mathscr{C}_1 \longrightarrow \mathscr{C}_2$, $G:\mathscr{C}_2 \longrightarrow \mathscr{C}_1$ are covariant functors such that G is an adjoint of the functor F, then G is left-exact and F is right-exact.*

PROOF. To prove that F is right-exact, let

$$0 \longrightarrow A' \xrightarrow{\alpha'} A \xrightarrow{\alpha''} A'' \longrightarrow 0 \qquad \text{(E)}$$

be an exact sequence in the category $\mathscr{C}_1$. We must show that the sequence

$$F(A') \xrightarrow{F(\alpha')} F(A) \xrightarrow{F(\alpha'')} F(A'') \longrightarrow 0$$

is exact. Let Y be an arbitrary object of the category $\mathscr{C}_2$. It is sufficient, in view of proposition 5.14, to prove that the sequence

$$0 \longrightarrow \operatorname{Hom}_{\mathscr{C}_2}(F(A''), Y) \longrightarrow \operatorname{Hom}_{\mathscr{C}_2}(F(A), Y) \longrightarrow \operatorname{Hom}_{\mathscr{C}_2}(F(A'), Y)$$

is exact. From the exactness of the sequence (E) it results that the sequence

$$0 \longrightarrow \operatorname{Hom}_{\mathscr{C}_1}(A'', G(Y)) \longrightarrow \operatorname{Hom}_{\mathscr{C}_1}(A, G(Y)) \longrightarrow \operatorname{Hom}_{\mathscr{C}_1}(A', G(Y))$$

is exact, while from the fact that G is an adjoint of F we infer that in the commutative diagram

$$\begin{array}{ccccccc} 0 \longrightarrow & \operatorname{Hom}_{\mathscr{C}_1}(A'', G(Y)) & \longrightarrow & \operatorname{Hom}_{\mathscr{C}_1}(A, G(Y)) & \longrightarrow & \operatorname{Hom}_{\mathscr{C}_1}(A', G(Y)) \\ & \uparrow & & \uparrow & & \uparrow \\ 0 \longrightarrow & \operatorname{Hom}_{\mathscr{C}_2}(F(A''), Y) & \longrightarrow & \operatorname{Hom}_{\mathscr{C}_2}(F(A), Y) & \longrightarrow & \operatorname{Hom}_{\mathscr{C}_2}(F(A'), Y) \end{array}$$

the vertical morphisms are isomorphisms, whence the exactness of the bottom row.

7. The Isomorphism Theorems in Abelian Categories

From the definition of an Abelian category it follows in particular that what is known as the 'first isomorphism theorem' in the theory of modules holds in such a category.

In the sequel we will show that actually all the three 'isomorphism theorems' in the theory of modules hold in any Abelian category.

To prove the second isomorphism theorem we need the following lemmas and propositions.

LEMMA 5.17. *Let A_1 and A_2 be two subobjects of A, $i_1 : A_1 \longrightarrow A$, $i_2 : A_2 \longrightarrow A$ the canonical injections. Suppose $i : A_1 \longrightarrow A_2$ is a monomorphism such that $i_2 i = i_1$. Denote by $p_1 : A \longrightarrow A/A_1$ the canonical surjection. Under these assumptions, the sequence*

$$0 \longrightarrow A_1 \xrightarrow{\ i\ } A_2 \xrightarrow{\ \alpha\ } A/A_1$$

is exact, where $\alpha = p_1 i_2$.

PROOF. Since the exactness in A_1 is clear, it remains to prove that the sequence is exact in A_2. We have $\alpha i = 0$, whence $\operatorname{im} i = A_1 \subset \ker \alpha$. We must show that we also have $\ker \alpha \subset \operatorname{im} i$. Denote by $\lambda : \ker \alpha \longrightarrow A_2$ the canonical injection. The relation $0 = \alpha\lambda = p_1(i_2\lambda)$ implies the existence of $\xi : \ker \alpha \longrightarrow A_1$ such that $i_1\xi = i_2\lambda$. But from this and from $i_2 i = i_1$ it follows that $i\xi = \lambda$, hence $\ker \alpha \subset \operatorname{im} i = A_1$.

LEMMA 5.18. *Let B_1 and B_2 be two quotient objects of B, $p_1 : B \longrightarrow B_1$, $p_2 : B \longrightarrow B_2$ the canonical projections. Suppose $p : B_2 \longrightarrow B_1$ is an* *epimorphism such that* $pp_2 = p_1$. *Denote by* $i_1 : B \backslash B_1 \longrightarrow B$ *the canonical injection. Under these assumptions, the sequence*

$$0 \longleftarrow B \xleftarrow{p} B_2 \xleftarrow{\beta} B \backslash B_1$$

is exact, where $\beta = p_2 i_1$.

Lemma 5.18 is a dual to lemma 5.17, so we need not prove it.

PROPOSITION 5.19. *Assume* $A_1, A_2, A, i_1, i_2, i, p_1, p_2, \alpha$ *have the same meaning as in lemma 5.17. Then the sequence*

$$0 \longrightarrow A_1 \xrightarrow{i} A_2 \xrightarrow{\alpha} A/A_1 \xrightarrow{j} A/A_2 \longrightarrow 0$$

is exact, where j *is the unique morphism which makes the following diagram commutative:*

$$\begin{array}{ccccc} A_1 & \xrightarrow{i_1} & A & \xrightarrow{p_1} & A/A_1 \\ {\scriptstyle i}\downarrow & & \downarrow{\scriptstyle 1_A} & & \downarrow{\scriptstyle j} \\ A_2 & \xrightarrow[i_2]{} & A & \xrightarrow[p_2]{} & A/A_2. \end{array}$$

PROOF. The exactness in A_1 and A_2 results from lemma 5.17 whereas the exactness in A/A_1 and A/A_2 results from lemma 5.18 by setting: $B_1 = A/A_2$, $B_2 = A/A_1$, $B = A$ and taking into account the canonical projections.

COROLLARY 5.20. (*The second isomorphism theorem*). *Assume* A_1, A_2, A, j *have the same meaning as in proposition 5.19. Then the sequence*

$$0 \longrightarrow A_2/A_1 \xrightarrow{\rho} A/A_1 \xrightarrow{j} A/A_2 \longrightarrow 0$$

is exact, where ρ *is the unique morphism which makes the following diagram commutative:*

$$\begin{array}{ccccc} A_1 & \xrightarrow{i} & A_2 & \longrightarrow & A_2/A_1 \\ {\scriptstyle 1_{A_1}}\downarrow & & \downarrow{\scriptstyle i_2} & & \downarrow{\scriptstyle \rho} \\ A_1 & \xrightarrow[i_1]{} & A & \longrightarrow & A/A_1. \end{array}$$

PROOF. According to proposition 5.19, $\ker j = \operatorname{im} \alpha$. On the other hand, the category being Abelian, $\operatorname{im} \alpha$ is isomorphic to $\operatorname{coim} \alpha$. But, by definition $\operatorname{coim} \alpha = \operatorname{coker}(\ker \alpha)$. By proposition 5.19, $\ker \alpha = i$ and therefore $\operatorname{coker} i = A_2/A_1$ whence the corollary.

COROLLARY 5.21. (*The dual of the second isomorphism theorem*). *Assume* B_1, B_2, B, p_1, p_2, p *have the same meaning as in lemma 5.18. Then the sequence*

$$0 \longleftarrow B_2 \backslash B_1 \longleftarrow B \backslash B_1 \longleftarrow B \backslash B_2 \longleftarrow 0$$

is exact, where the morphisms are defined in an obvious manner.

As an application of the second isomorphism theorem we shall prove the so-called '3 × 3 lemma', which will be used in the theory of derived functors. We first prove two auxiliary lemmas.

LEMMA 5.22. *If the sequences*

$$0 \xrightarrow{\beta} B \xrightarrow{\beta'} A \longrightarrow B'$$

$$0 \xrightarrow{\gamma} C \xrightarrow{\gamma'} A \longrightarrow C'$$

in the Abelian category $\mathscr{C}$ *are exact and if* $\gamma'\beta$ *is a monomorphism then* $\beta'\gamma$ *is also a monomorphism.*

LEMMA 5.23. *If the sequences*

$$B \longrightarrow A \xrightarrow{\beta} B' \xrightarrow{\beta'} 0$$

$$C \longrightarrow A \xrightarrow{\gamma} C' \xrightarrow{\gamma'} 0$$

in the Abelian category $\mathscr{C}$ *are exact and if* $\gamma'\beta$ *is an epimorphism then* $\beta'\gamma$ *is also an epimorphism.*

Since lemma 5.23 is dual to lemma 5.22 we indicate only the following:

PROOF OF LEMMA 5.22. Clearly lemma 5.22 holds in the particular category $\mathscr{Ab}$ of Abelian groups. This implies its validity in any Abelian category. Indeed, if X is an arbitrary object of the category $\mathscr{C}$, then the following sequences in the category $\mathscr{Ab}$ are exact and fulfill the conditions stated in the lemma:

$$0 \longrightarrow \mathrm{Hom}_{\mathscr{C}}(X, B) \longrightarrow \mathrm{Hom}_{\mathscr{C}}(X, A) \longrightarrow \mathrm{Hom}_{\mathscr{C}}(X, B')$$

$$0 \longrightarrow \mathrm{Hom}_{\mathscr{C}}(X, C) \longrightarrow \mathrm{Hom}_{\mathscr{C}}(X, A) \longrightarrow \mathrm{Hom}_{\mathscr{C}}(X, C').$$

COROLLARY 5.24. *If the sequences*

$$0 \longrightarrow B \xrightarrow{\beta} A \xrightarrow{\beta'} B' \longrightarrow 0$$

$$0 \longrightarrow C \xrightarrow{\gamma} A \xrightarrow{\gamma'} C' \longrightarrow 0$$

in the Abelian category $\mathscr{C}$ *are exact and if* $\gamma'\beta$ *is an isomorphism then* $\beta'\gamma$ *is also an isomorphism.*

Proposition 5.25 (*the '3 × 3 lemma'*). *If the diagram*

$$\begin{array}{ccccccccc}
 & & 0 & & 0 & & 0 & & \\
 & & \downarrow & & \downarrow & & \downarrow & & \\
0 & \longrightarrow & A_1 & \xrightarrow{\lambda} & A & \xrightarrow{\mu} & A_2 & \longrightarrow & 0 \\
 & & \downarrow{\scriptstyle \chi_1} & & \downarrow{\scriptstyle \chi} & & \downarrow{\scriptstyle \chi_2} & & \\
0 & \longrightarrow & B_1 & \xrightarrow{\eta} & B & \xrightarrow{\zeta} & B_2 & \longrightarrow & 0 \\
 & & \downarrow{\scriptstyle \rho_1} & & \downarrow{\scriptstyle \rho} & & \downarrow{\scriptstyle \rho_2} & & \\
0 & \longrightarrow & C_1 & \xrightarrow{\alpha} & C & \xrightarrow{\beta} & C_2 & \longrightarrow & 0 \\
 & & \downarrow & & \downarrow & & \downarrow & & \\
 & & 0 & & 0 & & 0 & &
\end{array}$$

in the Abelian category $\mathscr{C}$ is commutative and if its columns and its last two rows are exact, then first row is also exact.

Proof. The exactness in A_1 follows from the fact that $\eta\chi_1$ is a monomorphism and $\eta\chi_1 = \chi\lambda$. Thus λ is necessarily a monomorphism. Let (N, ω) be the cokernel of λ. We have to show that there exists an isomorphism $\theta : N \longrightarrow A_2$ such that $\theta\omega = \mu$.

We apply the second isomorphism theorem to the objects A_1, B_1, B and the injections $\chi_1 : A_1 \longrightarrow B_1$, $\eta : B_1 \longrightarrow B$, $\eta\chi_1 : A_1 \longrightarrow B$. We get the following exact sequence:

$$0 \longrightarrow C_1 \xrightarrow{n_1} B/A_1 \xrightarrow{m_1} B_2 \longrightarrow 0$$

where n_1, m_2 satisfy the conditions: $n_1\rho_1 = \tau\eta$, $m_1\tau = \zeta$, where $\tau : B \longrightarrow B/A_1$ is the canonical surjection.

Now, we apply the same theorem to the objects A_1, A, B and the injections $\lambda : A_1 \longrightarrow A$, $\chi : A \longrightarrow B$, $\eta\chi_1 : A_1 \longrightarrow B$. We get the following exact sequence

$$0 \longrightarrow N \xrightarrow{n_2} B/A_1 \xrightarrow{m_2} C \longrightarrow 0$$

where n_2, m_2 satisfy the conditions: $n_2\omega = \tau\chi$, $m_2\tau = \rho$.

We now apply the dual of the second isomorphism theorem to the following objects and surjections: $B/A_1, C, C_2, m_2 : B/A_1 \longrightarrow C$, $\beta : C \longrightarrow C_2$, $\beta m_2 : B/A_1 \longrightarrow C_2$. We obtain the following exact sequence:

$$0 \longrightarrow N \xrightarrow{n_3} R \xrightarrow{m_3} C_1 \longrightarrow 0$$

where $(R, \sigma) = \ker(\beta m_2)$ and $\sigma n_3 = n_2$, $\alpha m_3 = m_2\sigma$.

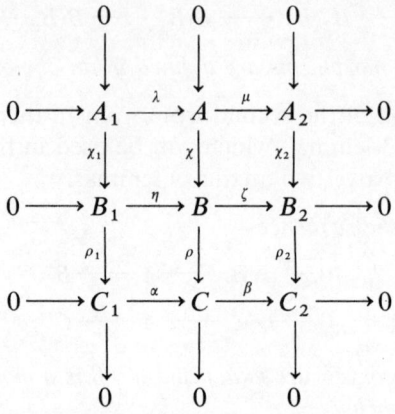

We apply the same theorem to the following objects and surjections: B/A_1, B_2, C_2, $m_1 : B/A_1 \longrightarrow B_2$, $\rho_2 : B_2 \longrightarrow C_2$, $\rho_2 m_1 = \beta m_2 : B/A_1 \longrightarrow C_2$. We obtain the following exact sequence:

$$0 \longrightarrow C_1 \xrightarrow{n_4} R \xrightarrow{m_4} A_2 \longrightarrow 0$$

with the conditions $\sigma n_4 = n_1$, $\chi_2 m_4 = m_1 \sigma$.

We assert that $m_3 n_4 = 1_{C_1}$. For, $\alpha m_3 n_4 = m_2 \sigma n_4 = m_2 n_1$ and $m_2 n_1 \rho_1 = m_2 \tau \eta = \rho \eta = \alpha \rho_1$, so that $m_2 n_1 = \alpha$. Hence $\alpha m_3 n_4 = \alpha$ and therefore $m_3 n_4 = 1_{C_1}$.

We are now in a position to apply corollary 5.24. We conclude that $\theta = m_4 n_3 : N \longrightarrow A_2$ is an isomorphism. On the other hand we have $\chi_2 \theta \omega = \chi_2 m_4 n_3 \omega = m_1 \sigma n_3 \omega = m_1 n_2 \omega = m_1 \tau \chi = \zeta \chi = \chi_2 \mu$ and thus $\theta \omega = \mu$. The proof is complete.

In the dual statement of the '3 × 3 lemma' the assumptions are that all columns and the first two rows are exact, and the conclusion is that the last row is also exact. This statement will also be called the '3 × 3 lemma'.

Direct and inverse images

Let $\mathscr{C}$ be an Abelian category and A an object of $\mathscr{C}$. We shall denote as above by $\mathscr{P}(A)$ the class of subobjects of A and by $\mathscr{Q}(A)$ the class of quotient objects of A. According to the second principle of duality (for Abelian categories) there exists a bijection $\varphi_A : \mathscr{P}(A) \longrightarrow \mathscr{Q}(A)$ which associates to the subobject (A', i) the quotient object (A'', j) such that $j = \text{coker } i$.

Let now $u : A \longrightarrow B$ be a morphism in $\mathscr{C}$. We propose to show that one associates to this morphism in a natural way the maps

$$\mathscr{P}(A) \longrightarrow \mathscr{P}(B)$$

$$\mathscr{Q}(A) \longrightarrow \mathscr{Q}(B)$$

$$\mathscr{P}(B) \longrightarrow \mathscr{P}(A)$$

$$\mathscr{Q}(B) \longrightarrow \mathscr{Q}(A)$$

which we denote somewhat improperly (to comply with a tradition) as follows: the first two by u, the last two by u^{-1}. The definitions of these maps are the following:

Let (A', i) be a subobject of A. Then by definition $u(A') = \text{im}(ui)$. Moreover, there exists a morphism $A' \longrightarrow u(A')$ defined to be the composition

$$A' \longrightarrow \text{coim}(ui) \longrightarrow \text{im}(ui)$$

If (A'', j) is a quotient object of A, then we denote by (A', i) the subobject of A which corresponds to it under the bijection φ i.e. such that $i = \ker j$. We put by definition $u(A'') = \operatorname{coker}(ui)$. Clearly there exists a morphism $A'' \longrightarrow u(A'')$ such that the diagram

$$\begin{array}{ccc} A & \xrightarrow{u} & B \\ \downarrow & & \downarrow \\ A'' & \longrightarrow & u(A'') \end{array}$$

is commutative.

We obtain by dualization the maps u^{-1}. The effective definition is as follows:

Let (B'', j) be a quotient object of B. By definition $u^{-1}(B'') = \operatorname{coim}(ju)$. If (B', i) is a subobject of B, we denote by (B'', j) the quotient object such that $j = \operatorname{coker} i$. We set by definition $u^{-1}(B') = \ker(ju)$. There exist the morphisms $u^{-1}(B') \longrightarrow B'$, $u^{-1}(B'') \longrightarrow B''$ such that the diagram

$$\begin{array}{ccc} u^{-1}(B') & \longrightarrow & B' \\ \downarrow & & \downarrow \\ A & \xrightarrow{u} & B \\ \downarrow & & \downarrow \\ u^{-1}(B'') & \longrightarrow & B'' \end{array}$$

is commutative.

We have remarked above that we can go over from the definition of the maps u to that of the maps u^{-1} and conversely by using the general principle of duality. Now, we are going to show that we can go over from one of the maps u to the other and analogously from one of the maps u^{-1} to the other by using the second duality principle for Abelian categories. Precisely, the following diagrams of classes and maps are commutative:

$$\begin{array}{ccc} \mathscr{Q}(A) & \xrightarrow{u} & \mathscr{Q}(B) \\ \uparrow{\scriptstyle \varphi_A} & & \uparrow{\scriptstyle \varphi_B} \\ \mathscr{P}(A) & \xrightarrow[u]{} & \mathscr{P}(B) \end{array} \qquad \begin{array}{ccc} \mathscr{Q}(B) & \xrightarrow{u^{-1}} & \mathscr{Q}(A) \\ \uparrow{\scriptstyle \varphi_B} & & \uparrow{\scriptstyle \varphi_A} \\ \mathscr{P}(B) & \xrightarrow[u^{-1}]{} & \mathscr{P}(A). \end{array}$$

To verify the commutativity of the first diagram, let (A', i) be a subobject of A. u associates to this subobject of A the subobject $\operatorname{im}(ui)$ of B and φ_B associates to this subobject of B its quotient object $\operatorname{coker}(\operatorname{im}(ui))$. Under φ_A the subobject (A', i) is carried onto $\operatorname{coker} i$, and the latter is carried under u onto coker(ui). Thus we must show that coker(im(ui)) = coker(ui). But we have

$$\operatorname{coker}(\operatorname{im}(ui)) = \operatorname{coker}(\ker(\operatorname{coker}(ui))) = \operatorname{coker}(ui)$$

The commutativity of the second diagram follows in a similar way from the relation

$$\ker(\operatorname{coim}(ju)) = \ker(ju).$$

These remarks enable us to reduce the study of the four maps described above to the study of one of them.

PROPOSITION 5.26. *Let (A', i) be a subobject of A and (A'', j) a quotient object of A. Also, let (B', k) be a subobject of B and (B'', l) a quotient object of B, and let $u: A \longrightarrow B$. Under these conditions the following formulas are valid:*

(i) $u^{-1}(u(A')) \supset A'$; (iii) $u(u^{-1}(B'')) \supset B''$;
(ii) $u^{-1}(u(A'')) \subset A''$; (iv) $u(u^{-1}(B')) \subset B'$.

PROOF. We first show that it is sufficient to prove only one of these relations, for instance relation (i). Indeed, relation (iii) then follows by duality. To deduce relation (ii) from relation (i) we argue as follows: Let A' be such that $\varphi_A(A') = A''$. Relation (i) implies $\varphi_A(u^{-1}(u(A))) \subset \varphi_A(A') = A''$. But $\varphi_A(u^{-1}(u(A'))) = u^{-1}(u(\varphi_A(A'))) = u^{-1}(u(A''))$. Formula (iv) results analogously from (iii).

To prove relation (i), consider the following commutative diagram:

$$\begin{array}{ccccc} & & A' & \xrightarrow{u'} & u(A') \\ & & \downarrow{\scriptstyle i} & & \downarrow{\scriptstyle \lambda} \\ u^{-1}(u(A')) = \ker(\mu u) & \longrightarrow & A & \longrightarrow & B \\ & & & & \downarrow{\scriptstyle \mu} \\ & & & & \operatorname{coker} \lambda. \end{array}$$

The diagram yields $\mu u i = \mu\lambda u' = 0$, whence the fact that i factorizes through $u^{-1}(u(A'))$, i.e. $A' \subset u^{-1}(u(A'))$.

LEMMA 5.27. *For any subobject (A', i) of A, the morphism $A' \xrightarrow{u'} u(A')$ is an epimorphism.*

PROOF. By definition u' equals the following composition:

$$A' \longrightarrow \operatorname{coim}(ui) \xrightarrow{\overline{ui}} \operatorname{im}(ui).$$

The first of these morphisms is surjective and the second is an isomorphism. Hence u' is an epimorphism.

PROPOSITION 5.28. *Let* $(A', i), (A'', j)$ *be a subobject (resp. a quotient object) of* $A, (C', k), (C'', l)$ *a subobject (resp. a quotient object) of* C, *and* $u : A \longrightarrow B, v : B \longrightarrow C$. *Then the following transitivity relations hold*:

$$vu(A') = v(u(A')) \qquad (vu)^{-1}(C'') = u^{-1}(v^{-1}(C''))$$
$$vu(A'') = v(u(A'')) \qquad (vu)^{-1}(C') = u^{-1}(v^{-1}(C')).$$

PROOF. As with proposition 5.26, it is sufficient to check only the first of these relations. The others follow by the two duality principles.

Consider the commutative diagram

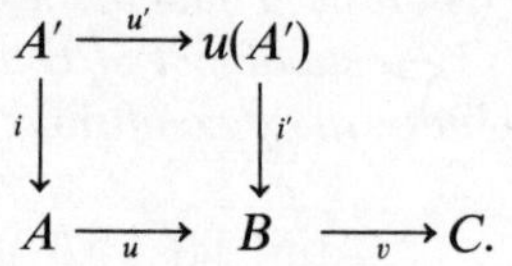

$$\begin{array}{ccc} A' & \xrightarrow{u'} & u(A') \\ i \downarrow & & \downarrow i' \\ A & \xrightarrow{u} B & \xrightarrow{v} C. \end{array}$$

By definition $(vu)(A') = \operatorname{im}(vui)$; therefore, by using the commutativity of the diagram, we obtain:

$$(vu)(A') = \operatorname{im}(vi'u') = \operatorname{im}((vi')u').$$

However, according to lemma 5.27, u' is surjective; thus $(vu)(A') = \operatorname{im}(vi') = v(u(A'))$.

In order to state and prove the third isomorphism theorem for Abelian categories we must introduce a few new notions.

Let $\mathscr{C}$ be an Abelian category and A an arbitrary object of $\mathscr{C}$. We have denoted above by $\mathscr{P}(A)$ (or $\mathscr{Q}(A)$) the class of subobjects (or quotient objects) of A. These two classes are provided with natural partial order relations.

PROPOSITION 5.29. *For any two subobjects* A_1, A_2 *of* A *there exist* $\sup(A_1, A_2)$ *and* $\inf(A_1, A_2)$ *in the sense of the order relation on* $\mathscr{P}(A)$.

PROOF. Let $\alpha_1 : A_1 \longrightarrow A$, $\alpha_2 : A_2 \longrightarrow A$ be the canonical injections of the two subobjects of A and let $(A_1 \amalg A_2, i_1, i_2)$ be their direct sum. Let α be such that the diagram

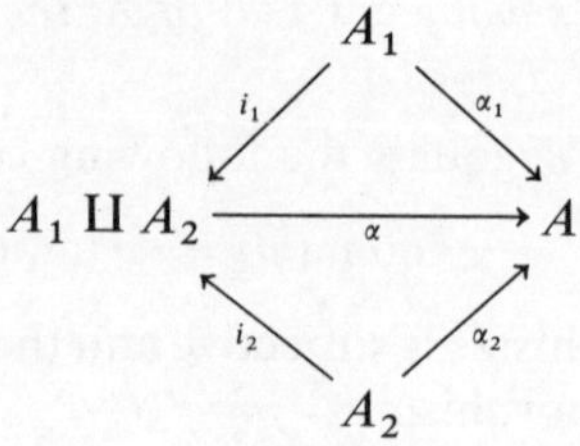

is commutative. We claim that $\operatorname{im}\alpha = \sup(A_1, A_2)$. First, it is clear that $\operatorname{im}\alpha \supset A_1$ and $\operatorname{im}\alpha \supset A_2$. Let now B be a subobject of A such that $A_1 \subset B$, $A_2 \subset B$. We must show that $\operatorname{im}\alpha \subset B$. To this end, let $\beta_1 : A_1 \longrightarrow B$, $\beta_2 : A_2 \longrightarrow B$, $\gamma : B \longrightarrow A$ be the respective canonical injections. There exists $\beta : A_1 \amalg A_2 \longrightarrow B$ such that the diagram

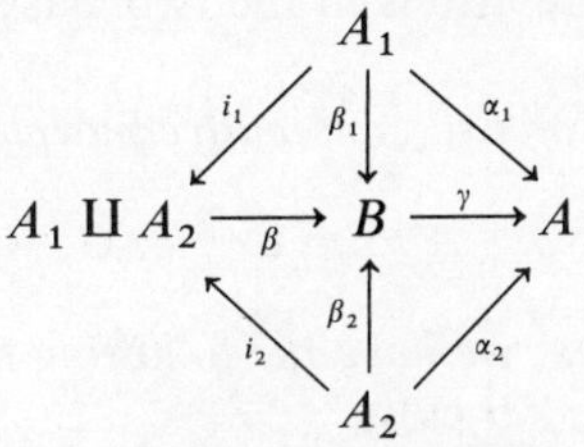

is commutative. Hence $\gamma\beta = \alpha$. But since γ is an injection it follows that $\ker\beta = \ker\alpha$, whence $\operatorname{coim}\beta = \operatorname{coim}\alpha$. Denote by $\bar{\alpha} : \operatorname{coim}\alpha \longrightarrow \operatorname{im}\alpha$ the isomorphism in the canonical factorization of α. It results that the diagram

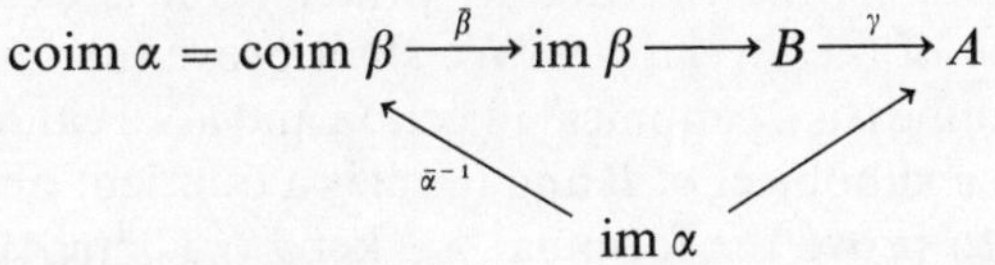

is commutative and therefore $\operatorname{im}\alpha \subset B$.

Now, let $\rho_1 : A \longrightarrow A/A_1$, $\rho_2 : A \longrightarrow A/A_2$ be the canonical surjections and let $(A/A_1 \prod A/A_2, p_1, p_2)$ be the direct product of the objects A/A_1, A/A_2. Let p be such that the diagram

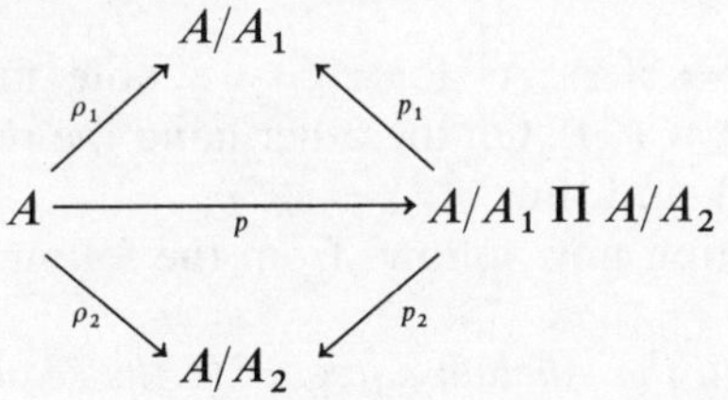

is commutative. We claim that $\ker p = \inf(A_1, A_2)$. It is clear that $\ker p \subset A_1$, $\ker p \subset A_2$. Now let $B \subset A_1$, $B \subset A_2$. If $\beta : B \longrightarrow A$ is the canonical injection, then clearly $p\beta = 0$ and thus $B \subset \ker p$.

Notation. We denote $\sup(A_1, A_2)$ by $A_1 \cup A_2$ and $\inf(A_1, A_2)$ by $A_1 \cap A_2$.

Remarks. 1. A statement dual to proposition 5.29 exists concerning the class $\mathcal{Q}(A)$.

2. When the first duality principle (which holds in any category) is used, the partial order relations in $\mathcal{P}(A)$ and $\mathcal{Q}(A)$ are preserved. When the second duality principle (which holds only in Abelian categories) is used, the partial order relations in the two classes are reversed.

PROPOSITION 5.30. *For two consecutive morphisms*

$$A \xrightarrow{u} B \xrightarrow{v} C$$

in an Abelian category $\mathscr{C}$, *we have the following relations.*

(i) $\operatorname{Im} u \cap \ker v = u(\ker vu)$;
(ii) $\operatorname{Im} u \cup \ker v = v^{-1}(\operatorname{im} vu)$;
(iii) $\operatorname{coim} v \cap \operatorname{coker} u = v^{-1}(\operatorname{coker} vu)$;
(iv) $\operatorname{coim} v \cup \operatorname{coker} u = u(\operatorname{coim} vu)$.

PROOF. In view of the two duality principles it is clearly sufficient to prove only one of these relations. We shall prove (ii).

First, we assume u is a canonical injection and v is a canonical surjection, that is (A, u) is a subobject of B and (C, v) is a quotient object of B. In this case we have to prove the formula $A \cup \ker v = v^{-1}(v(A))$.

To do this, notice first that according to proposition 5.26 we have $v^{-1}(v(A)) \supset A$. It is also obvious that $v^{-1}(v(A)) \supset v^{-1}(0) = \ker v$ whence $v^{-1}(v(A)) \supset A \cup \ker v$. To deduce from this the desired relation, it is sufficient to show that, if C is a subobject of B such that $C \supset A$, $C \supset \ker v$ and $v^{-1}(v(A)) \supset C$, then $C = v^{-1}(v(A))$. To this end we prove that $v(C) = v(v^{-1}(v(A)))$.

But $v(v^{-1}(v(A))) = v(A)$. As for $v(C)$ we note first that the relation $C \supset A$ implies $v(C) \supset v(A)$. On the other hand the relation $v^{-1}(v(A)) \supset C$ implies $v(A) \supset v(C)$, and thus $v(A) = v(C)$.

The desired relation now follows from the following:

LEMMA 5.31. *If in the Abelian category* $\mathscr{C}$ *the sequence*

$$0 \longrightarrow X' \xrightarrow{\xi'} X \xrightarrow{\xi''} X'' \longrightarrow 0$$

is exact, ξ' *is a canonical injection and* ξ'' *a canonical surjection and if* (X_1, ξ_1), (X_2, ξ_2) *are subobjects of* X *satisfying the relation* $X' \subset X_1 \subset X_2$, *then in order that* $X_1 = X_2$ *it is necessary and sufficient that* $\xi''(X_1) = \xi''(X_2)$.

PROOF. By the second isomorphism theorem we have the commutative diagram

$$\begin{array}{ccccccccc} 0 & \longrightarrow & X_1/X' & \longrightarrow & X/X' & \longrightarrow & X/X_1 & \longrightarrow & 0 \\ & & \downarrow & & \downarrow & & \downarrow & & \\ 0 & \longrightarrow & X_2/X' & \longrightarrow & X/X' & \longrightarrow & X/X_2 & \longrightarrow & 0 \end{array}$$

where the rows are exact. The hypothesis implies that the first vertical morphism is an isomorphism. Since the second one is the identity, it follows that the third morphism is also an isomorphism. Hence $X_1 = X_2$.

Now, to obtain the general formula (ii), we consider, in addition to the morphisms in the proposition, the following consecutive morphisms.

$$\text{im } u \xrightarrow{\mu} B \xrightarrow{\nu} \text{im } v.$$

We have $\text{im } u = \text{im } \mu$, $\ker v = \ker \nu$, $\text{im } \nu\mu = \text{im } vu$ and $\nu^{-1}(\text{im } \nu\mu) = v^{-1}(\text{im } vu)$. The relation (ii) now follows immediately.

PROPOSITION 5.32. (*The third isomorphism theorem*).

Let $v: A \longrightarrow C$ be a morphism in an Abelian category, and M and N two subobjects of A. Then we have the following isomorphisms

$$v(M) \approx M/(M \cap \ker v) \approx (M \cup \ker v)/\ker v$$

$$M/M \cap N \approx M \cup N/N.$$

PROOF. According to proposition 5.30, $M \cup \ker v = v^{-1}(v(M))$. This implies $M \cup \ker v/\ker v \approx v(M)$. Similarly $M \cap \ker v = u(\ker vu)$, where $u: M \longrightarrow A$ is the canonical injection of M. Hence it follows that $M \cap \ker v$ is the kernel of the surjection $M \longrightarrow v(M)$ and therefore $v(M) \approx M/(M \cap \ker v)$. To obtain the last isomorphism we consider the morphism $v: A \longrightarrow A/N$ and we use the relations already proved.

8. The Conditions AB3, AB4, AB5

In order to obtain precise results, we are often obliged to impose additional conditions to an Abelian category $\mathscr{C}$. The most manageable such conditions considered until now are those introduced by Grothendieck under the name of conditions AB3, AB4, AB5, and their duals which are usually denoted by AB3*, AB4*, AB5*. We next formulate these conditions.

AB3. Any family $(A_i)_{i \in I}$ of objects of the category $\mathscr{C}$ possesses at least one direct sum.

An immediate consequence of this condition is the following:

Let $(A_i, u_i)_{i \in I}$ be a family of subobjects of the object A. Then there exists $\sup_{i \in I} A_i$.

For, it is sufficient to consider the image of the morphism

$$\bigoplus_{i \in I} A_i \longrightarrow A$$

induced by the morphisms u_i.

If a category $\mathscr{C}$ satisfies condition AB3 and if $(u_i)_{i \in I}$ is a family of monomorphisms $u_i : A_i \longrightarrow B_i$, then the induced morphism $u : \bigoplus_{i \in I} A_i \longrightarrow \bigoplus_{i \in I} B_i$ is not in general a monomorphism. This leads to the formulation of the following condition:

AB4. The condition AB3 is valid and, in addition, the direct sum of a family of monomorphisms is also a monomorphism.

The condition AB5 is stronger than AB4 and is stated as follows:

AB5. The condition AB3 is valid and, in addition, for any directed family $(A_i)_{i \in I}$ of subjects of A and for any subobject B of A, the relation

$$(\sup_{i \in I} A_i) \cap B = \sup_{i \in I} (A_i \cap B)$$

holds.

We mention two important categories which satisfy all of these conditions: The category of left (or right) Λ-modules, where Λ is a ring, and the category of sheaves of Λ-modules on an arbitrary topological space.

It is easy to verify that condition AB5 can be formulated in the following equivalent form:

Condition AB3 is valid and if $(A_i)_{i \in I}$ is a directed family of subobjects of A and if $u_i : A_i \longrightarrow B$ are morphisms such that the condition $A_i \subset A_j$ implies that the restriction of u_j to A_i coincides with u_i, then there exists a unique morphism $u : \sup_{i \in I} A_i \longrightarrow B$ such that the restriction of u to A_i coincides with u_i for any $i \in I$.

Also, the condition AB3 is equivalent to the existence of arbitrary inductive limits, as follows from proposition 3.4. Also, AB5 is equivalent to the fact that there exist inductive limits and the inductive limits over directed families of indices are exact, i.e. if I is a directed set and

$$A_i \longrightarrow B_i \longrightarrow C_i$$

is an exact sequence for any $i \in I$, then

$$\varinjlim A_i \longrightarrow \varinjlim B_i \longrightarrow \varinjlim C_i$$

is an exact sequence. We shall use condition AB5 mostly under this form.

9. Generators

Definition. Let $\mathscr{C}$ be a category and $(U_i)_{i \in I}$ a family of objects of $\mathscr{C}$. The family $(U_i)_{i \in I}$ is said to be a *family of generators* of the category $\mathscr{C}$ if for any object A of $\mathscr{C}$ and any subobject B of A distinct from A there exists at least an index $i \in I$ and a morphism $u : U_i \longrightarrow A$ which cannot be factorized through the canonical injection $i : B \longrightarrow A$ *of* B into A.

An object U of $\mathscr{C}$ is said to be a *generator* of the category $\mathscr{C}$ if the family (U) is a family of generators of the category $\mathscr{C}$.

Example. The object Λ of ${}_{\Lambda}\mathscr{C}$ is a generator for the category ${}_{\Lambda}\mathscr{C}$. For, let $N \subset M$, $N \neq M$ and $c \in M - N$. Consider then the morphism $u : \Lambda \longrightarrow M$ defined by $u(x) = xc$. This morphism satisfies the required conditions.

PROPOSITION 5.33. *Let $\mathscr{C}$ be an Abelian category satisfying condition* AB3. *Then $(U_i)_{i \in I}$ is a family of generators for $\mathscr{C}$ if and only if the object $\bigoplus_{i \in I} U_i$ is a generator of $\mathscr{C}$.*

PROOF. Let $B \subset A$, $B \neq A$ and $u_i : U_i \longrightarrow A$ such that u_i cannot be factorized through $\beta : B \longrightarrow A$. Then the morphism of components u_i and 0 form $\bigoplus_{i \in I} U_i$ *to* A cannot be factorized through $\beta : B \longrightarrow A$.

Conversely, if $u : \bigoplus_{i \in I} U_i \longrightarrow A$ cannot be factorized through β then there exists at least one index $i \in I$ such that $u_i : U_i \longrightarrow A$ cannot be factorized through β.

PROPOSITION 5.34. *Let $\mathscr{C}$ be an Abelian category satisfying condition* AB3. *If U is a generator of the category $\mathscr{C}$ then any object of $\mathscr{C}$ is isomorphic to a quotient of an object of the form $U^{(I)}$, where I is a set.*

PROOF. Let A be in $\mathscr{C}$ and $I = \operatorname{Hom}_{\mathscr{C}}(U, A)$.

Consider the morphism $u : U^I \longrightarrow A$ defined as follows: For each $i \in I$, $u_i = i$. u is obviously a surjection because U is a generator.

PROPOSITION 5.35. *Let $\mathscr{C}$ be an Abelian category which possesses a generator U. Then the class of all subobjects of an object A of $\mathscr{C}$ is a set.*

PROOF. Notice first that if the category $\mathscr{C}$ possesses a generator U and if B_1, B_2 are two distinct subobjects of A, then there exists at least one morphism $u : U \longrightarrow A$ which can be factorized through $\beta_2 : B_2 \longrightarrow A$ but cannot be factorized through $\beta_1 : B_1 \longrightarrow A$. For, consider the object

$B_1 \cap B_2$. It defines a subobject of A, a subobject of B_1 and a subobject of B_2 distinct from B_2. It follows that there exists a morphism $\mu : U \longrightarrow B_2$ which cannot be factorized through $B_1 \cap B_2$. The morphism $\beta_2\mu$ can be factorized through β_2 but cannot be factorized through β_1. For, otherwise we would have $\beta_2\mu = \beta_1\nu$. Thus we would have the following commutative diagram

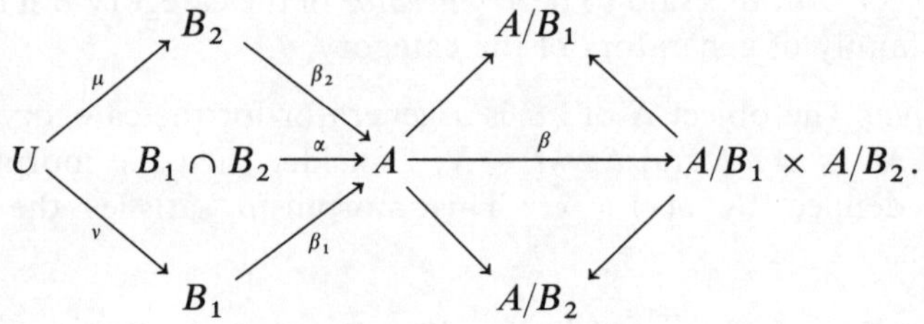

But this implies that $\beta_2\mu$ can be factorized through $B_1 \cap B_2$. For, $(B_1 \cap B_2, \alpha)$ is the kernel of the morphism β and from the commutativity of the diagram it follows that $\beta\mu\beta_2 = 0$, which is absurd.

Now, consider the map which associates to each subobject B of A the subset of $\text{Hom}_{\mathscr{C}}(U, A)$ consisting of those morphisms which can be factorized through $B \xrightarrow{\beta} A$, where β is the canonical injection of the subobject B. From the above considerations it follows that this map is one-to-one.

The dual of the notion of a generator is that of a cogenerator. We invite the reader to dualized proposition 5.33, 5.34, and 5.35.

In homological algebra and in algebraic geometry there appear a very important type of categories, namely the Grothendieck categories.

Definitions. An Abelian category which satisfies the condition AB5 and which possesses a family of generators is said to be a *Grothendieck category*.

A category $\mathscr{C}$ such that $\mathscr{O}\mathscr{C}\mathscr{C}$ is a set is said to be a *small* category.

Examples. The category ${}_{\Lambda}\mathscr{C}$, the category of sheaves of Abelian groups over a topological space are Grothendieck categories.

As follows from chapter 6, the Grothendieck categories are most adequate for homological algebra. In the following paragraph we will show that any small Abelian category is a full Abelian subcategory of a Grothendieck category.

10. Full Embedding of a Small Abelian Category into a Grothendieck Category

Definition. Let $\mathscr{C}$, $\mathscr{C}'$ be two Abelian categories and $T:\mathscr{C} \longrightarrow \mathscr{C}'$ a functor. We say that T realizes an *embedding* of $\mathscr{C}$ into $\mathscr{C}'$ if:

(a) T is an exact functor;

(b) T is a faithful functor, i.e. if $f \in \text{Hom}_{\mathscr{C}}(X, Y)$ is such that $Tf = 0$, then $f = 0$, for any objects X, Y of $\mathscr{C}$.

The functor T is said to realize a *full embedding* of $\mathscr{C}$ into $\mathscr{C}'$ if it is an embedding of $\mathscr{C}$ into $\mathscr{C}'$ and T is full, i.e. for any $g \in \text{Hom}_{\mathscr{C}'}(TX, TY)$ there exists $f \in \text{Hom}_{\mathscr{C}}(X, Y)$ such that $Tf = g$ for any objects X, Y of $\mathscr{C}$.

We shall prove in the following that any small Abelian category can be fully embedded into a Grothendieck category. This result is due to P. Gabriel.

Let $\mathscr{C}$ be a small Abelian category; we denote by $\mathscr{C}^{\wedge}$ the category of all contravariant additive functors from $\mathscr{C}$ to $\mathscr{A}b$. We denote by $h:\mathscr{C} \longrightarrow \mathscr{C}^{\wedge}$ the functor which associates to each object X of $\mathscr{C}$ the functor h_X defined by

$$h_X(Y) = \text{Hom}_{\mathscr{C}}(Y, X).$$

By an argument similar to that of theorem 1.6 it is proved that h is fully faithful and, moreover, by proposition 3.11, h is left-exact.

Let $\mathscr{S}$ be the full subcategory of $\mathscr{C}^{\wedge}$ consisting of the left-exact functors, and $I:\mathscr{S} \longrightarrow \mathscr{C}^{\wedge}$ the inclusion functor, which is obviously fully faithful. We shall prove that I possesses a left-adjoint functor R^0 which is exact.

Before giving the effective construction we make some observations about the category $\mathscr{C}$.

$\mathscr{C}^{\wedge}$ is a category of diagrams with values in $\mathscr{A}b$, hence it is an Abelian category satisfying condition AB5. To see this, it is sufficient to remark that to deal with diagrams it is sufficient to work on the components. The objects h_X form a family of generators for the category $\mathscr{C}^{\wedge}$. In conclusion, $\mathscr{C}^{\wedge}$ is a Grothendieck category.

Let X be an object of $\mathscr{C}$. We denote by $E(X)$ all the epimorphisms in $\mathscr{C}$ which end in X; an element of $E(X)$ is a couple (A, p) where A is an object of $\mathscr{C}$ and $p:A \longrightarrow X$ an epimorphism. For any object X of $\mathscr{C}$, $E(X)$ satisfies the conditions:

(a) $(X, 1_X) \in E(X)$.

(b) If $(A, p) \in E(X)$ and $(B, q) \in E(A)$, then $(B, pq) \in E(X)$.

(c) If $(A, p) \in E(X)$ and $f:X' \longrightarrow X$ is an arbitrary morphism in $\mathscr{C}$, then in the commutative diagram

$$\begin{array}{ccc} A' = A \amalg_X X' & \xrightarrow{p'} & X' \\ \downarrow & & \downarrow f \\ A & \xrightarrow[p]{} & X \end{array}$$

which is canonically constructed, $(A', p') \in E(X')$.

Remark. The assignment $X \rightsquigarrow E(X)$ defines a Grothendieck topology on $\mathscr{C}$; this is the so called canonical topology. From the following, it follows that $\mathscr{S}$ is the category of sheaves of Abelian groups on $\mathscr{C}$ relatively to this Grothendieck topology.

For each object X of $\mathscr{C}$, we define the category $\mathscr{E}(X)$ as follows:

(i) $\mathcal{O}b(\mathscr{E}(X)) = E(X)$

(ii) a morphism from (A, p) to (A', p') is a morphism $u: A \longrightarrow A'$ such that $p'u = p$.

Let $f: X \longrightarrow X'$ be an epimorphism in $\mathscr{C}$. Then f induces a functor $f^*: \mathscr{E}(X) \longrightarrow \mathscr{E}(X')$ defined as follows: $f^*(A, p) = (A, f \circ p)$. f induces also a functor $f_*: \mathscr{E}(X') \longrightarrow \mathscr{E}(X)$ defined as follows: for $(A', p') \in E(X')$, $f_*(A', p') = (A' \amalg_{X'} X, p)$ where $p: A' \amalg_{X'} X \longrightarrow X$ is the canonical morphism. We infer that for any $(A', p') \in \mathscr{E}(X')$, there is a canonical morphism $f^*f_*(A', p') \longrightarrow (A', p')$, which shows that the image of the set $E(X)$ under f^* is cofinal in $\mathscr{E}(X')$. We shall use later these remarks.

We now construct a functor $H: \mathscr{C}^\wedge \longrightarrow \mathscr{C}^\wedge$ as follows: Let F be an object of $\mathscr{C}^\wedge$ and X an object of $\mathscr{C}$. For any $(A, p) \in E(X)$ we denote by $m_p: \ker p \longrightarrow A$, the kernel of p. We also denote by $F(A, p) = \ker F(m_p)$; that is, we have the exact sequence:

$$0 \longrightarrow F(A, p) \longrightarrow F(A) \xrightarrow{F(m_p)} F(\ker p).$$

If $u: (A, p) \longrightarrow (A', p')$ is a morphism of $\mathscr{E}(X)$, then it is easily seen that there exists a unique morphism $F(p', p): F(A', p') \longrightarrow F(A, p)$ (which is independent of u) such that the diagram

$$\begin{array}{ccc} & & F(A', p') \\ & \nearrow^{F(X, p')} & \\ F(X) & & \downarrow F(p, p') \\ & \searrow_{F(X, p)} & \\ & & F(A, p) \end{array}$$

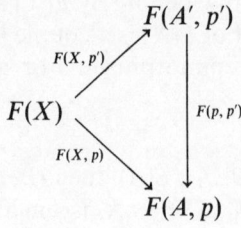

is commutative, where the morphisms $F(X,p):F(X)\longrightarrow F(A,p)$ are constructed in a canonical way from the definition of $F(A,p)$.

We set

$$HF(X) = \varinjlim_{(A,p)\in \mathscr{E}(X)} F(A,p).$$

This inductive limit is taken in $\mathscr{Ab}$. Using the particular properties of the category $\mathscr{E}(X)$ it is easy to see that in fact the above inductive limit is an inductive limit over a directed set.

If $f:X\longrightarrow Y$, then $f_*:E(Y)\longrightarrow E(X)$ and therefore for any $(B,q)\in E(Y)$ there exists a unique morphism $F(B,q)\longrightarrow F(f_*(B,q))$. These morphisms define a unique morphism $HF(f):HF(Y)\longrightarrow HF(X)$. If we denote by $\psi_X:F(X)\longrightarrow HF(X)$ the unique morphism induced by the morphisms $F(X,p)$, then the diagram, which is constructed in a canonical manner

$$\begin{array}{ccc} F(Y) & \xrightarrow{\psi_Y} & HF(Y) \\ {\scriptstyle F(f)}\downarrow & & \downarrow{\scriptstyle HF(f)} \\ F(X) & \xrightarrow[\psi_X]{} & HF(X) \end{array}$$

is commutative.

In conclusion, the assignment $X \rightsquigarrow HF(X)$, $f\rightsquigarrow HF(f)$ is a contravariant additive functor from $\mathscr{C}$ to $\mathscr{Ab}$.

PROPOSITION 5.36. *For any object F of $\mathscr{C}^\wedge$, HF is a functor which carries epimorphisms into monomorphisms.*

The assignment $F \rightsquigarrow HF$ defines a left-exact functor from $\mathscr{C}^\wedge$ to $\mathscr{C}^\wedge$.

PROOF. Let $f:X\longrightarrow X'$ be an epimorphism in $\mathscr{C}$ and $(A,p)\in E(X)$. Then we have the commutative diagram in $\mathscr{C}$

$$\begin{array}{ccccccccc} 0 & \longrightarrow & \ker p & \xrightarrow{m_p} & A & \xrightarrow{p} & X & \longrightarrow & 0 \\ & & \downarrow & & \downarrow{\scriptstyle 1_A} & & \downarrow{\scriptstyle f} & & \\ 0 & \longrightarrow & \ker f\circ p & \xrightarrow{m_{f\circ p}} & A & \xrightarrow{f\circ p} & X' & \longrightarrow & 0 \end{array}$$

whence we deduce the commutative diagram:

$$\begin{array}{ccccccc} 0 & \longrightarrow & F(A,p) & \longrightarrow & F(A) & \longrightarrow & F(\ker p) \\ & & {\scriptstyle F(f\circ p,\,p)}\uparrow & & \uparrow{\scriptstyle 1_{F(A)}} & & \uparrow \\ 0 & \longrightarrow & F(A,f\circ p) & \longrightarrow & F(A) & \longrightarrow & F(\ker (f\circ p)) \end{array}$$

It follows that $F(f\circ p,p)$ is a monomorphism. But $(A,f\circ p) = f^*(A,p)$ and the image of $E(X)$ under f^* is cofinal in $E(X')$. Hence, $HF(X')$ may be obtained as an inductive limit taken over the image of f^*. Since $\mathscr{Ab}$

satisfies AB5, we deduce that $HF(f)$ is a monomorphism, as an inductive limit of monomorphisms.

Let

$$0 \longrightarrow F' \longrightarrow F \longrightarrow F''$$

be an exact sequence of objects in $\mathscr{C}^{\wedge}$ and X an object of $\mathscr{C}$. For any $(A, p) \in E(X)$ we have the commutative diagram

$$\begin{array}{ccccccc}
0 \longrightarrow & F'(A) & \longrightarrow & F(A) & \longrightarrow & F''(A) \\
& \downarrow{\scriptstyle F'(m_p)} & & \downarrow{\scriptstyle F(m_p)} & & \downarrow{\scriptstyle F''(m_p)} \\
0 \longrightarrow & F'(\ker p) & \longrightarrow & F(\ker p) & \longrightarrow & F''(\ker p)
\end{array}$$

which yields the exact sequence:

$$0 \longrightarrow F'(A, p) \longrightarrow F(A, p) \longrightarrow F''(A, p).$$

Thus, by taking the inductive limit over the category $\mathscr{E}(X)$ we get the exact sequence

$$0 \longrightarrow HF'(X) \longrightarrow HF(X) \longrightarrow HF''(X)$$

(the inductive limits are taken in $\mathscr{Ab}$ which satisfies AB5).

PROPOSITION 5.37. *Let F be an object of $\mathscr{C}^{\wedge}$ such that for any epimorphism $f: X \longrightarrow X'$ in $\mathscr{C}$, $F(f)$ is a monomorphism. Then HF is a left-exact functor.*

PROOF. Let $(A, p) \in E(X)$; then we have the commutative diagram in $\mathscr{C}$.

$$\begin{array}{ccccccccc}
& & & & 0 & & 0 & & \\
& & & & \downarrow & & \downarrow & & \\
0 & \longrightarrow & \ker p & \xrightarrow{m_{p'}} & A \,\Pi_X \ker f & \xrightarrow{p'} & \ker f & \longrightarrow & 0 \\
& & \downarrow{\scriptstyle 1_{\ker p}} & & \downarrow & & \downarrow{\scriptstyle f'} & & \\
0 & \longrightarrow & \ker p & \xrightarrow{m_p} & A & \xrightarrow{p} & X & \longrightarrow & 0 \\
& & & & \downarrow{\scriptstyle f \circ p} & & \downarrow{\scriptstyle f} & & \\
& & & & X' & \xrightarrow{1_X} & X' & &
\end{array}$$

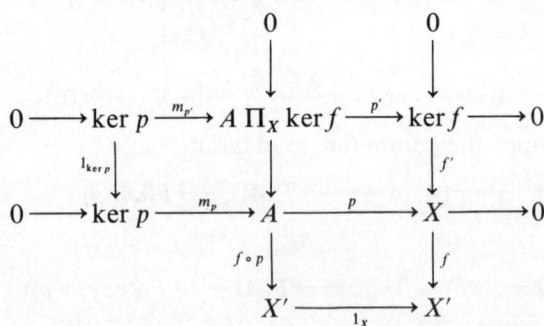

where $(A \,\Pi_X \ker f, p') = f'_*(A, p)$ and $(A, f \circ p) = f^*(A, p)$. From this diagram we deduce the following commutative diagram which has exact rows:

$$\begin{array}{ccccccc} & & & & 0 & & 0 \\ & & & & \downarrow & & \downarrow \\ 0 & \longrightarrow & F(A, f \circ p) & \longrightarrow & F(A, p) & \longrightarrow & F(A \Pi_X \ker f, p') \\ & & \downarrow & & \downarrow & & \downarrow \\ 0 & \longrightarrow & F(A, f \circ p) & \longrightarrow & F(A) & \longrightarrow & F(A \Pi_X \ker f). \end{array}$$

Since $f^*(E(X))$ is cofinal in $E(X')$, we obtain by taking the inductive limit over $\mathscr{E}(X)$ the exact sequence

$$0 \longrightarrow HF(X') \longrightarrow HF(X) \longrightarrow \varinjlim_{(A,p)\in \mathscr{E}(X)} F(f^*(A, p))$$

To finish the proof, we remark that if F is a functor which carries epimorphisms into monomorphism, then, for any $u:(A, p) \longrightarrow (A', p')$, $F(p', p)$ is a monomorphism. As a consequence of this fact it follows (since $\mathscr{A}b$ satisfies AB5) that the canonical morphism

$$\varinjlim_{(A,p)\in \mathscr{E}(X)} F(f'_*(A, p)) \longrightarrow HF(\ker f)$$

is a monomorphism, which completes the proof.

Let F be an object of $\mathscr{C}^\wedge$ and $\psi_F : F \longrightarrow HF$ the morphism constructed above, which is in fact a functorial morphism from the identity functor of the category $\mathscr{C}^\wedge$ to H. Proposition 5.36 implies that HF is a functor which carries epimorphisms into monomorphisms hence HHF is a left-exact functor according to proposition 5.37. Furthermore, $H\psi \circ \psi$ defines a functorial morphism from the identity functor of $\mathscr{C}^\wedge$ to the functor HH. We denote by $R^0 = HH$ and by $\rho = H\psi \circ \psi$. Thus for any object F of $\mathscr{C}^\wedge$ there exists an object R^0F of $\mathscr{S}$ and a functorial morphism $\rho_F : F \longrightarrow R^0F$ such that for each object G of $\mathscr{S}$ and any morphism $\alpha : F \longrightarrow G$ there exists a unique morphism $\alpha_0 : R^0F \longrightarrow G$ such that $\alpha_0 \rho_F = \alpha$. The construction and uniqueness of α_0 is carried out in a canonical manner starting from the definition of R^0F. Thus the assignment $F \rightsquigarrow R^0F$ is a functor from $\mathscr{C}^\wedge$ to $\mathscr{S}$ which is a left-adjoint of the functor I.

We now prove that the category $\mathscr{S}$ is an Abelian category and R^0 is an exact functor. We notice that $\mathscr{S}$ is an additive category and moreover, for any morphism $\alpha : F \longrightarrow G$ of $\mathscr{S}$, Ker α defined as being the functor $X \rightsquigarrow \ker \alpha_X$ for any X in $\mathscr{C}$, is a left-exact functor.

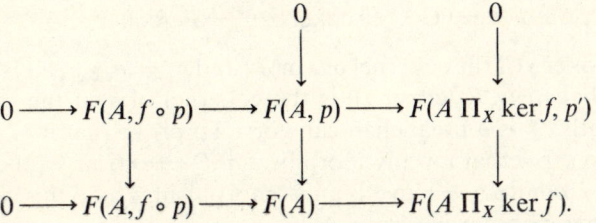

We define the cokernel of the morphism α by means of the diagram:

$$F \xrightarrow{\alpha} G \xrightarrow{\varphi} \operatorname{coker} \alpha \xrightarrow{\rho_1} R^0(\operatorname{coker} \alpha),$$

where $(\varphi, \operatorname{coker} \alpha)$ is the cokernel of α in $\mathscr{C}^\wedge$ and $p_1 = \rho_{\operatorname{coker} \alpha}$. It is immediate to check that $(\rho_1\varphi, R^0(\operatorname{coker} \alpha))$ is the cokernel of α in the category $\mathscr{S}$. Consequently $\mathscr{S}$ is a preabelian category. To prove that it is Abelian we have still to show that for any morphism $\alpha : F \longrightarrow G$ in $\mathscr{S}$, the canonical morphism $\alpha : \operatorname{coim} \alpha \longrightarrow \operatorname{im} \alpha$ is an isomorphism. To do this, we consider the following diagram in $\mathscr{C}^\wedge$:

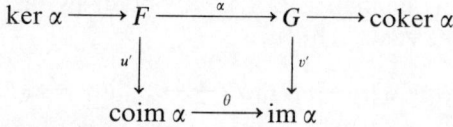

$$\begin{array}{ccccccc}
\ker \alpha & \longrightarrow & F & \xrightarrow{\alpha} & G & \longrightarrow & \operatorname{coker} \alpha \\
 & & \Big\downarrow{\scriptstyle u'} & & \Big\downarrow{\scriptstyle v'} & & \\
 & & \operatorname{coim} \alpha & \xrightarrow{\theta} & \operatorname{im} \alpha & &
\end{array}$$

where θ is an isomorphism. We infer from the definition of R^0F that if F is a left-exact functor, then $\rho_F : F \longrightarrow R^0F$ is an isomorphism. From the above diagram we obtain the diagram

$$\begin{array}{ccccccc}
\ker \alpha \longrightarrow G & & \xrightarrow{\quad\quad\alpha\quad\quad} & & F & \xrightarrow{\varphi} & \operatorname{coker} \alpha \\
\Big\downarrow{\scriptstyle u'} & \searrow^{u} & & \nearrow^{v} & \Big\uparrow{\scriptstyle v'} & \searrow^{p'} & \Big\downarrow{\scriptstyle \rho_1} \\
 & & R^0(\operatorname{coim} \alpha) \xrightarrow{\bar{\alpha}} R^0(\operatorname{im} \alpha) & & & & R^0(\operatorname{coker} \alpha) \\
 & \nearrow_{\mu} & & \nwarrow_{\nu} & & & \\
\operatorname{coim} \alpha & & \xrightarrow{\quad\quad\theta\quad\quad} & & \operatorname{im} \alpha & &
\end{array}$$

where $\mu = \rho_{\operatorname{coim} \alpha}$, $\nu = \rho_{\operatorname{im} \alpha}$. All the morphisms are constructed in a canonical way so that all the diagrams be commutative. In particular, $\bar{\alpha} = R^0\theta$. $R^0(\operatorname{im} \alpha)$ and $R^0(\operatorname{coim} \alpha)$ play the role of image and coimage for α in the category $\mathscr{S}$. Indeed, it is sufficient to prove that $R^0(\operatorname{im} \alpha)$ together with v is the kernel of $p' = \rho_1\varphi$. To do this, we notice that v is a monomorphism in the category $\mathscr{S}$ and $(p', R^0(\operatorname{coker} \alpha))$ is the cokernel of v. Thus $\mathscr{S}$ is an Abelian category.

The functor R^0 is left-exact as shown by proposition 5.37. It is also right-exact since it possesses a right-adjoint functor. Thus we obtain:

PROPOSITION 5.38. *Let $\mathscr{C}$ be a small Abelian category. We denote by $\mathscr{C}^\wedge$ the category of contravariant additive functors from $\mathscr{C}$ to $\mathscr{Ab}$ and by $\mathscr{S}$ the full subcategory of $\mathscr{C}^\wedge$ consisting of left-exact functors. $\mathscr{S}$ is an Abelian category and the inclusion functor $I : \mathscr{S} \longrightarrow \mathscr{C}^\wedge$ possesses an exact left-adjoint functor.*

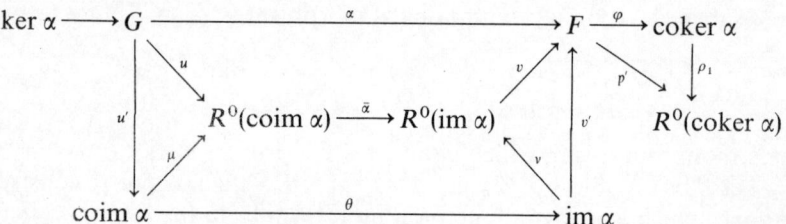

We use this opportunity to introduce the following notion.

Definition. Let $\mathscr{C}$, $\mathscr{C}'$ be two Abelian categories, $T:\mathscr{C} \longrightarrow \mathscr{C}'$ an exact functor and S a right-adjoint functor of T which is fully faithful. We say that the couple (T, S) defines the category $\mathscr{C}'$ as a *quotient category of* $\mathscr{C}$.

Remark. Our notion of a quotient category is less general than that of P. Gabriel.

The most important properties of the category $\mathscr{C}$ are inherited by the category $\mathscr{C}'$, as shown by the following

PROPOSITION 5.39. *Let $\mathscr{C}$ and $\mathscr{C}'$ be two Abelian categories, $T:\mathscr{C} \longrightarrow \mathscr{C}'$ a functor and $S:\mathscr{C}' \longrightarrow \mathscr{C}$ a right-adjoint of T such that the couple (T, S) defines $\mathscr{C}'$ as a quotient category of $\mathscr{C}$. Then, if $\mathscr{C}$ satisfies one of the axioms* AB3, AB4, AB5, AB3*, *then $\mathscr{C}'$ satisfies the same axiom. If $\mathscr{C}$ satisfies* AB4* *or* AB5* *and S is exact, then $\mathscr{C}'$ satisfies the same axiom. If $\mathscr{C}$ possesses a family of generators, then so does $\mathscr{C}'$.*

PROOF. We verify, for instance, that if $\mathscr{C}$ satisfies AB3, then so does $\mathscr{C}'$. Let $(X_i')_{i\in I}$ be a family of objects of $\mathscr{C}'$. Consider the family of objects of $\mathscr{C}$ $(S(X_i'))_{i\in I}$ and their direct sum $\bigoplus_{i\in I} S(X_i') = Y$; we denote by $\alpha_i:S(X_i') \longrightarrow Y$ the canonical injections. Then $T(Y)$ together, with the morphisms $T(\alpha_i)$ is a direct sum of the family $(T(S(X_i'))_{i\in I}$; since the functor S is fully faithful, it follows that $TS(X_i')$ is isomorphic with X_i' for any $i \in I$. The other conditions are verified in a similar manner.

If $(U_i)_{i\in I}$ is a family of generators for the category $\mathscr{C}$, then $(T(U_i))_{i\in I}$ is a family of generators for the category $\mathscr{C}'$. The verification of this assertion is straightforward.

It follows from proposition 5.39 that, under the hypothesis of proposition 5.38, the category $\mathscr{S}$ is a quotient category of $\mathscr{C}^{\wedge}$ and is therefore a Grothendieck category. To complete the proof of the programme proposed at the end of the preceding paragraph, we now prove the following:

THEOREM 5.40. *Under the hypothesis and notations of proposition 5.38, the functor $R^0 \circ h$, where $h:\mathscr{C} \longrightarrow \mathscr{C}^{\wedge}$ is the functor which associates to X the functor $\mathrm{Hom}_{\mathscr{C}}(h_X, X)$, is a full embedding of $\mathscr{C}$ into $\mathscr{S}$, that is, any small Abelian category can be fully embedded into a Grothendieck category.*

PROOF. For any object X of $\mathscr{C}$, the functor h_X is left-exact, hence $R^0 h_X$ is isomorphic under ρ_{h_X} with h_X. This implies that for any objects X, Y of $\mathscr{C}$

$$R^0(h_X, h_Y):\mathrm{Hom}_{\mathscr{C}^{\wedge}}(h_X, h_Y) \longrightarrow \mathrm{Hom}_{\mathscr{S}}(R^0 h_X, R^0 h_Y)$$

is an isomorphism. Consequently, $R^0 \circ h$ is a fully faithful functor. It remains to prove that it is exact.

As a composition of two left-exact functors, $R^0 \circ h$ is left-exact. We have still to show that it is right-exact and to this end it is sufficient to prove that it preserves epimorphisms. Let $f: Z \longrightarrow Y$ be an epimorphism in $\mathscr{C}$. $R^0 h(f)$ is an epimorphism if and only if $R^0(\operatorname{coker} h(f)) = 0$ or $H(\operatorname{coker} h(f)) = 0$, where H is the functor constructed at the beginning of this paragraph. Let X be an object of $\mathscr{C}$; according to the definition of H,

$$H(\operatorname{coker} h(f)(X)) = \varinjlim_{(A,p) \in E(X)} \operatorname{coker} h(f)(A, p).$$

Consider the diagram:

$$\begin{array}{ccccccc}
h_Z(X) & \xrightarrow{h(f)(X)} & h_Y(X) & \longrightarrow & \operatorname{coker} h(f)(X) & \longrightarrow & 0 \\
{\scriptstyle h_Z(p)}\downarrow & & {\scriptstyle h_Y(p)}\downarrow & & \downarrow & & \\
h_Z(A) & \longrightarrow & h_Y(A) & \longrightarrow & \operatorname{coker} h(f)(A) & \longrightarrow & 0.
\end{array}$$

Let $a \in h_Y(A)$; then $a: A \longrightarrow Y$ is a morphism of $\mathscr{C}$. From the commutative diagram:

$$\begin{array}{ccc}
Z & \xrightarrow{f} & Y \\
{\scriptstyle a'}\uparrow & & \uparrow{\scriptstyle a} \\
Z \prod_Y A & \xrightarrow[f']{} & A
\end{array}$$

constructed in a canonical manner, we deduce $h(f')(a) = h(f)(a)$, which shows, according to the construction of $H(\operatorname{coker} h(f))$, that $H(\operatorname{coker} h(f)(X)) = 0$, using the properties of inductive limits for Abelian groups. Thus $R^0(\operatorname{coker} h(f)) = 0$, which completes the proof.

We have assumed above that $\mathscr{C}$ is a small category because we have used—to construct certain functors—inductive limits, and we know how to construct these limits only when the indices form a set. Certain foundational difficulties arise in the case of an arbitrary Abelian category. A method to circumvent these difficulties is furnished by the following general result.

PROPOSITION 5.41. *Let $\mathscr{C}$ be an Abelian category and $(A_i)_{i \in I}$ a set of objects of $\mathscr{C}$. Then there exists in $\mathscr{C}$ a small Abelian full subcategory $\mathscr{A}$ such that A_i is an object of $\mathscr{A}$ for any $i \in I$.*

PROOF. Denote by $\mathscr{A}_0$ the full subcategory of $\mathscr{C}$ consisting of the objects A_i. $\mathscr{A}_0$ is a small category. We denote by $\mathscr{A}_1$ the full subcategory of $\mathscr{C}$

formed by all the objects of $\mathscr{A}_0$, all the kernels and cokernels of the morphisms in $\mathscr{A}_0$ (for any morphism in $\mathscr{A}_0$, we choose a single kernel and a single cokernel) and the finite direct sums formed with the objects of $\mathscr{A}_0$. $\mathscr{A}_1$ is a small subcategory of $\mathscr{C}$. By iteration, denote by $\mathscr{A}_{n+1} = (\mathscr{A}_n)_1$ $(n = 1, 2, \cdots)$. For any n, $\mathscr{A}_n$ is a full small subcategory of $\mathscr{C}$. Obviously $\mathscr{A}_n \subset \mathscr{A}_{n+1}$. Denote $\mathscr{A} = \bigcup_{n=0,1,2,\cdots} \mathscr{A}_n$, i.e. the smallest full subcategory of $\mathscr{C}$ which contains simultaneously $\mathscr{A}_n$ for $n = 0, 1, 2, \cdots$. Clearly $\mathscr{A}$ is a small subcategory of $\mathscr{C}$. $\mathscr{A}$ is Abelian; for, let $f: A \longrightarrow B$ be a morphism in $\mathscr{A}$. There exists an n such that A and B are objects of $\mathscr{A}_n$. Then $\ker f$ and $\operatorname{coker} f$ are objects in $\mathscr{A}_{n+1}$ together with the canonical morphisms, etc.

Proposition 5.41 enables us to study the Abelian categories locally. An application of this proposition is to be found in chapter 6.

CHAPTER 6

Injective and Projective Objects in Abelian Categories

1. The Notion of an Injective (Projective) Object and its General Properties

Definition. Let $\mathscr{C}$ be an Abelian category and Q an object of $\mathscr{C}$. We say that Q is *injective* if for any monomorphism $\alpha: A' \longrightarrow A$ and any morphism $u: A' \longrightarrow Q$ there exists a morphism $v: A \longrightarrow Q$ such that $v\alpha = u$. Using the language of diagrams this means that any diagram of the form

$$\begin{array}{ccccc} 0 & \longrightarrow & A' & \longrightarrow & A \\ & & \downarrow & & \\ & & Q & & \end{array}$$

where the row $0 \longrightarrow A' \longrightarrow A$ is exact can be imbedded into a commutative diagram of the form,

$$\begin{array}{ccccc} 0 & \longrightarrow & A' & \longrightarrow & A \\ & & \downarrow & \swarrow & \\ & & Q & & \end{array}$$

PROPOSITION 6.1. *Let $\mathscr{C}$ be an Abelian category and $(Q_i)_{i \in I}$ a family of objects of $\mathscr{C}$ such that the direct product $Q = \prod_{i \in I} Q_i$ exists. Then Q is injective if and only if each Q_i is injective.*

PROOF. Let $\pi_i: Q \longrightarrow Q_i$ be the canonical projections and $\alpha_i: Q_i \longrightarrow Q$ morphisms such that $\pi_i \alpha_i = 1_{Q_i}$. If Q is injective and $f: X' \longrightarrow X$ is a monomorphism in $\mathscr{C}$ and $g: X' \longrightarrow Q_i$ is a morphism in $\mathscr{C}$, then $\alpha_i \circ g: X' \longrightarrow Q$. There exists $h: X \longrightarrow Q$ with $h \circ f = \alpha_i \circ g$ and therefore $\pi_i \circ h \circ f = \pi_i \circ \alpha_i \circ g = g$. Hence Q_i is injective.

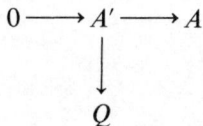

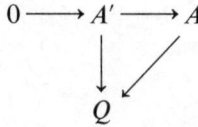

Conversely, assume that Q_i is injective for any $i \in I$, and let $g: X' \longrightarrow Q$ be a morphism in $\mathscr{C}$. Let $h_i: X \longrightarrow Q_i$ be such that $\pi_i \circ g = h_i \circ f$. From the definition of direct products it follows that there exists a unique morphism $h:X \longrightarrow Q$ such that $\pi_i \circ h = h_i$. Clearly $h \circ f = g$, so that Q is injective.

PROPOSITION 6.2. *If the subobject Q of C is injective, then Q is a direct summand of C.*

PROOF. Let $i:Q \longrightarrow C$ be the canonical injection of Q in C. It is sufficient to show that there exists $p:C \longrightarrow Q$ such that $pi = 1_Q$. However, if we consider the diagram

$$\begin{array}{ccc} Q & \xrightarrow{i} & C \\ {\scriptstyle 1_Q}\downarrow & & \\ Q & & \end{array}$$

the fact that Q is injective yields a morphism $p:C \longrightarrow Q$ such that $pi = 1_Q$.

PROPOSITION 6.3. *Let $\mathscr{C}_1$ and $\mathscr{C}_2$ be two Abelian categories. Assume that the functor $G:\mathscr{C}_2 \longrightarrow \mathscr{C}_1$ is the adjoint of the functor $F:\mathscr{C}_1 \longrightarrow \mathscr{C}_2$. Under these conditions, if the functor F is exact, then G transforms any injective object of $\mathscr{C}_2$ into an injective object of $\mathscr{C}_1$.*

PROOF. Let Q be an injective object of $\mathscr{C}_2$. We have to prove that $G(Q)$ is an injective object of the category $\mathscr{C}_1$. To do this, it is sufficient to show that for any exact sequence

$$0 \longrightarrow X' \longrightarrow X \longrightarrow X'' \longrightarrow 0$$

in the category $\mathscr{C}_1$, the sequence of Abelian groups and homomorphisms of Abelian groups

$$\begin{aligned} 0 \longrightarrow \mathrm{Hom}_{\mathscr{C}_1}(X'', G(Q)) &\longrightarrow \mathrm{Hom}_{\mathscr{C}_1}(X, G(Q)) \\ &\longrightarrow \mathrm{Hom}_{\mathscr{C}_1}(X', G(Q)) \longrightarrow 0 \end{aligned} \qquad (*)$$

is exact. But the hypothesis implies that the sequence

$$0 \longrightarrow F(X') \longrightarrow F(X) \longrightarrow F(X'') \longrightarrow 0$$

is exact. Hence, Q being injective, the sequence

$$0 \longrightarrow \mathrm{Hom}_{\mathscr{C}_2}(F(X''), Q) \longrightarrow \mathrm{Hom}_{\mathscr{C}_2}(F(X), Q) \longrightarrow \mathrm{Hom}_{\mathscr{C}_2}(F(X'), Q) \longrightarrow 0$$

is exact. Since we have the commutative diagram

$$\begin{array}{ccccccccc}
0 & \longrightarrow & \mathrm{Hom}_{\mathscr{C}_2}(F(X''),Q) & \longrightarrow & \mathrm{Hom}_{\mathscr{C}_2}(F(X),Q) & \longrightarrow & \mathrm{Hom}_{\mathscr{C}_2}(F(X'),Q) & \longrightarrow & 0 \\
 & & \downarrow & & \downarrow & & \downarrow & & \\
0 & \longrightarrow & \mathrm{Hom}_{\mathscr{C}_1}(X'',G(Q)) & \longrightarrow & \mathrm{Hom}_{\mathscr{C}_1}(X,G(Q)) & \longrightarrow & \mathrm{Hom}_{\mathscr{C}_1}(X',G(Q)) & \longrightarrow & 0
\end{array}$$

where the vertical arrows are isomorphisms, it follows that the sequence (*) is exact.

The notion dual to that of injective object is called projective object. We leave it to the reader to formulate the propositions dual to propositions 6.1, 6.2, and 6.3.

Definition. Let $\mathscr{C}$ be an Abelian category. We say that $\mathscr{C}$ *is with sufficiently many injectives* if any object of $\mathscr{C}$ is a subobject of an injective object of $\mathscr{C}$.

The Abelian categories with sufficiently many injectives represent the best setting for 'doing homological algebra'. Before studying in detail a few such categories, we prove the following proposition which is in a sense a converse of proposition 6.3.

PROPOSITION 6.4. *Let $\mathscr{C}$, $\mathscr{C}'$ be two Abelian categories, $T:\mathscr{C} \longrightarrow \mathscr{C}'$ an additive functor and $S:\mathscr{C}' \longrightarrow \mathscr{C}$ a right adjoint of T. If $\mathscr{C}$ is a category with sufficiently many injectives and S preserves the injective objects, then T is exact.*

PROOF. It is sufficient to prove that T preserves monomorphisms, i.e. if $i:X \longrightarrow Y$ is a monomorphism in $\mathscr{C}$ then $T(i):T(X) \longrightarrow T(Y)$ is a monomorphism in $\mathscr{C}'$. Let $\Psi:1_{\mathscr{C}} \longrightarrow ST$ and $\Phi:TS \longrightarrow 1_{\mathscr{C}'}$ be functorial morphisms such that $\Phi(T(A))T(\Psi(A)) = 1_{T(A)}$ and $S(\Phi(B))\Psi(S(B)) = 1_{S(B)}$ for any A in $\mathscr{C}$ and B in $\mathscr{C}'$ (cf. proposition 1.13), and let $i:X \longrightarrow Y$ be a monomorphism in $\mathscr{C}$. We have in $\mathscr{C}'$ the diagram

$$\begin{array}{ccc}
 & & T(Y) \\
 & \overset{T(i)}{\nearrow} & \\
T(X) & & \\
 & \underset{f}{\searrow} & \\
 & & Q
\end{array}$$

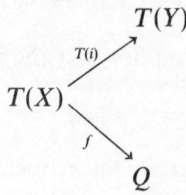

where f is a monomorphism and Q is an injective object. From this diagram we obtain the following commutative diagram

$$\begin{array}{ccccc} & & X & \xrightarrow{i} & Y \\ & & \downarrow{\scriptstyle \Psi(X)} & & \downarrow{\scriptstyle \Psi(Y)} \\ S(Q) & \xleftarrow{S(f)} & ST(X) & \xrightarrow{ST(i)} & ST(Y). \end{array}$$

Let $\alpha : Y \longrightarrow S(Q)$ be a morphism in $\mathscr{C}$ such that $\alpha \circ i = S(f) \circ \Psi(X)$. By applying the functor T to this diagram and taking into account the relations:

$$\Phi(T(X)) \circ T(\Psi(X)) = 1_{T(X)}, \qquad f \circ \Phi(T(X)) = \Phi(Q) \circ TS(f),$$

we get: $f = \Phi(Q) \circ TS(f) \circ T\Psi(X)$. But $T(\alpha) \circ T(i) = TS(f) \circ T\Psi(X)$, hence $f = \Phi(Q) \circ T(\alpha) \circ T(i)$, whence we conclude that $T(i)$ is a monomorphism.

We shall now indicate a few important categories with sufficiently many injectives. More precisely, we shall prove the following.

PROPOSITION 6.5. *Let Λ be a ring with a unity element. The category ${}_\Lambda\mathscr{C}$ is with sufficiently many injectives.*

PROOF. We shall prove the proposition in several steps.

(a) In the category $\mathscr{Ab}$ an object Q is injective if and only if Q is a divisible group.

For, let Q be an injective object in the category $\mathscr{Ab}$, $y_0 \in Q$ and $n \in Z$, $n \neq 0$. We must produce an element $y \in Q$ such that $ny = y_0$. To do this, consider the homomorphism of Abelian groups $\varphi : Z \longrightarrow Q$ defined by

$$\varphi(m) = my_0 \qquad \text{for any} \quad m \in Z.$$

Consider the diagram

$$\begin{array}{ccccc} 0 & \longrightarrow & Z & \xrightarrow{\psi} & Z \\ & & \downarrow{\scriptstyle \varphi} & & \\ & & Q & & \end{array}$$

where ψ is defined by $\psi(m) = nm$. Since Q is by hypothesis injective, there exists $u : Z \longrightarrow Q$ such that $u\psi = \varphi$. Let $y = u(1)$. We have $y_0 = \varphi(1) = u(\psi(1)) = u(n) = nu(1) = ny$.

Conversely, let Q be a divisible group. Let A' be a subgroup of A and $u : A' \longrightarrow Q$ a homomorphism of Abelian groups. We have to show that u can be extended to A. Consider the set of pairs (V, v) where V is a subgroup of A which contains A' and v is a homomorphism of V into Q which extends u. We introduce an order relation into this set as follows: $(V', v') \leq (V'', v'')$ if and only if $V' \subset V''$ and v'' extends v'. The Zorn Lemma shows that there exists a maximal element (V_0, v_0) in this set. We show that $V_0 = A$. Indeed, assume that there exists $x \in A$ such that $x \notin V_0$. Consider now the subgroup Zx of A. Two cases may arise:

I. $Zx \cap V_0 \neq \{0\}$.
II. $Zx \cap V_0 = \{0\}$.

Since case II is easily seen to lead to an absurdity, we deal only with case I. Let n_0 be the least positive integer such that $n_0 x \in V_0$. Clearly $Zx \cap V_0$ coincides with the subset of elements of the form $mn_0 x$ where m runs through the set Z. Consider the subgroup $V_0 + Zx = V$. We show that there exists $v: V \longrightarrow Q$ such that v extends v_0, which will complete the proof. Since $n_0 x \in V_0$ it follows that $v_0(n_0 x)$ is defined. Let $y_0 = v_0(n_0 x)$. Since Q is divisible, there exists $y \in Q$ such that $n_0 y = y_0$. Let $z \in V_0 + Zx$, i.e. $z = a + nx$, $a \in V_0$, $n \in Z$. Set $v(z) = v_0(a) + ny$. $v(z)$ does not depend on the representation of z in the form $z = a + nx$. For, if $z = b + mx$, $b \in V_0$, $m \in Z$, then $a - b = (m - n)\, x \in V_0 \cap Zx$ and therefore $m - n = n'n_0$. Hence

$$v_0(a - b) = v_0(a) - v_0(b) = v_0(n'n_0 x) = n'v_0(n_0 x)$$
$$= n'y_0 = n'n_0 y = (m - n)y.$$

Hence $v_0(a) + ny = v_0(b) + my$. Clearly v is a homomorphism which extends v_0. Thus Q is injective.

(b) Let $\varphi : \Lambda \longrightarrow \Gamma$ be a homomorphism of rings with a unity element ($\varphi(1) = 1$). If the category ${}_\Lambda\mathscr{C}$ is with sufficiently many injectives then the category ${}_\Gamma\mathscr{C}$ is also with sufficiently many injectives.

Proof. Consider the functors

$$F : {}_\Gamma\mathscr{C} \longrightarrow {}_\Lambda\mathscr{C}, \qquad G : {}_\Lambda\mathscr{C} \longrightarrow {}_\Gamma\mathscr{C}$$

defined by

$$F(A) = A_{[\varphi]}, \qquad G(B) = \mathrm{Hom}_\Lambda(\Gamma_{[\varphi]}, B).$$

We recall that the left Γ-module structure of $G(B)$ is defined as follows: if $u : \Gamma_{[\varphi]} \longrightarrow B$ then by definition $(\gamma u)(x) = u(x\gamma)$ for any $x \in \Gamma_{[\varphi]} = \Gamma$. The functor G is an adjoint of the functor F, and F is an exact functor. According to proposition 6.3 G then transforms an injective Λ-module into an injective Γ-module. Let now B be an arbitrary Γ-module. $B_{[\varphi]}$ is a Λ-module. By hypothesis there exists an injective Λ-module Q and an imbedding $\mu: B_{[\varphi]} \longrightarrow Q$. G being left-exact we get the imbedding (of Γ-modules) $G(\mu): G(B_{[\varphi]}) \longrightarrow G(Q)$. But $G(Q)$ is injective. We show that we can imbed B into $G(Q)$. It is obviously sufficient to imbed B into $G(B_{[\varphi]})$. This is achieved as follows: To each $b \in B = B_{[\varphi]}$ we associate the homomorphism $\hat{b} \in \mathrm{Hom}_\Lambda(\Gamma_{[\varphi]}, B_{[\varphi]})$ defined by $\hat{b}(\gamma) = \gamma b$. If we take into account the homomorphism $\mathrm{Hom}_\Lambda(\Gamma_{[\varphi]}, B_{[\varphi]}) \longrightarrow B_{[\varphi]}$ which associates to each $u \in \mathrm{Hom}_\Lambda(\Gamma_{[\varphi]}, B_{[\varphi]})$ the value $u(1)$ it follows that with the map $b \longrightarrow \hat{b}$ we have realized an imbedding of B into $G(B_{[\varphi]})$, whence our assertion.

(c) The category $\mathscr{Ab}$ is with sufficiently many injectives.

Proof. In view of (a) we must prove that any Abelian group can be imbedded into a divisible group. This is clear for the case of the group Z and hence for any free group. Since any factor group of a divisible group is divisible and the direct sum of a family of divisible groups is a divisible group, we infer that any Abelian group can be imbedded into a divisible group.

(d) For any ring Λ with a unity element, the category ${}_\Lambda\mathscr{C}$ is with sufficiently many injectives.

Proof. It follows from (b) and (c) by considering the ring homomorphism $\varphi: Z \longrightarrow \Lambda$ which sends the unity of Z into the unity of Λ.

2. Essential Extensions

Proposition 6.6. *Let $\mathscr{C}$ be an Abelian category and M and P two objects of $\mathscr{C}$ such that $P \supset M$. Under these conditions the following assertions are equivalent:*

(i) *For any subobject Q of P, $Q \cap M = 0$ implies $Q = 0$.*

(ii) *Any morphism $u: P \longrightarrow A$ which induces a monomorphism on M is a monomorphism.*

Proof. (i) $\Rightarrow$ (ii). Let $Q = \ker u$. By hypothesis $u\alpha: M \longrightarrow A$, where $\alpha: M \longrightarrow P$ is the canonical injection of M into P, is a monomorphism. We have to prove that $\ker u = 0$. It is sufficient to show that $Q \cap M = 0$.

Otherwise, we would have $0 \neq Q \cap M \subset \ker(u\alpha)$, as seen from the following commutative diagram, where $(M \cap Q, \mu)$ is the kernel of $\beta\alpha$:

$$\begin{array}{ccccccc} M \cap Q & \xrightarrow{\mu} & M & \xrightarrow{\alpha} & P & \xrightarrow{\beta} & P/Q: \\ & & & & \Big\downarrow \mu & \swarrow p & \\ & & & & A & & \end{array}$$

But this contradicts the fact that $u\alpha$ is a monomorphism.

(ii) $\Rightarrow$ (i). Let $Q \subset P$, $Q \cap M = 0$. Consider the morphism $P \xrightarrow{\beta} P/Q$. The kernel of the morphism $\beta\alpha$ is $Q \cap M = 0$. Hence $\beta\alpha$ is a monomorphism. It follows that β is a monomorphism and thus $Q = 0$.

Definition. The pair (P, M) where M is a subobject of P is said to be an *essential extension* of M if the equivalent conditions (i), (ii) are satisfied. We will also say that $\mu : M \longrightarrow P$ is an *essential monomorphism* if μ is a monomorphism and if the pair (P, M) is an essential extension of M. Still simpler, we will say that P is an essential extension of M.

It is sometimes interesting to consider also the dual definition, which we explicitate for the convenience of the reader.

Definition. An epimorphism $f : M \longrightarrow M'$ is said to be *essential* if for any morphism $g : X \longrightarrow M$ such that fg is an epimorphism, g is an epimorphism.

Examples.

PROPOSITION 6.7. *Let A be a commutative ring with unity element, and let ${}_A\mathscr{C}$ be the category of left A-modules. In order that the A-module P be an essential extension of its submodule M it is necessary and sufficient that for any $\xi \in P$, $\xi \neq 0$, we have $A\xi \cap M \neq \{0\}$.*

PROOF. The necessity is evident, since $A\xi \neq \{0\}$. For the sufficiency, let Q be a submodule of P such that $Q \cap M = \{0\}$. We have to show that $Q = \{0\}$. Assume $0 \neq \xi \in Q$. Then $A\xi \cap M \neq \{0\}$, which is absurd in view of $A\xi \cap M \subset Q \cap M$.

Example 1. Let A be a domain of integrity, considered as an object of the category ${}_A\mathscr{C}$. Let B be the field of quotients of A, considered also as an object of ${}_A\mathscr{C}$. It follows immediately from proposition 6.7 that B is an essential extension of F.

2. Let A be a local ring of maximal ideal m, and let k be the residual field A/m. Consider the epimorphism $f : A \longrightarrow A/m$. (All is conceived,

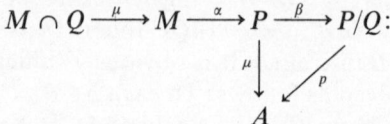

of course, in the category ${}_A\mathscr{C}$). f is an essential epimorphism. Indeed, let X be an A-module and $g: X \longrightarrow A$ such that fg is surjective. Thus $g(X) + m = A$. From this it follows by the Lemma of Nakayama that $g(X) = A$, so that g is an epimorphism.

Definition. An essential extension (I, M) of the object M where I is an injective object is called an *injective envelope* of the object M.

We shall see (corollary 6.19) that in a Grothendieck category any object possesses at least one injective envelope and that two injective envelopes are isomorphic.

Examples of injective envelopes

1. Consider in the category $\mathscr{Ab}$ of Abelian groups the additive group of integers Z. Then clearly the additive group of rationals Q is an injective envelope of Z.

2. Let p be a prime. Consider in the category $\mathscr{Ab}$ the multiplicative group Z_{p^∞} of all complex numbers which satisfy an equation of the form

$$z^{p^n} = 1, \qquad n > 0.$$

Let m be an arbitrary positive integer. Clearly $Z_{p^m} \subset Z_{p^\infty}$. We prove that Z_{p^∞} is an injective envelope of Z_{p^m}.

(a) Z_{p^∞} is injective. It is sufficient to show that Z_{p^∞} is infinitely divisible. Let $\zeta \in Z_{p^\infty}$. We have to check that for any integer $n > 0$ there exists $\zeta_1 \in Z_{p^\infty}$ such that $\zeta_1^n = \zeta$. We may obviously prove this only for prime numbers.

If $n = p$, this is evidently the case. Assume now n is prime with p and $\zeta^{p^r} = 1$. We have

$$p^r s + nq = 1.$$

Hence

$$\zeta = \zeta^{p^r s + nq} = \zeta^{p^r s}\zeta^{nq} = (\zeta^q)^n.$$

Thus Z_{p^∞} is infinitely divisible and therefore injective.

(b) The extension (Z_{p^∞}, Z_{p^m}) is essential for any $m > 0$. Since $Z_{p^\infty} = \sup_{m' \geq m} Z_{p^{m'}}$, it is sufficient to prove that the extensions $(Z_{p^{m''}}, Z_{p^m})$, $m' > m$, are essential. This results from the fact that, as easily seen, the extension $(Z_{p^{m+1}}, Z_{p^m})$ is essential.

3. Properties of Injective Envelopes

LEMMA 6.8. *If $M_1 \subset P_1$, $M_2 \subset P_2$ are essential extensions, then $M_1 \oplus M_2 \subset P_1 \oplus P_2$ is an essential extension.*

PROOF. Let Q be a subobject of $P_1 \oplus P_2$. We must prove that if $Q \neq 0$ then $Q \cap (M_1 \oplus M_2) \neq 0$. Let p_1, p_2 be the canonical projections of $P_1 \oplus P_2$ onto P_1 and P_2. From the hypothesis it follows that either $p_1(Q) \neq 0$ or $p_2(Q) \neq 0$. Assume $p_1(Q) \neq 0$. Since $p_1(Q) \subset P_1$ it follows, by the fact that $M_1 \subset P_1$ is an essential extension, that $p_1(Q) \cap M_1 \neq 0$. Thus we have $Q \cap p_1^{-1}(M_1) = Q_1 \neq 0$.

Two cases are possible: either $p_2(Q_1) \neq 0$ or $Q_1 \subset p_2^{-1}(M_2)$. In the first case it follows that $p_2(Q_1) \cap M_2 \neq 0$ and therefore $Q_1 \cap p_2^{-1}(M_2) \neq 0$. But $Q_1 \cap p_2^{-1}(M_2) \subset M_1 \oplus M_2$. In the second case it follows that $Q_1 \subset M_1 \oplus M_2$.

PROPOSITION 6.9. *If I_1 is the injective envelope of M_1 and I_2 is the injective envelope of M_2 then, $I_1 \oplus I_2$ is the injective envelope of $M_1 \oplus M_2$.*

The proof is obvious in view of lemma 6.8.

PROPOSITION 6.10. *Let $\mathscr{C}$ be an Abelian category and X an object of $\mathscr{C}$. If Q_1 and Q_2 are two injective envelopes of X, then Q_1 and Q_2 are isomorphic.*

PROOF. Let $f_i : X \longrightarrow Q_i$ $(i = 1, 2)$ be the canonical monomorphisms. There exists $\alpha : Q_1 \longrightarrow Q_2$ such that $\alpha f_1 = f_2$. α is a monomorphism. For, if $Y = \ker \alpha$, then $Y \cap X \neq 0$ and therefore $\ker(\alpha \circ f_1) \neq 0$, which contradicts the fact that f_2 is a monomorphism.

α is an epimorphism, since otherwise $\operatorname{im} \alpha$ is a direct summand of Q_2 which contains $\operatorname{im} f_2$ and this contradicts the fact that Q_2 is an essential extension.

In what follows we prove a proposition which shows the existence of injective envelopes in a very large variety of categories.

PROPOSITION 6.11. *Let Λ be a ring with a unity element. Then in ${}_\Lambda\mathscr{C}$ any object possesses an injective envelope.*

PROOF. We first prove the following:

LEMMA. *A module I is injective if and only if I has no proper essential extension.*

PROOF. Assume I is injective; let $I \longrightarrow B$ be an essential extension of I. According to proposition 6.2, I is then a direct summand of B, so that I is isomorphic with B.

Conversely, assume I has no proper essential extension. We prove that any extension $I \longrightarrow B$ is split, which in view of propositions 6.5 and 6.1 shows that I is injective. Consider the set $\mathscr{M}$ of all submodules X of B such that $X \cap I = 0$; $\mathscr{M}$ is an inductive set. For, if $(X_\alpha)_{\alpha \in I}$ is subset of $\mathscr{M}$ which is linearly ordered by inclusion, then the union $X = \bigcup_{\alpha \in I} X_\alpha$ is a submodule of B with $X \cap I = 0$, hence X belongs to $\mathscr{M}$. According to Zorn's lemma, $\mathscr{M}$ possesses a maximal element M. The composition

$$I \xrightarrow{\ i\ } B \xrightarrow{\ p\ } B/M$$

is an essential extension of I, which is immediately seen. But then $p \circ i$ is by hypothesis an isomorphism. It follows that $B = I \oplus M$.

To prove proposition 6.11, let A be a left Λ-module and $A \longrightarrow Q_0$ a monomorphism with Q_0 injective, which exists by proposition 6.5. Let $\mathscr{N}$ be the set of all submodules X of Q_0 such that X is an essential extension of A; $\mathscr{N}$ is an inductive set, since if $(X_\alpha)_{\alpha \in I}$ is a subset of $\mathscr{N}$ which is linearly ordered by inclusion, then $\bigcup_{\alpha \in I} X_\alpha$ is an essential extension of A. According to Zorn's lemma, the set $\mathscr{N}$ possesses a maximal element Q and Q is an essential extension of A.

Let now Q_1 be an essential extension of Q. Then there exists a monomorphism $\alpha : Q_1 \longrightarrow Q_0$ such that the diagram

$$\begin{array}{ccc} Q & \longrightarrow & Q_0 \\ \downarrow & \nearrow_{\alpha} & \\ Q_1 & & \end{array}$$

is commutative, where $Q \longrightarrow Q_0$ is the canonical monomorphism which defines Q as a subobject of Q_0. But this implies that Q is isomorphic with Q_1, since otherwise the maximality of Q in $\mathscr{N}$ would be contradicted. Thus Q has no proper essential extensions. Consequently Q is injective by the lemma.

Remark. Let $\mathscr{C}$ be an Abelian category satisfying condition AB5 and X an object of $\mathscr{C}$. If $X \subset Q_0$ and Q_0 is an injective object of $\mathscr{C}$, then we may paraphrase the above proof to show that X possesses an injective envelope.

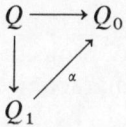

4. Projective Objects

The notion of a projective object is dual to that of an injective object, as shown by the following:

Definition. Let $\mathscr{C}$ be an Abelian category and P an object of $\mathscr{C}$. We say that P is projective if for any epimorphism $\alpha: A \longrightarrow A'$ and any morphism $u: P \longrightarrow A'$ there exists a morphism $v: P \longrightarrow A$ such that $\alpha v = u$.

Using the language of diagrams this means that any diagram of the form

$$\begin{array}{ccccc} 0 & \longleftarrow & A' & \longleftarrow & A \\ & & \uparrow & & \\ & & P & & \end{array}$$

where the row $0 \longleftarrow A' \longleftarrow A$ is exact can be imbedded into a commutative diagram of the form

$$\begin{array}{ccccc} 0 & \longleftarrow & A' & \longleftarrow & A. \\ & & \uparrow & \nearrow & \\ & & P & & \end{array}$$

We may dualize for projective objects propositions 6.1 and 6.2 in an evident manner. Also, we may dualize propositions 6.3 and 6.4, the dualization consisting in interchanging the words 'right-adjoint' and 'left-adjoint,' and 'injective' and 'projective.'

The following proposition characterizes the projective objects in the category ${}_{\Lambda}\mathscr{C}$.

PROPOSITION 6.12. *In order that a Λ-module P be projective it is necessary and sufficient that P be a direct summand of a free Λ-module.*

PROOF. Let F_P be the Λ-module consisting of all formal finite sums $\Sigma \lambda_i x_i$, where $\lambda_i \in \Lambda$, $x_i \in P$. F_P is a free Λ-module. The map

$$1x \longrightarrow x$$

extends to a homomorphism $\varphi: F_P \longrightarrow P$. Consider the exact sequence

$$0 \longrightarrow \ker \varphi \longrightarrow F_P \xrightarrow{\varphi} P \longrightarrow 0.$$

Now, if P is projective, then there exists a morphism: $\psi: P \longrightarrow F_P$ such that $\varphi\psi = 1_P$. This means that the sequence splits and P is a direct summand of the free Λ-module F_P.

We have still to prove that each direct summand of a free Λ-module is projective. According to the dual of proposition 6.1 it is sufficient to show that every free Λ-module is projective. Let then F be a free Λ-module with base $(x_i)_{i \in I}$. Let $g: A \longrightarrow B$ be an epimorphism and let $f: F \longrightarrow B$. For each $i \in I$ select an element $y_i \in A$ with the property that $g(y_i) = f(x_i)$. There exists a morphism $h: F \longrightarrow A$ such that $h(x_i) = y_i$ for each $i \in I$. We have $gh = f$, so that F is projective.

COROLLARY 6.13. *Each Λ-module A is a quotient of a projective Λ-module* (*The dual of proposition* 6.5).

PROOF. It is sufficient to consider the epimorphism $F_A \longrightarrow A$ defined in the proof of proposition 6.12.

It is not true that any object of ${}_\Lambda\mathscr{C}$ possesses a projective envelope.

In chapter 7 we indicate examples of categories in which any object possesses a projective envelope.

5. Localization in Rings

We propose to describe in the following all the quotient categories of the category $\mathscr{C}_\Lambda$ of right Λ-modules, where Λ is a ring with a unity element. This problem is closely related to the notion of an injective envelope.

We begin with a few general considerations on the quotient categories defined in chapter 5.

PROPOSITION 6.14. *Let $\mathscr{C}, \mathscr{C}'$ be Abelian categories, $T: \mathscr{C} \longrightarrow \mathscr{C}'$ an exact functor and S a right-adjoint functor of T which is fully faithful* (*i.e. $\mathscr{C}'$ is a quotient category of $\mathscr{C}$*). *Denote by* $\ker T$ *the full subcategory of $\mathscr{C}$ consisting of the objects X of $\mathscr{C}$ such that $T(X) = 0$. The subcategory* $\ker T$ *satisfies the following conditions*:

(a) *If in the exact sequence*

$$0 \longrightarrow A' \longrightarrow A \longrightarrow A'' \longrightarrow 0$$

of objects and morphisms in $\mathscr{C}$, A is an object of $\ker T$, *then A' and A'' are objects of* $\ker T$ *and conversely.*

(b) *If $D: \mathscr{D} \longrightarrow \ker T$ is a functor where $\mathscr{D}$ is a small category, such that $\varinjlim D$ exists in $\mathscr{C}$, then $\varinjlim D$ is an object of* $\ker T$.

PROOF. To prove (a), it is sufficient to use the fact that T is an exact functor. To prove (b), we use the fact that T preserves inductive limits, since it possesses a right-adjoint functor.

The full subcategory $\ker T$ of $\mathscr{C}$ is said to be the *localizant subcategory* associated to the quotient category $\mathscr{C}'$. We shall see that for the category $\mathscr{C}_\Lambda$ there exists a correspondence between the localizant subcategories and the quotient categories.

PROPOSITION 6.15. *Under the hypothesis and the notations of proposition 6.14, the following assertions are equivalent for the object X of $\mathscr{C}$:*

(a) *If $f:P \longrightarrow Q$ is a morphism in $\mathscr{C}$ such that* $\ker f$ *and* $\operatorname{coker} f$ *are objects of* $\ker T$, *then the map* $\operatorname{Hom}_{\mathscr{C}}(f, X)$ *is a bijection.*

(b) *X does not contain non-null objects of* $\ker T$ *and any exact sequence*

$$0 \longrightarrow X \longrightarrow A \longrightarrow B \longrightarrow 0$$

where B is an object of $\ker T$, *is split.*

PROOF. (a) $\Rightarrow$ (b) is evident.

(b) $\Rightarrow$ (a). Let $f:P \longrightarrow Q$ be a morphism in $\mathscr{C}$ such that $\ker f$ and $\operatorname{coker} f$ are objects of $\ker T$. Let $p:Q \longrightarrow X$ be a morphism such that $p \circ f = 0$. Then $p = 0$, since X has no subobject from $\ker T$. Thus $\operatorname{Hom}_{\mathscr{C}}(f, X)$ is a monomorphism. If $p:P \longrightarrow X$ is a morphism of $\mathscr{C}$ then we have the commutative diagram

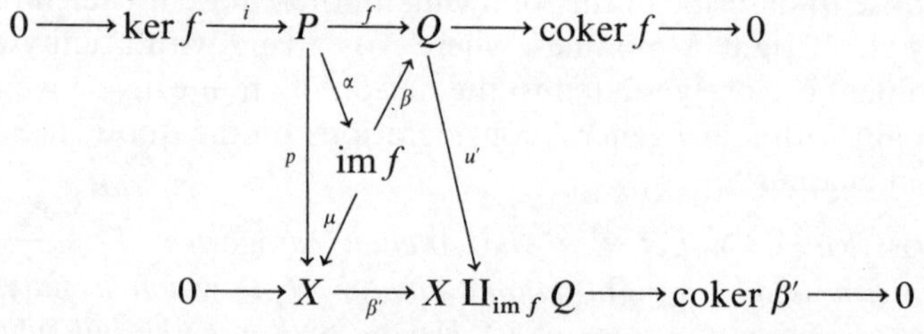

$p \circ i = 0$, since X has no subobject of $\ker T$, and α is the cokernel of i. $\operatorname{coker} \beta'$ is an object of $\ker T$, since $\operatorname{coker} f$ is so. There exists a morphism $\gamma: X \amalg_{\operatorname{im} f} Q \longrightarrow X$ such that $\gamma\beta' = 1_X$. But then $\gamma u' f = \gamma u' \beta \alpha = \gamma \beta' u \alpha = p$ which shows that $\operatorname{Hom}_{\mathscr{C}}(f, X)$ is an epimorphism.

Definition. Under the hypothesis and notations of proposition 6.14, an object X of $\mathscr{C}$ is said to be $\ker T$*-closed* (or simply closed if no confusion is possible), if it satisfies one of the equivalent assertions of proposition 6.15.

We now prove a proposition which characterizes the closed objects.

PROPOSITION 6.16. *Under the hypothesis and notations of proposition 6.14, an object X of $\mathscr{C}$ is* ker T*-closed if and only if it is isomorphic with an object of the form $S(X')$, where X' is an object of $\mathscr{C}'$.*

PROOF. Let X' be an object of $\mathscr{C}'$ and $f:P \longrightarrow Q$ a morphism in $\mathscr{C}$ such that $\ker f$ and $\operatorname{coker} f$ are objects of $\ker T$. Then Tf is an isomorphism in $\mathscr{C}'$ and we have the commutative diagram

$$\begin{array}{ccc} \operatorname{Hom}_{\mathscr{C}}(Q, SX') & \xrightarrow{\operatorname{Hom}_{\mathscr{C}}(f, SX')} & \operatorname{Hom}_{\mathscr{C}}(P, SX') \\ \downarrow \wr & & \downarrow \wr \\ \operatorname{Hom}_{\mathscr{C}'}(TQ, X') & \xrightarrow[\operatorname{Hom}_{\mathscr{C}'}(Tf, X')]{} & \operatorname{Hom}_{\mathscr{C}'}(TP, X') \end{array}$$

where the vertical maps are adjunction isomorphisms. This implies that $\operatorname{Hom}_{\mathscr{C}}(f, SX')$ is an isomorphism. To complete the proof, let $\Psi:1_{\mathscr{C}} \longrightarrow ST$ and $\Phi:TS \longrightarrow 1_{\mathscr{C}'}$, be the functorial morphisms such that $\Phi(T(A))T(\Psi(A)) = 1_{T(A)}$ and $S(\Phi(B))\Psi(S(B)) = 1_{S(B)}$ for any A in $\mathscr{C}$ and B in $\mathscr{C}'$ (cf. proposition 1.13; Φ and Ψ are said to be adjunction arrows quasi-inverse to each other). It is sufficient to prove that for any object X of $\mathscr{C}$ $\ker \Psi(X)$ and $\operatorname{coker} \Psi(X)$ are objects of $\ker T(\Psi(X):X \longrightarrow STX)$. Since Φ is an isomorphism, from the relation $\Phi(T(X))T(\Psi(X)) = 1_{T(X)}$ we deduce that $T\Psi(X)$ is an isomorphism, which completes the proof, T being an exact functor.

Let X be an injective object of $\mathscr{C}$, which has no subobject in $\ker T$. Then it follows from proposition 6.15(b), that X is $\ker T$-closed. It is readily seen that TX is an injective object of $\mathscr{C}'$.

COROLLARY 6.17. *Under the hypothesis and the notations of propositions 6.14, if $\mathscr{C}$ is a category with injective envelopes, then so is $\mathscr{C}'$.*

PROOF. Let X' be an object of $\mathscr{C}'$. Consider the object SX' of $\mathscr{C}$ and let $i:SX' \longrightarrow Q$ be an injective envelope of SX'. Q has no non-null subobject in $\ker T$ and therefore it is closed. TQ is injective in $\mathscr{C}'$ and $Ti:TSX' \longrightarrow TQ$ is an essential extension.

We now prove a proposition which will be used in the sequel.

PROPOSITION 6.18. *Under the hypothesis and notations of proposition 6.14, we denote by $\mathscr{C}_1$ the full subcategory of $\mathscr{C}$ consisting of the* ker T*-closed objects. $\mathscr{C}_1$ can be canonically organized as an Abelian category and S establishes an equivalence between this category and $\mathscr{C}$.*

PROOF. Let $\Psi:1_{\mathscr{C}} \longrightarrow ST$ be an adjunction arrow of S with T and $\Phi:TS \longrightarrow 1_{\mathscr{C}'}$ a quasi-inverse of Ψ. Let X, Y be objects of $\mathscr{C}$ which are ker T-closed, i.e. according to proposition 6.16, $\Psi(X)$ and $\Psi(Y)$ are isomorphisms. If $f:X \longrightarrow Y$ is a morphism in $\mathscr{C}_1$, we consider the commutative diagram

$$\begin{array}{ccccccccc} 0 \longrightarrow & \ker f & \xrightarrow{i} & X & \xrightarrow{f} & Y & \xrightarrow{p} & \operatorname{coker} f & \longrightarrow 0 \\ & \downarrow{\scriptstyle \Psi(\ker f)} & & \downarrow{\scriptstyle \Psi(X)\,\wr} & & \downarrow{\scriptstyle \Psi(Y)\,\wr} & & \downarrow{\scriptstyle \Psi(\operatorname{coker} f)} & \\ & ST(\ker f) & \xrightarrow{STi} & STX & \xrightarrow{STf} & STY & \xrightarrow{STp} & ST(\operatorname{coker} f) & \end{array}$$

where the top row is exact in $\mathscr{C}$. The kernel of f in $\mathscr{C}_1$ is the couple $(ST(\ker f), \Psi(X)^{-1} \circ STi)$, and the cokernel of f in $\mathscr{C}_1$ is the couple $(ST(\operatorname{coker} f), STp \circ \Psi(Y))$. We leave it to the reader to check that $\mathscr{C}_1$ is Abelian.

A morphism f in $\mathscr{C}_1$ is a monomorphism if and only if it is a monomorphism in $\mathscr{C}$; f is an epimorphism in $\mathscr{C}_1$ if and only if coker f is an object of ker T. The fact that S establishes an equivalence between $\mathscr{C}_1$ and $\mathscr{C}$ is deduced from the fact that S is fully faithful and any object X of $\mathscr{C}_1$ is isomorphic with STX.

As announced at the beginning of this paragraph, we now study briefly the localization in rings, i.e. we study the quotient categories of the category $\mathscr{C}_1$.

Let Λ be a ring with a unity element and $\mathscr{C}_\Lambda$ the category of unitary right Λ-modules. Let $\mathscr{C}'$ be a quotient category of $\mathscr{C}_\Lambda$, i.e. an Abelian category together with two functors: $T:\mathscr{C}_\Lambda \longrightarrow \mathscr{C}'$, which is exact, and $S:\mathscr{C}' \longrightarrow \mathscr{C}_\Lambda$, a fully faithful right-adjoint of T. We shall prove in the following that the category ker T determines the category $\mathscr{C}'$ up to an equivalence. Precisely, let $\mathscr{C}', \mathscr{C}''$ be two quotient categories of $\mathscr{C}_\Lambda$ and T', S', T'', S'' the respective functors. We say that $\mathscr{C}'$ and $\mathscr{C}''$ are equivalent quotient categories if there exists an equivalence $H:\mathscr{C}' \longrightarrow \mathscr{C}''$ such that $H \circ T' = T''$. In this case, the localizant categories ker T' and ker T'' coincide in a canonical way. We shall understand by a quotient category of the category $\mathscr{C}_\Lambda$ a 'class of equivalent quotient categories.'

Definition. A full subcategory $\mathscr{A}$ of the category $\mathscr{C}_\Lambda$ is said to be localizant if it satisfies the following conditions:

(a) If in the exact sequence

$$0 \longrightarrow A' \longrightarrow A \longrightarrow A'' \longrightarrow 0$$

in $\mathscr{C}_\Lambda$ A is an object of $\mathscr{A}$, then A', A'' are objects of $\mathscr{A}$ and conversely.

(b) An inductive limit of objects of $\mathscr{A}$ taken in $\mathscr{C}_\Lambda$ is an object of $\mathscr{A}$.

The localizant subcategories have already appeared with the occasion of proposition 6.14; however, the above notion is not dependent on any notion of quotient category, but, as will be seen in the sequel, we will define quotient categories by starting from localizant categories. To do this we shall use the following notion:

Definition. Let Λ be a ring with a unity element and let $\mathscr{F}$ be a system of right ideals of Λ. We say that $\mathscr{F}$ is a *topologizing system of ideals* on Λ if the following axioms are satisfied:

1. $\Lambda \in \mathscr{F}$
2. If $\mathfrak{a} \in \mathscr{F}$ and $x \in \Lambda$, then the ideal
$$\{\mathfrak{a}:x\} = \{y | y \in \Lambda, xy \in \mathfrak{a}\}$$
is an element of $\mathscr{F}$.
3. If $\mathfrak{a}$ and $\mathfrak{b}$ are two right ideals of Λ, such that $\mathfrak{a} \in \mathscr{F}$ and for any $x \in \mathfrak{a}$, $\{\mathfrak{b}:x\} \in \mathscr{F}$, then $\mathfrak{b} \in \mathscr{F}$.

The justification of the concept just introduced consists in the following propositions, as well as in the subsequent constructions.

PROPOSITION 6.19. *Let Λ be a ring with a unity element. There exists a one-to-one correspondence between the localizant subcategories of $\mathscr{C}_\Lambda$ and the topologizing systems of ideals in Λ.*

PROOF. Let U be the set of topologizing systems of right-ideals of Λ and V the class of localizant subcategories of $\mathscr{C}_\Lambda$. We define the map $\varphi: U \longrightarrow V$ as follows: if $\mathscr{F}$ is an element of U, then $\varphi(\mathscr{F})$ is the full subcategory of $\mathscr{C}_\Lambda$ formed by the modules A such that for any $x \in A$, $\operatorname{Ann} x = \{\lambda | \lambda \in \Lambda, x\lambda = 0\}$ is a right-ideal of $\mathscr{F}$. It is easy to verify that $\varphi(\mathscr{F})$ is a localizant subcategory of $\mathscr{C}_\Lambda$.

We now define a map $\psi: V \longrightarrow U$ as follows: if $\mathscr{A}$ is a localizant subcategory of $\mathscr{C}_\Lambda$, we denote by $\psi(\mathscr{A})$ the system of right-ideals of Λ which are the annulators for the elements of the objects of $\mathscr{A}$. It is straightforward to see that $\psi(\mathscr{A})$ is a topologizing system and that $\varphi \circ \psi = 1$ and $\psi \circ \varphi = 1$ (where 1 is the suitable identity map).

We now consider the category $\mathscr{C}_\Lambda$ and a localizant subcategory $\mathscr{A}$ of $\mathscr{C}_\Lambda$. Denote by $\mathscr{F}$ the topologizing system associated to $\mathscr{A}$ by the preceding proposition. We shall define a functor $H: \mathscr{C}_\Lambda \longrightarrow \mathscr{C}_\Lambda$ as follows: for any object X of $\mathscr{C}_\Lambda$, we denote by X_1 the greatest subobject of X in $\mathscr{A}$ (the existence of X_1 is assured by the fact that $\mathscr{A}$ is closed under inductive limits). Consider the Abelian group

$$H(X) = \varinjlim_{\mathfrak{a} \in \mathscr{F}} \mathrm{Hom}_\Lambda(\mathfrak{a}, X/X_1),$$

which we make into a right Λ-module as follows: let $m \in H(X)$ and $\lambda \in \Lambda$. Denote by $\bar{\lambda} \in \Lambda \longrightarrow \Lambda$ the map defined by $x \longrightarrow \lambda x$. Let now $u:\mathfrak{a} \longrightarrow X/X_1$ be a morphism of $\mathscr{C}_\Lambda$ which represents the element m of the inductive limit. We consider the element $\{\mathfrak{a}:\lambda\} \in \mathscr{F}$ and we denote by $\bar{\lambda}_1$ the restriction of $\bar{\lambda}$ to $\{\mathfrak{a}:\lambda\}$. The image of $\bar{\lambda}_1$ is contained in $\mathfrak{a}$. We set by definition $m\lambda$ as being the image in the inductive limit $H(X)$ of the morphism $u\bar{\lambda}_1$. Thus $H(X)$ becomes a Λ-module.

Let now $f:X \longrightarrow Y$ be a morphism in $\mathscr{C}_\Lambda$. It is obvious that $f(X_1)$ is contained in Y_1. Thus f induces a unique morphism $f':X/X_1 \longrightarrow Y/X_1$. f' induces in turn a unique morphism $H(f):H(X) \longrightarrow H(Y)$ which is a homomorphism of right Λ-modules. This concludes the construction of the functor H.

The topologizing system $\mathscr{F}$ ordered by inclusion is a left directed set. From the fact that $\mathscr{Ab}$ satisfies AB5 and from the construction of H we deduce that H is a left-exact functor. Moreover, if X is an object of $\mathscr{A}$, $H(X) = 0$, since $X_1 = X$.

If α_Λ is the canonical morphism of $\mathrm{Hom}_\Lambda(\Lambda, X/X_1)$ into the inductive limit $H(X)$, $\beta:X/X_1 \longrightarrow \mathrm{Hom}_\Lambda(\Lambda, X/X_1)$ is the canonical isomorphism, and $p:X \longrightarrow X/X_1$ is the canonical epimorphism, then $\psi(X) = \alpha_\Lambda\beta p$ is a Λ-homomorphism from X into $H(X)$ which defines a functorial morphism ψ from $1_{\mathscr{C}_\Lambda}$ to H.

Concerning the functorial morphism ψ we prove the following.

PROPOSITION 6.20. *For any object X of $\mathscr{C}_\Lambda$,* ker $\psi(X)$ *and* coker $\psi(X)$ *are objects of $\mathscr{A}$.*

PROOF. Clearly ker $\psi(X)$ contains X_1 since H is a left-exact functor. We shall prove that the canonical morphism $\alpha_\Lambda\beta:X/X_1 \longrightarrow HX$ is a monomorphism, or, equivalently, that if X has no non-null subobject in $\mathscr{A}$, then $\psi(X)$ is a monomorphism. Indeed, assume X has no non-null subobject in $\mathscr{A}$ and let $x \in X$ be such that $\psi(X)(x) = 0$. Then there exists an element $\mathfrak{a} \in \mathscr{F}$ such that denoting by $\bar{x}:\Lambda \longrightarrow X$ the Λ-homomorphism defined by $\bar{x}(\lambda) = x\lambda$, and by $i:\mathfrak{a} \longrightarrow \Lambda$ the canonical inclusion, $\bar{x} \circ i = 0$ (by using the properties of inductive limits). But this implies that $\mathfrak{a}$ is contained in the annulator of X, which shows that $x = 0$. (We deduce immediately that if $\mathfrak{b}$ is a right-ideal of Λ which contains an ideal in $\mathscr{F}$, then $\mathfrak{b} \in \mathscr{F}$). Thus, ker $\psi(X) = X_1$.

Let now $x \in H(X)$. We shall prove that the right-ideal $\{\mathrm{im}\,\psi(X):x\} = \{\lambda/\lambda \in \Lambda, x\lambda \in \mathrm{im}\,\psi(X)\}$ is an element of $\mathscr{F}$. For simplicity, we admit that $\psi(X)$ is a monomorphism, which does not reduce the generality, and we identify X with $\mathrm{im}\,\psi(X)$. Let $u:\mathfrak{a} \longrightarrow X$ be a Λ-homomorphism which represents x in the inductive limit $H(X)$.

From the preceding considerations, for any $\lambda \in \Lambda$, $x\lambda$ is defined as being the image in the inductive limit of the Λ-homomorphism $u \circ \bar{\lambda}_1$, where $\bar{\lambda}_1:\{\mathfrak{a}:\lambda\} \longrightarrow \mathfrak{a}\bar{\lambda}_1(y) = \lambda y, y \in \{\mathfrak{a}:\lambda\}$.

In order that $x\lambda \in X$, (i.e. $\lambda \in \{X:x\}$), it is necessary that $x\lambda = \psi(X)(x')$, where $x' \in X$. In this case, we can identify the morphism $\psi(X)$ with the canonical morphism $\alpha_\Lambda:\mathrm{Hom}_\Lambda(\Lambda, X) \longrightarrow H(X)$. With this observation, the relation $x\lambda = \psi(X)(x')$ is equivalent with the existence of the element $\mathfrak{b} \in \mathscr{F}$ contained in $\{\mathfrak{a}:\lambda\}$ such that the diagram

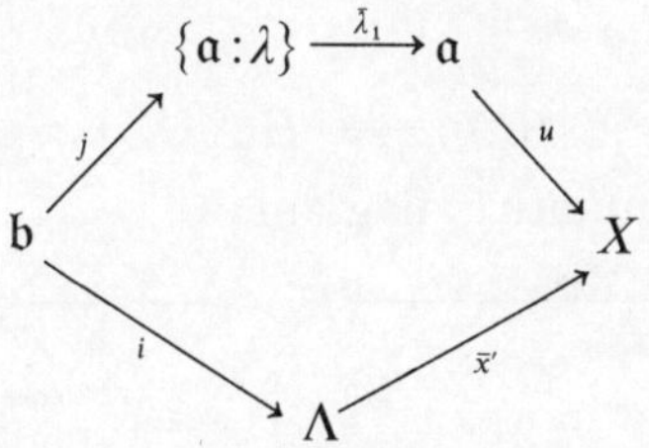

is commutative, where i,j are canonical monomorphisms and $\bar{x}'$ is defined as follows: $x'(y) = \bar{x}'y, y \in \Lambda$. We denote by $\mathfrak{c}$ the ideal $\{X:x\}$, and let $y \in \mathfrak{a}$. We shall prove that $\{\mathfrak{c}:y\} \in \mathscr{F}$. The element $z \in \Lambda$ is in $\{\mathfrak{c}:y\}$ if $yz \in \mathfrak{c}$, i.e. if we have a diagram of the above type where λ is replaced by yz. But a diagram of this type is valid for any element of $\mathfrak{a}$, hence $\{\mathfrak{c}:y\}$ contains $\mathfrak{a}$. Thus $\{\mathfrak{c}:y\} \in \mathscr{F}$ for any $y \in \mathfrak{a}$. From the definition of the topologizing system (3) it follows that $\mathfrak{c} \in \mathscr{F}$. The proof is complete.

Definition. The Λ-module X is said to be closed relatively to the localizant subcategory $\mathscr{A}$ if the homomorphism $\psi(X)$ is an isomorphism.

From the above construction we derive the following remarks:

1. For any object X of $\mathscr{C}_\Lambda$ the object $H(X)$ has no non-null subobject in $\mathscr{A}$. This results from the construction of $H(X)$.

2. If X is a closed object of $\mathscr{C}_\Lambda$ then any exact sequence

$$0 \longrightarrow X \xrightarrow{\;i\;} Y \longrightarrow A \longrightarrow 0$$

in $\mathscr{C}_\Lambda$, where A is an object of $\mathscr{A}$, is split. To see this, consider the commutative diagram:

$$\begin{array}{ccccccccc}
0 & \longrightarrow & X & \xrightarrow{i} & Y & \longrightarrow & A & \longrightarrow & 0 \\
 & & \downarrow{\scriptstyle \psi(X)} & & \downarrow{\scriptstyle \psi(Y)} & & \downarrow{\scriptstyle \psi(A)} & & \\
0 & \longrightarrow & H(X) & \xrightarrow[H(i)]{} & H(Y) & \longrightarrow & H(A) & &
\end{array}$$

where the bottom row is exact and $H(A) = 0$. We deduce that $H(i)$ is an isomorphism. Hence $\psi(X)^{-1} \circ H(i)^{-1} \circ \psi(Y) \circ i = 1_X$.

3. For any object X of $\mathscr{C}_\Lambda$, the object $H(X)$ is closed.

To see this, notice in the construction of the functor H that for any object X of $\mathscr{C}_\Lambda$ the canonical epimorphism $p: X \longrightarrow X/X_1$ induces the isomorphism $H(p)$. More precisely, we have the commutative diagram

$$\begin{array}{ccc}
X & \xrightarrow{p} & X/X_1 \\
\downarrow{\scriptstyle \psi(X)} & & \downarrow{\scriptstyle \psi(X/X_1)} \\
H(X) & \xrightarrow{H(p)} & H(X/X_1).
\end{array}$$

We consider the commutative diagram

$$\begin{array}{ccccccccc}
0 & \longrightarrow & X/X_1 & \xrightarrow{H(p)^{-1} \circ \psi(X/X_1) = \alpha} & H(X) & \longrightarrow & \operatorname{coker} \psi(X) & \longrightarrow & 0 \\
 & & \downarrow{\scriptstyle \psi(X/X_1)} & & \downarrow{\scriptstyle \psi(H(X))} & & \downarrow{\scriptstyle \psi(\operatorname{coker} \psi(X))} & & \\
0 & \longrightarrow & H(X/X_1) & \xrightarrow{H(\alpha)} & H(H(X)) & \longrightarrow & H(\operatorname{coker} \psi(X)). & &
\end{array}$$

The top row is exact owing to the commutativity of the preceding diagram and to the fact that p is an epimorphism. The bottom row is also exact and $H(\operatorname{coker} \psi(X)) = 0$ according to proposition 6.20, and the fact that H is zero on the objects of $\mathscr{A}$. Thus $H(\alpha)$ is an isomorphism. By applying the functor H to the preceding diagram we obtain that $H(\psi(X))$ is an isomorphism.

We denote by $\bar{\mathscr{A}}$ the full subcategory of $\mathscr{C}$ consisting of the objects which are closed relatively to $\mathscr{A}$ (or simply closed).

$\bar{\mathscr{A}}$ is an Abelian category; for, the monomorphisms in $\bar{\mathscr{A}}$ are those morphisms in $\bar{\mathscr{A}}$ which are monomorphisms in $\mathscr{A}$, the epimorphisms in $\bar{\mathscr{A}}$ are those morphisms $f: X \longrightarrow Y$ in $\bar{\mathscr{A}}$ such that $\operatorname{Coker} f$ considered in $\mathscr{C}_\Lambda$ is an object in $\mathscr{A}$. If $f: X \longrightarrow Y$ is a morphism in $\bar{\mathscr{A}}$, the kernel of f is defined by means of the canonical commutative diagram

$$\begin{array}{ccccccc}
0 & \longrightarrow & \ker f & \xrightarrow{i} & X & \xrightarrow{f} & Y \\
 & & \downarrow{\scriptstyle \psi(\ker f)} & & \downarrow{\scriptstyle \psi(X)} & & \downarrow{\scriptstyle \psi(Y)} \\
0 & \longrightarrow & H(\ker f) & \xrightarrow{H(i)} & H(X) & \longrightarrow & H(Y)
\end{array}$$

where $(\ker f, i)$ is the kernel of f in $\mathscr{C}_\Lambda$. The kernel of f is the couple $(H(\ker f), \psi(X)^{-1} \circ H(i))$. Likewise, from the commutative diagram

$$\begin{array}{ccccccc} X & \xrightarrow{f} & Y & \xrightarrow{p} & \operatorname{coker} f & \longrightarrow & 0 \\ & & \downarrow{\scriptstyle \psi(Y)} & & \downarrow{\scriptstyle \psi(\operatorname{coker} f)} & & \\ & & H(Y) & \xrightarrow{H(p)} & H(\operatorname{coker} f) & & \end{array}$$

where $(\operatorname{coker} f, p)$ is the cokernel of f in the category $\mathscr{C}_\Lambda$, we deduce that $(H(\operatorname{coker} f), H(p) \circ \psi(Y))$ is a cokernel of f in $\bar{\mathscr{A}}$. It is left to the reader to verify that the canonical homomorphism from $\operatorname{coim} f$ to $\operatorname{im} f$ is an isomorphism in $\bar{\mathscr{A}}$.

We define a functor $T:\mathscr{C}_\Lambda \longrightarrow \bar{\mathscr{A}}$ as follows: if X is an object of $\mathscr{C}_\Lambda$, $T(X) = H(X)$, and if $f:X \longrightarrow Y$ is a Λ-homomorphism, $Tf = Hf$. The functor T is a right-adjoint of the inclusion functor $S:\bar{\mathscr{A}} \longrightarrow \mathscr{C}_\Lambda$. An adjunction arrow of S with T is furnished by the functorial morphism $\Psi:1_{\mathscr{C}_\Lambda} \longrightarrow H = ST$ constructed above. Moreover, the functor T is exact and S is fully faithful. Thus, the couple (T, S) defines $\bar{\mathscr{A}}$ as a quotient category of $\mathscr{C}_\Lambda$. Thus the following proposition is true:

PROPOSITION 6.21. *Let $\mathscr{A}$ be a localizant subcategory of $\mathscr{C}_\Lambda$. There exists a quotient category $\bar{\mathscr{A}}$ of $\mathscr{C}_\Lambda$, and the canonical functors, $T:\mathscr{C}_\Lambda \longrightarrow \bar{\mathscr{A}}$, $S:\bar{\mathscr{A}} \longrightarrow \mathscr{C}_\Lambda$ (T exact, S fully faithful and right-adjoint to T) such that* $\mathscr{A} = \ker T$.

PROOF. It is sufficient by the above considerations to prove that, if $TX = 0$, then X is an object of $\mathscr{A}$, and this follows from the definition of H and the properties of inductive limits.

It follows from proposition 6.21 that there exists a one-to-one correspondence between the quotient categories of $\mathscr{C}_\Lambda$ and the localizant subcategories of $\mathscr{C}_\Lambda$.

By using proposition 5.39, we deduce that any quotient category of $\mathscr{C}_\Lambda$ is a Grothendieck category, with direct products and from corollary 6.17 and proposition 6.11, we deduce that any quotient category of $\mathscr{C}_\Lambda$ is with injective envelopes. Thus we have

COROLLARY 6.22. *Let $\mathscr{C}'$ be a quotient category of $\mathscr{C}_\Lambda$. Then $\mathscr{C}'$ is a Grothendieck category with direct products and injective envelopes.*

6. Characterization of Grothendieck Categories

In the present paragraph we shall prove that any Grothendieck category is equivalent with a quotient category of a category $\mathscr{C}_\Lambda$ with a suitable ring Λ. This result is due to P. Gabriel and N. Popescu [1].

Before doing this, we make some preliminary considerations which have also an intrinsic importance.

Let $\mathscr{C}$ be an Abelian category and U an object of $\mathscr{C}$. We denote by Λ the ring $\mathrm{Hom}_{\mathscr{C}}(U, U)$, and by $S:\mathscr{C} \longrightarrow \mathscr{C}_\Lambda$ the functor defined by

$$S(X) = \mathrm{Hom}(U, X)$$

$$S(f) = \mathrm{Hom}(U, f),$$

where X is an object of $\mathscr{C}$ and f a morphism of $\mathscr{C}$, and $S(X)$ has a right Λ-module structure defined as follows: for $\lambda \in \Lambda, \lambda: U \longrightarrow U$ being a morphism in $\mathscr{C}$, and $x \in S(X)$, $x: U \longrightarrow X$ being a morphism in $\mathscr{C}$, $x\lambda$ is the composition of the morphisms x and λ in $\mathscr{C}$. Obviously Sf is a Λ-homomorphism for any morphism f in $\mathscr{C}$. S is a left-exact functor.

PROPOSITION 6.23. *Let $\mathscr{C}$ be an Abelian category and U an object of $\mathscr{C}$ such that for any set $(U_i)_{i \in I}$ of objects in $\mathscr{C}$ where U_i is isomorphic with U for any $i \in I$, there exists the direct sum $\bigoplus_{i \in I} U_i$. Denote by Λ the ring $\mathrm{Hom}_{\mathscr{C}}(U, U)$ and by $S:\mathscr{C} \longrightarrow \mathscr{C}_\Lambda$ the functor defined above. Under these conditions the functor S possesses a left adjoint functor $T:\mathscr{C}_\Lambda \longrightarrow \mathscr{C}$.*

PROOF. Let M be an object of $\mathscr{C}_\Lambda$. We can construct an exact sequence in $\mathscr{C}_\Lambda$:

$$\bigoplus_{j \in J} \Lambda_j \xrightarrow{g} \bigoplus_{i \in I} \Lambda_i \xrightarrow{p} M \longrightarrow 0 \tag{*}$$

where I and J are sets and $\Lambda_i \approx \Lambda_j \approx \Lambda$ for any $i \in I, j \in J$. (Cf. [3], chapter I, section 2.)

Consider the direct sums in the category $\mathscr{C}: \bigoplus_{j \in J} U_j, \bigoplus_{i \in I} U_i$ where $U_i \approx U_j \approx U$ for any $i \in I, j \in J$. We have for any object X of $\mathscr{C}$, according to proposition 3.7:

$$\mathrm{Hom}_\Lambda(\bigoplus_{i \in I} \Lambda_i, S(X)) = \prod_{i \in I} \mathrm{Hom}_{\mathscr{C}}(\Lambda_i, S(X)) = \prod_{i \in I} S(X)_i$$

where $S(X)_i \approx S(X)$ for any $i \in I$.

But we have $S(X) = \mathrm{Hom}_{\mathscr{C}}(U, X)$ and also $S(X)_i \approx \mathrm{Hom}_{\mathscr{C}}(U_i, X)$, so that $\prod_{i \in I} S(X)_i \approx \prod_{i \in I} \mathrm{Hom}_{\mathscr{C}}(U_i, X) \approx \mathrm{Hom}_{\mathscr{C}}(\bigoplus_{i \in I} U_i, X)$.

As easily seen, all these equalities are in fact functorial in X (i.e. they are 'independent of X').

Finally, we obtain a functorial isomorphism

$$\psi_0(X):\mathrm{Hom}_\Lambda\Big(\bigoplus_{i\in I}\Lambda_i, S(X)\Big)\xrightarrow{\approx}\mathrm{Hom}_{\mathscr{C}}\Big(\bigoplus_{i\in I}U_i, X\Big)$$

and analogously, for any object X of $\mathscr{C}$ a functorial isomorphism

$$\psi_1(X):\mathrm{Hom}_\Lambda\Big(\bigoplus_{j\in J}\Lambda_j, S(X)\Big)\xrightarrow{\approx}\mathrm{Hom}_{\mathscr{C}}\Big(\bigoplus_{j\in J}U_j, X\Big).$$

The morphism q enables us to construct the functorial morphism

$$\alpha=\psi_0\circ\mathrm{Hom}_{\mathscr{C}}(q,\ \ )\circ\psi_1^{-1}:\mathrm{Hom}_{\mathscr{C}}\Big(\bigoplus_{j\in J}U_j,\ \ \Big)\longrightarrow\mathrm{Hom}_{\mathscr{C}}\Big(\bigoplus_{i\in I}U_i,\ \ \Big).$$

However, by corollary 1.7, there exists a unique morphism $q':\bigoplus_{j\in J}U_j\longrightarrow\bigoplus_{i\in I}U_i$ such that $\mathrm{Hom}_{\mathscr{C}}(q',\ \ )=\alpha$. We define $TM=\operatorname{coker} q'$.

We leave it to the reader to check that TM does not depend on the free modules from the sequence (*) and T so defined is indeed a functor from $\mathscr{C}_\Lambda$ to $\mathscr{C}$.

From the commutative diagram

$$\begin{array}{ccccccc}
0\longrightarrow & \mathrm{Hom}_\Lambda(M,SX) & \longrightarrow & \mathrm{Hom}_\Lambda\big(\bigoplus_{i\in I}\Lambda_i,SX\big) & \longrightarrow & \mathrm{Hom}_\Lambda\big(\bigoplus_{j\in J}\Lambda_j,SX\big) \\
 & \downarrow & & \downarrow{\scriptstyle\psi_0(X)} & & \downarrow{\scriptstyle\psi_1(X)} \\
0\longrightarrow & \mathrm{Hom}_{\mathscr{C}}(TM,X) & \longrightarrow & \mathrm{Hom}_{\mathscr{C}}\big(\bigoplus_{i\in I}U_i,X\big) & \longrightarrow & \mathrm{Hom}_{\mathscr{C}}\big(\bigoplus_{j\in J}U_j,X\big)
\end{array}$$

where the top row is induced by the exact sequence (*), the bottom row by the exact sequence

$$\bigoplus_{j\in J}U_j\longrightarrow\bigoplus_{i\in I}U_i\longrightarrow TM\longrightarrow 0$$

and the dotted arrow is constructed so that the diagram be commutative, we deduce that T is left-adjoint for S, the morphism represented by the dotted arrow being in fact functorial.

PROPOSITION 6.24. *Let $\mathscr{C}$ be an Abelian category and U an object of $\mathscr{C}$. Consider the ring $\Lambda=\mathrm{Hom}_{\mathscr{C}}(U,U)$ and the functor $S:\mathscr{C}\longrightarrow\mathscr{C}_\Lambda$ defined at the beginning of this paragraph. U is a generator of $\mathscr{C}$ if and only if S is faithful.*

Proof. Assume S is faithful; if $i:X' \longrightarrow X$ is a monomorphism of $\mathscr{C}$ which is not an isomorphism, and $p:X \longrightarrow X/X'$ is the cokernel of i, then, if Si is an isomorphism we deduce $Sp = 0$, since S is left-exact. Thus $p = 0$, which is a contradiction.

Conversely, assume U is a generator of $\mathscr{C}$ and let $f:X \longrightarrow Y$ be a morphism in $\mathscr{C}$ such that $Sf = 0$. If $f = i \circ p$ is the canonical decomposition of f with i monomorphism and p epimorphism, then $Sp = 0$ and if $j:\ker f \longrightarrow X$ is the kernel of p, we infer that Sj is an isomorphism. Since U is a generator, j is an isomorphism and therefore $p = 0$, hence $f = 0$.

Theorem 6.25. (P. Gabriel–N. Popescu). *Let $\mathscr{C}$ be an Abelian category satisfying condition* AB5 *and let U be an object of $\mathscr{C}$. We denote by Λ the ring* $\mathrm{Hom}_{\mathscr{C}}(U, U)$ *and by $S:\mathscr{C} \longrightarrow \mathscr{C}_\Lambda$ the functor considered in proposition 6.24. Let $T:\mathscr{C}_\Lambda \longrightarrow \mathscr{C}$ be a left-adjoint of S and $\Phi : TS \longrightarrow 1_{\mathscr{C}}$ an adjunction arrow of T with S. The following assertions are equivalent:*

(a) *U is a generator of $\mathscr{C}$.*
(b) *S is a fully faithful functor.*
(c) *Φ is a functorial isomorphism and T is exact.*
(d) *T is exact and induces an equivalence between $\mathscr{C}$ and* $\ker T$, *the quotient category of $\mathscr{C}_\Lambda$ relatively to the localizant subcategory* $\ker T$.

Proof. The equivalence (c) $\Leftrightarrow$ (d) is immediate according to propositions 6.18 and 6.21. Also, by corollary 6.22, the induction (c) $\Rightarrow$ (b) $\Rightarrow$ (a) is clear. It remains to show that (a) $\Rightarrow$ (c) and to this end we will first show that (a) $\Rightarrow$ (b).

Indeed, let X, Y be objects of $\mathscr{C}$ and $S(X, Y):\mathrm{Hom}_{\mathscr{C}}(X, Y) \longrightarrow \mathrm{Hom}_{\mathscr{C}_\Lambda}(SX, SY)$ the canonical map: $f \longrightarrow Sf$, $f \in \mathrm{Hom}_{\mathscr{C}}(X, Y)$. According to proposition 6.24, $S(X, Y)$ is a monomorphism. We shall prove that $S(X, Y)$ is also an epimorphism. To do this, let $\varphi : SX \longrightarrow SY$ be a morphism of $\mathscr{C}_\Lambda$. We denote by I the set $\mathrm{Hom}_{\mathscr{C}}(U, X)$ and let $(U_f)_{f \in I}$ be a family of objects of $\mathscr{C}$, each U_f being isomorphic with U. Denote by $p: \bigoplus_{f \in I} U_f \longrightarrow X$ the canonical morphism which induces f on the summand U_f, i.e. such that the diagram

$$\begin{array}{ccc} U_f & & \\ {\scriptstyle i_f}\downarrow & \searrow{\scriptstyle f} & \\ \bigoplus_{f \in I} U_f & \xrightarrow{p} & X \end{array}$$

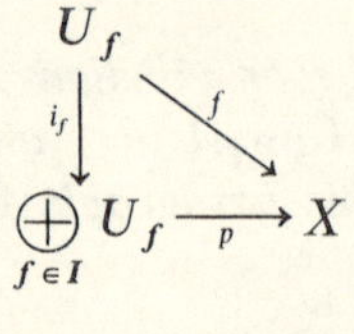

is commutative for any $f \in I$, i_f being the canonical injections. Let also $q: \bigoplus_{f \in I} U_f \longrightarrow Y$ be the canonical morphism such that the diagram

$$\begin{array}{ccc} U_f & & \\ {\scriptstyle i_f}\downarrow & \searrow^{\varphi(f)} & \\ \bigoplus_{f \in I} U_f & \xrightarrow{q} & Y \end{array}$$

is commutative for any $f \in I$.

If we prove that $\ker q$ contains $\ker p$, then we deduce that $q = \psi \circ p$ and therefore $\varphi = S\psi$ (since p is an epimorphism as follows from proposition 5.34). Let J be a finite subset of I, and K_J the kernel of the canonical morphism $p_j: \bigoplus_{f \in J} U_f \longrightarrow X$ induced by p. Since $\mathscr{C}$ satisfies AB5, $\ker p = \sup_J K_J$, J running through the family of finite subsets of I ordered by inclusion. To finish, it is sufficient to prove that, for any J, K_J is contained in $\ker q$. To do this, let $h: K_J \longrightarrow \bigoplus_{f \in J} U_f$ be the canonical monomorphism (which defines K_J as a subobject of $\oplus U_f$) and $\alpha: U \longrightarrow K_J$ a morphism in $\mathscr{C}$. Let also π_f be the canonical projection $\pi_f: \bigoplus_{f \in J} U_f \longrightarrow U_f$; obviously, preserving the notation $i_f: U_f \longrightarrow \bigoplus_{f \in J} U_f$ for the canonical injections, we have $\sum_{f \in J} i_f \circ \pi_f = 1$, where 1 is the identity of $\bigoplus_{f \in J} U_f$. We have:

$$p_J \circ h \circ \alpha = p_J \circ \sum_{f \in I} (i_f \circ \pi_f \circ h \circ \alpha) = 0 = \sum_{f \in I} f \circ \lambda_f$$

where $\lambda_f = \pi_f \circ h \circ \alpha$ is an element of Λ.

This implies

$$\begin{aligned} q_J \circ h \circ \alpha &= \sum_{f \in J} (q_J \circ i_f \circ \pi_f \circ h \circ \alpha) = \sum_{f \in J} \varphi(f) \circ \lambda_f \\ &= \sum_{f \in J} \varphi(f \circ \lambda_f) = \varphi(\sum_{f \in J} f \circ \lambda_f) = 0 \end{aligned}$$

for any $\alpha: U \longrightarrow K_J$. We conclude that $q_J \circ h = 0$, and therefore K_J is contained in $\ker q$.

In order to prove that (a) $\Rightarrow$ (c), we shall prove that S induces an equivalence between $\mathscr{C}$ and the full subcategory of $\mathscr{C}_\Lambda$ consisting of closed objects relatively to a localizant subcategory $\mathscr{A}$ of $\mathscr{C}_\Lambda$, abelianized in conformity with the considerations preceding proposition 6.21 (remark 3). To do this, we now introduce a few definitions and lemmas.

We shall say that the object N of $\mathscr{C}_\Lambda$ is negligible if for any morphism $f:\Lambda \longrightarrow N$ and any object X of $\mathscr{C}$ the canonical morphism

$$\mathrm{Hom}_\Lambda(\Lambda, SX) \longrightarrow \mathrm{Hom}_\Lambda(\ker f, SX)$$

is an isomorphism.

Lemma 6.26. *A Λ-module N is negligible if and only if for any morphism $f:M \longrightarrow N$ in $\mathscr{C}_\Lambda$ and any object X of $\mathscr{C}$ the canonical map*

$$\mathrm{Hom}_\Lambda(M, SX) \longrightarrow \mathrm{Hom}_\Lambda(\ker f, SX)$$

is an isomorphism.

Proof. Assume N is negligible and $f:M \longrightarrow N$ a morphism in $\mathscr{C}_\Lambda$. Consider the commutative diagram

$$\begin{array}{ccccccc} & & \ker(f\circ p\circ\psi_i) & \xrightarrow{\gamma_i} \Lambda_i \xrightarrow{\psi_i} & \bigoplus_{i\in I}\Lambda_i & & \\ & & \downarrow{\scriptstyle\delta_i} & & \downarrow{\scriptstyle p} & & \\ 0 & \longrightarrow & \ker f & \xrightarrow{\qquad\varphi\qquad} & M & \xrightarrow{f} & N \end{array}$$

where the bottom row is exact, the couple $(\ker(f\circ p\circ\psi_i), \gamma_i)$ is the kernel of $f\circ p\circ\psi_i$, p is an epimorphism, $\Lambda_i \approx \Lambda$ for any $i\in I$, ψ_i are the canonical injections, and δ_i are constructed in a canonical way. Let X be an object of $\mathscr{C}$ and $u:M \longrightarrow SX$ a Λ-homomorphism such that $u\circ\varphi = 0$. Then $u\circ\varphi\circ\delta_i = 0$ for any $i\in I$. Since N is negligible, we deduce that $u\circ p\circ\psi_i = 0$ for any i, and therefore $u\circ p = 0$, which implies $u = 0$ since p is an epimorphism. Thus the canonical map induced by φ

$$\mathrm{Hom}_\Lambda(M, SX) \longrightarrow \mathrm{Hom}_\Lambda(\ker f, SX)$$

is a monomorphism.

Let now $u:\ker f \longrightarrow SX$ be a Λ-homomorphism. Then $u\circ\delta_i:\ker(f\circ p\circ\psi_i) \longrightarrow SX$. Let $\lambda_i:\Lambda_i \longrightarrow SX$ be such that $\lambda_i\circ\gamma_i = u\circ\delta_i$. The morphisms λ_i induce the morphism $\lambda: \bigoplus_{i\in I}\Lambda_i \longrightarrow SX$ such that $\lambda\circ\psi_i = \lambda_i$. To finish the proof, it remains to show that λ is zero on $\ker p$ and to this end one proves that it is zero on $\ker(f\circ p)$, by using an argument similar to that in the above proof of the implication (a) $\Rightarrow$ (b).

Lemma 6.27. *Let $\mathscr{A}$ be the full subcategory of $\mathscr{C}_\Lambda$ consisting of all negligible objects. $\mathscr{A}$ is a localizant subcategory of $\mathscr{C}_\Lambda$.*

PROOF. Let N be a negligible Λ-module, $N' \xrightarrow{i} N$ a submodule of N and $f:\Lambda \longrightarrow N'$ a Λ-homomorphism. Clearly $\ker(i \circ f)$ is canonically isomorphic with $\ker f$, whence it follows that N' is negligible.

Let $p:N \longrightarrow N''$ be an epimorphism. If $f:\Lambda \longrightarrow N''$ is a morphism in $\mathscr{C}_\Lambda$, then there exists a morphism $h:\Lambda \longrightarrow N$ such that $p \circ h = f$, since Λ has a base reduced to the unity. Consider the commutative diagram

$$\begin{array}{ccccccc}
0 \longrightarrow & \ker f' & \xrightarrow{i'} & \Lambda\Pi_{N''}N & \xrightarrow{f'} & N \\
 & \downarrow{\scriptstyle p''} & & \downarrow{\scriptstyle p'} & & \downarrow{\scriptstyle p} \\
0 \longrightarrow & \ker f & \xrightarrow[i]{} & \Lambda & \xrightarrow[f]{} & N'' \\
 & & & & & \downarrow \\
 & & & & & 0
\end{array}$$

where the rows are exact and p'' is an isomorphism. Let X be an object of $\mathscr{C}$ and $u:\Lambda \longrightarrow SX$ a Λ-homomorphism such that $u \circ i = 0$. Then $u \circ i \circ p'' = 0$ and therefore $u \circ p' \circ i' = 0$, which means that $u \circ p' = 0$; thus $u = 0$ since p' is an epimorphism.

There exists a Λ-homomorphism $q:\Lambda \longrightarrow \Lambda\Pi_{N''}N$ such that $p' \circ q = 1_\Lambda$. Then $q \circ i \circ p'' = i'$. For, if $x \in \ker f'$, from $p' \circ i'(x) = 0$ it follows that $x = 0$ which shows that $\operatorname{im} i'$ is contained in $\operatorname{im} q$.

If $u:\ker f \longrightarrow SX$ is a Λ-homomorphism, then there exists $\alpha:\Lambda\Pi_{N''}N \longrightarrow SX$ such that $\alpha \circ i' = u \circ p''$. Then $\alpha \circ q:\Lambda \longrightarrow SX$ and $\alpha \circ q \circ i \circ p'' = \alpha \circ i = u \circ p''$, i.e. $\alpha \circ q \circ i = u$.

We have thus proved that a quotient object of a negligible object is a negligible object.

We now consider the exact sequence in $\mathscr{C}_\Lambda$

$$0 \longrightarrow N' \xrightarrow{i} N \xrightarrow{p} N'' \longrightarrow 0$$

where N', N'' are objects of $\mathscr{A}$. Let $f:\Lambda \longrightarrow N$ be a Λ-homomorphism. We consider the commutative diagram

$$\begin{array}{ccccccc}
 & & & \ker f & & & \\
 & & \swarrow{\scriptstyle k} & \downarrow{\scriptstyle u} & & & \\
 & \ker(pf) & \xrightarrow[j]{} & \Lambda & & & \\
 & \downarrow{\scriptstyle h} & & \downarrow{\scriptstyle f} & & & \\
0 \longrightarrow & N' & \xrightarrow[i]{} & N & \xrightarrow[p]{} & N'' & \longrightarrow 0
\end{array}$$

where $(\ker f, k)$ is the kernel of h, $(\ker f, u)$ is the kernel of f, $(\ker(p \circ f), j)$

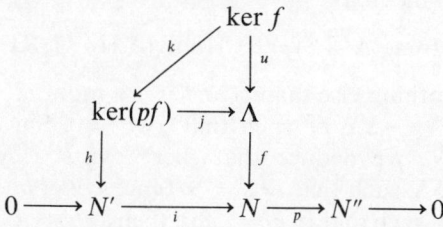

is the kernel of $p \circ f$, and k, h, j are constructed in a canonical manner. Let X be an object of $\mathscr{C}$ and $\alpha : \Lambda \longrightarrow SX$ a homomorphism such that $\alpha \circ u = 0$. Then $\alpha \circ j \circ k = 0$ and hence $\alpha \circ j = 0$, i.e. $\alpha = 0$ since N' and N'' are negligible. If $\alpha : \ker f \longrightarrow SX$ is a Λ-homomorphism, there exists $\beta : \ker(p \circ f) \longrightarrow SX$ such that $\beta \circ k = \alpha$ and also there exists $\gamma : \Lambda \longrightarrow SX$ such that $\gamma \circ j = \beta$. Obviously, $\gamma \circ u = \gamma \circ j \circ k = \beta \circ k = \alpha$.

We have still to show that an inductive limit of negligible modules is a negligible module and to do this it is sufficient to prove that a direct sum of negligible modules is negligible. This in turn reduces to the fact that a finite sum of negligible modules is negligible, which is proved by induction.

We shall say that a morphism $u : M \longrightarrow N$ in $\mathscr{C}_\Lambda$ is covering if Coker u is a negligible object.

LEMMA 6.28. *A morphism $u : M \longrightarrow N$ in $\mathscr{C}_\Lambda$ is covering if and only if for any Λ-homomorphism $f : \Lambda \longrightarrow N$ and any object X of $\mathscr{C}$ the sequence*

$$0 \longrightarrow \operatorname{Hom}_\Lambda(\Lambda, SX) \xrightarrow{\varphi} \operatorname{Hom}_\Lambda(\Lambda \,\Pi_N\, M, SX) \longrightarrow \operatorname{Hom}_\Lambda(\ker u, SX)$$

induced by u and f is exact.

PROOF. Consider the commutative diagram

$$
\begin{array}{ccccccccc}
 & & & & & \ker(p\circ f) & & & \\
 & & & & \overset{q}{\nearrow} & & \overset{h}{\searrow} & & \\
0 & \longrightarrow & \ker u' & \xrightarrow{i'} & \Lambda \,\Pi_N\, M & \xrightarrow{u'} & \Lambda & & \\
 & & \downarrow{\scriptstyle v} & & \downarrow{\scriptstyle f'} & & \downarrow{\scriptstyle f} & & \\
0 & \longrightarrow & \ker u & \longrightarrow & M & \xrightarrow{u} & N & \xrightarrow{p} & \operatorname{coker} u \longrightarrow 0
\end{array}
$$

where the rows are exact, v is an isomorphism and $(\ker(p \circ f), h)$ is the kernel of $p \circ f$, and q is an epimorphism.

Let X be an object of $\mathscr{C}$ and $\alpha : \Lambda \longrightarrow SX$ a Λ-homomorphism with $\alpha u' = 0$. Then $\alpha \circ h \circ q = 0$, hence $\alpha \circ h = 0$, hence $\alpha = 0$, since coker u is negligible. Thus the map

$$\operatorname{Hom}_\Lambda(\Lambda, SX) \longrightarrow \operatorname{Hom}_\Lambda(\Lambda \,\Pi_N\, M, SX)$$

induced by u' is nothing else than φ and it is a monomorphism.

Let $\alpha : A \,\Pi_N\, M \longrightarrow SX$ be such that $\alpha \circ i' = 0$. Since $(\ker(p \circ f), q)$ is the cokernel of i', we deduce that there exists a Λ-homomorphism $t : \ker(p \circ f) \longrightarrow SX$ such that $t \circ q = \alpha$. Since coker u is negligible, there exists $l : \Lambda \longrightarrow SX$ such that $l \circ h = t$. But then $l \circ u' = l \circ h \circ q = t \circ q = \alpha$, which completes the proof.

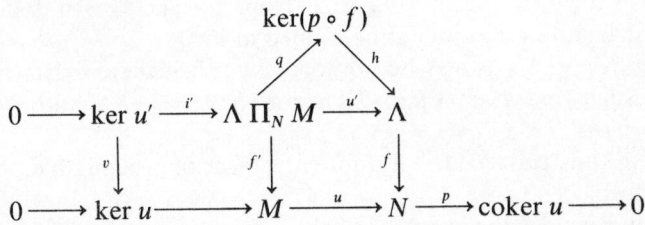

Let $\bar{\mathscr{A}}$ be the full subcategory of $\mathscr{C}_\Lambda$ consisting of the Λ-modules which are closed relatively to the localizant subcategory $\mathscr{A}$. $\bar{\mathscr{A}}$ is an Abelian category according to the considerations preceding proposition 6.21 (remark 3). Clearly for any object X of $\mathscr{C}$ SX is a closed object. Thus, S induces a functor $S_1 : \mathscr{C} \longrightarrow \bar{\mathscr{A}}$.

LEMMA 6.29. *The functor S_1 is exact.*

PROOF. S_1 is obviously left-exact. It remains to show that if $u : X \longrightarrow Y$ is an epimorphism in $\mathscr{C}$, then the morphism Su of $\mathscr{C}_\Lambda$ is covering, i.e. according to lemma 6.28, that for any object Z of $\mathscr{C}$ we have the exact sequence:

$$\begin{aligned} 0 \longrightarrow \mathrm{Hom}_\Lambda(\Lambda, SZ) \longrightarrow \mathrm{Hom}_\Lambda(\Lambda \,\Pi_{SY}\, SX, SZ) \\ \longrightarrow \mathrm{Hom}_\Lambda(\ker Su, SZ), \end{aligned} \tag{1}$$

induced by any Λ-homomorphism $f : \Lambda \longrightarrow SY$. Since S is fully faithful and $\Lambda = SU$, we deduce that there exists a morphism $g : U \longrightarrow Y$ in $\mathscr{C}$ such that $f = Sg$. Thus we have the commutative diagram

$$\begin{array}{ccccccccc} 0 & \longrightarrow & \ker u & \xrightarrow{i} & X \,\Pi_Y\, U & \xrightarrow{u} & U & \longrightarrow & 0 \\ & & & & \downarrow & & \downarrow g & & \\ & & & & X & \xrightarrow[f]{} & Y & \longrightarrow & 0 \end{array} \tag{2}$$

where the rows are exact. Since S is left-exact, we have the exact sequence

$$0 \longrightarrow \ker Su \longrightarrow SX \,\Pi_{SY}\, \Lambda \longrightarrow \Lambda.$$

The exactness of (1) is deduced from the exactness of the top row of (2) and the fact that S is fully faithful.

LEMMA 6.30. *Let X be an object of $\mathscr{C}$ and $(X_i)_{i \in I}$ a family of subobjects of X directed by the inclusion relation such that $X = \sup_{i \in I} X_i$. Then $S_1 X = \sup_{i \in I} S_1 Z_i$.*

PROOF. It is sufficient to prove that the canonical morphism

$$\sup_{i \in I} SX_i \longrightarrow SX$$

is covering, which can be deduced as in lemma 6.29 by using the relation

$$(\sup_{i \in I} X_i) \,\Pi_X\, U = \sup_{i \in I} (X_i \,\Pi_X\, U),$$

valid for any morphism $f : U \longrightarrow X$ (since $\mathscr{C}$ satisfies AB5).

Lemma 6.30 implies that S_1 commutates with direct sums indexed by sets. In particular, any direct sum $\bigoplus_{i \in I} \Lambda_i$, where $\Lambda_i \approx \Lambda$ for any $i \in I$, is of the form $S_1(\bigoplus_{i \in I} U_i)$, where $U_i \approx U$ for any $i \in I$.

Let now M be an object of $\mathscr{A}$. Then M can be inserted into the exact sequence

$$SY \xrightarrow{Sf} SX \longrightarrow M \longrightarrow 0$$

where $f: Y \longrightarrow X$ is a morphism in $\mathscr{C}$. Clearly M is isomorphic with $S_1(\operatorname{coker} f)$. Thus S_1 is a fully faithful functor, and any object M of $\bar{\mathscr{A}}$ is isomorphic with an object of the form $S_1 M'$, where M' is an object of $\mathscr{C}$. This means that S_1 is an equivalence according to proposition 1.15.

Thus theorem 6.25 is completely proved.

Corollary 6.31. *Any Grothendieck category is a category with direct products.*

Proof. The proof follows from theorem 6.25 and proposition 5.39.

Corollary 6.32. *Any Grothendieck category is a category with injective envelopes and cogenerators.*

Proof. The first part follows from theorem 6.25 and corollary 6.17.

Let U be a generator of the Grothendieck category $\mathscr{C}$ and $(U_i)_{i \in I}$ the family of all subobjects of U. Let $Y_i = U/U_i$, and let Q_i be an injective envelope of $Y_i (i \in I)$. Then $W = \prod_{i \in I} Q_i$ is a cogenerator of $\mathscr{C}$, which is of course injective as a product of injective objects.

Here is a theorem which will be used in the sequel to study some properties of Abelian categories.

Theorem 6.33 (B. Mitchell). *Let $\mathscr{C}$ be an Abelian category and U a projective generator of $\mathscr{C}$. Let $\Lambda = \operatorname{Hom}_{\mathscr{C}}(U, U)$ and $S: \mathscr{C} \longrightarrow \mathscr{C}_\Lambda$ be the canonical functor defined by $S(X) = \operatorname{Hom}_{\mathscr{C}}(U, X)$. If X and Y are two quotient objects of U, then $S(X, Y): \operatorname{Hom}_{\mathscr{C}}(X, Y) \longrightarrow \operatorname{Hom}_{\mathscr{C}}(SX, SY)$ is an isomorphism.*

Proof. S is an exact and faithful functor (proposition 6.24). To finish the proof, let $f: SX \longrightarrow SY$ be a Λ-homomorphism. Then we have the diagram

$$\begin{array}{ccccccccc} 0 & \longrightarrow & SK & \xrightarrow{Si} & \Lambda & \xrightarrow{Sp} & SX & \longrightarrow & 0 \\ & & & & & & \downarrow{\scriptstyle f} & & \\ & & & & \Lambda & \xrightarrow[Sq]{} & SY & \longrightarrow & 0 \end{array}$$

where the rows are exact and constructed according to the hypothesis of the theorem. Let $g:\Lambda \longrightarrow \Lambda$ be a Λ-homomorphism such that $Sq \circ g = f \circ Sp$. But $g = Sg'$, where $g' \in \mathrm{Hom}_{\mathscr{C}}(U, U)$. From $q \circ g' \circ i = 0$, we obtain in $\mathscr{C}$ a morphism $f':X \longrightarrow Y$ such that $f'p = qg'$. Hence $Sf' \circ Sp = Sq \circ g$, hence $Sf' = f$ since Sp is an epimorphism.

COROLLARY 6.34. *Let $\mathscr{C}$ be a small Abelian category. There exists a ring Λ and a full embedding $T:\mathscr{C} \longrightarrow \mathscr{C}_\Lambda$.*

PROOF. According to proposition 5.40 there exists a full embedding $T_1:\mathscr{C} \longrightarrow \mathscr{C}_1$ where $\mathscr{C}_1$ is a Grothendieck category. It remains to notice in the constructions preceding proposition 5.40 that $\mathscr{C}_1$ contains a generator U such that for any object X of $\mathscr{C}$, TX is a quotient object of U. Thus, $\mathscr{C}_1$ contains by corollary 6.32 an injective cogenerator V such that TX is a subobject of V. The category $\mathscr{C}_1^0$, the dual of $\mathscr{C}_1$, contains a projective generator V^0 (the object 'dual' to V). According to theorem 6.33 there exists an exact functor $T_2:\mathscr{C}_1^0 \longrightarrow \mathscr{C}_\Lambda$, where Λ is a suitable ring, such that, if $D:\mathscr{C}_1 \longrightarrow \mathscr{C}_1^0$ is the functor of passage to the dual, then for any objects X, Y of $\mathscr{C}$, $T_2(DTX, DTY)$ is an isomorphism. In conclusion the composition of functors $T_2 \circ D \circ T \circ A$, where $A:\mathscr{C}^0 \longrightarrow \mathscr{C}$ is the functor of passage to the dual, is a full embedding of $\mathscr{C}^0$ into $\mathscr{C}_\Lambda$, which completes the proof.

COROLLARY 6.35. *Let $\mathscr{C}$ be an Abelian category, and consider the following diagram in $\mathscr{C}$:*

$$\begin{array}{ccccccc} & & X' & \longrightarrow & X & \longrightarrow & X'' & \longrightarrow & 0 \\ & & \big\downarrow{\scriptstyle f'} & & \big\downarrow{\scriptstyle f} & & \big\downarrow{\scriptstyle f''} & & \\ 0 & \longrightarrow & Y' & \longrightarrow & Y & \longrightarrow & Y'' & & \end{array}$$

where the rows are exact. Then there exists a morphism $\theta:\ker f'' \longrightarrow \operatorname{coker} f'$ such that we have the exact sequence:

$$\ker f' \longrightarrow \ker f \longrightarrow \ker f'' \xrightarrow{\theta} \operatorname{coker} f' \longrightarrow \operatorname{coker} f \longrightarrow \operatorname{coker} f''.$$

PROOF. According to proposition 5.41, there exists a small Abelian full subcategory $\mathscr{A}$ of $\mathscr{C}$, which contains the diagram in the corollary. Let $T:\mathscr{A} \longrightarrow \mathscr{C}_\Lambda$ be a full embedding. It is well known that the corollary is valid for the category $\mathscr{C}_\Lambda$ ([3], chapter 3). To finish the proof, it is sufficient to notice that a sequence:

$$X \longrightarrow Y \longrightarrow Z$$

in $\mathscr{A}$ is exact if and only if the sequence

$$TX \longrightarrow TY \longrightarrow TZ$$

in $\mathscr{C}_\Lambda$ is exact.

Remark. By using the characterization of Grothendieck categories given in proposition 6.25, it can be shown that any category of diagrams on a Grothendieck category is a Grothendieck category, which generalizes a result due to Grothendieck.

7. The Theorem of Krull–Remak–Schmidt

Let $\mathscr{C}$ be an Abelian category and M an object of $\mathscr{C}$.

Definition. The subobject N of M is said to be *irreducible* in M if the following conditions are satisfied:

(a) $N \neq M$.
(b) $P \cap Q = N$ implies that either $P = N$ or $Q = N$.

If 0 is irreducible in M then M is said to be *coirreducible*.

Definition. An object X of an Abelian category $\mathscr{C}$ is said to be *indecomposable* if from $X = X_1 \oplus X_2$ we deduce $X_1 = 0$ or $X_2 = 0$.

In the following we shall assume that $\mathscr{C}$ is a category with injective envelopes or that $\mathscr{C}$ is a Grothendieck category.

PROPOSITION 6.36. *The following assertions are equivalent:*
(a) *M is a coirreducible object.*
(b) *The injective envelope $E(M)$ is indecomposable.*

PROOF. (a) $\Rightarrow$ (b). Let M be coirreducible. Assume that $E(M) = Q_1 \oplus Q_2$, $Q_1 \neq 0 \neq Q_2$. Hence $Q_1 \cap M \neq 0 \neq Q_2 \cap M$. Clearly $Q_1 \cap M \subset M$, $Q_2 \cap M \subset M$ and $(Q_1 \cap M) \cap (Q_2 \cap M) = 0$, which contradicts the fact that M is coirreducible.

(b) $\Rightarrow$ (a). Let $E(M)$ be indecomposable.
Assume that two subobjects P, Q of M exist such that

$$M = P \cup Q, \qquad P \neq 0 \neq Q$$

Then $E(P) \subset E(H)$ and consequently $E(P) = E(M)$ since $E(M)$ is indecomposable. We have $E(P) \cap Q = E(M) \cap Q = Q \neq 0$ and therefore $(E(P) \cap Q) \cap P = Q \cap P$. I claim that $Q \cap P$. Indeed if $P \cap Q = 0$ then $(E(P) \cap Q) \cap P = 0$ hence $E(P) \cap Q = 0$.

COROLLARY 6.37. *If $N_1, N_2, \cdots N_r$ are irreducible subobjects of M and if $0 = N_1 \cap N_2 \cap \cdots \cap N_r$ and 0 is not the intersection of a smaller number of objects N_i, then*

$$E(M) = E(M/N_1) \oplus E(M/N_2) \oplus \cdots \oplus E(M/N_r)$$

and the objects $E(M/N_i)$ are indecomposable.

PROPOSITION 6.38. *Let I be an injective object. Then the following assertions are equivalent:*

(a) *I is indecomposable;*

(b) *The ring $\mathrm{Hom}_{\mathscr{C}}(I, I)$ is a local ring.*

PROOF. (a) $\Rightarrow$ (b). Notice first that $u \in \mathrm{Hom}_{\mathscr{C}}(I, I)$ is invertible if and only if $\ker u = 0$. For, $\ker u = 0$ implies, since I is injective, that $u(I) = I$ by condition (a). It is sufficient to prove that if $\ker u \neq 0$ and $\ker v \neq 0$, then $\ker(u + v) \neq 0$. But we have $\ker(u + v) \supset \ker u \cap \ker v$, so that otherwise we would have $0 = \ker u \cap \ker v$, which contradicts proposition 6.36.

(b) $\Rightarrow$ (a). If I were not indecomposable, we would have $I = I_1 \oplus I_2$. But then i_1p_1, i_2p_2 are not invertible and their sum is invertible, which contradicts (b).

Remark. In proving the implication (b) $\Rightarrow$ (a) we have not used the hypothesis that I is injective. We shall give below conditions under which the implication (a) $\Rightarrow$ (b) is also valid without assuming that I is injective. The conditions will then refer to the category $\mathscr{C}$.

Henceforth we shall again assume that the category $\mathscr{C}$ satisfies axiom AB5.

LEMMA. 6.39. *Let $M = M_1 \oplus M_2$ and denote by i_1, i_2, p_1, p_2 the canonical injections and projections. If $P \xrightarrow{\alpha} M$ is a subobject of M such that $p_1\alpha : P \longrightarrow M_1$ is an isomorphism then $M = P \oplus M_2$. The canonical injections are $i_1p_1\alpha$, i_2 and the canonical projections $(p_1\alpha)^{-1}p_1, p_2$.*

The proof is immediate.

Definition. An object M of an Abelian category $\mathscr{C}$ is said to be *noetherian* if any sequence

$$M_1 \subset M_2 \subset \cdots \subset M_i \subset \cdots$$

of subobjects of M is stationary, i.e. there exists an integer n_0 such that $M_i = M_{i+1}$ for any $i \geq n_0$.

LEMMA 6.40. *Any subobject of a noetherian object is the intersection of a finite number of irreducible subobjects.*

PROOF. Assume there exist objects which have not the property in the lemma; let $\mathfrak{M}\mathscr{C}$ the set formed by these objects. From the definition of a noetherian object it results that $\mathfrak{M}\mathscr{C}$ has at least one maximal element. Let N be such an object. N cannot be irreducible. But then $N = N_1 \cap N_2$ and N_1, N_2 have the property in the lemma, which is absurd.

PROPOSITION 6.41. *The injective envelope of any noetherian object is the direct sum of a finite number of indecomposable injective objects.*

PROOF. Let M be such an object. By lemma 6.40 $0 = N_1 \cap N_2 \cap \cdots \cap N_v$, where $N_1, N_2, \cdots, N_v$ are irreducible subobjects of M. We may obviously assume that 0 is not the intersection of a smaller number of N_i's. Thus we are in the conditions of corollary 6.37. It follows that

$$E(M) = E(M/N_1) \oplus \cdots \oplus E(M/N_v)$$

and $E(M/N_i)$ are indecomposable injective objects.

We shall show in the sequel that the decomposition given by proposition 6.41 is essentially unique. The proof of the uniqueness theorem will be preceded by a few lemmas.

LEMMA 6.42. *Let* $M = \bigoplus_{i \in I} M_i$ *and* $a, b : M \longrightarrow M$ *be such that* $a + b = 1_M$. *Suppose that for each* $i \in I$ *the ring of endomorphisms of* M_i *is a local ring. Under these conditions for any finite system* $i_1, i_2, \cdots, i_s$ *of elements of* I, *there exist subobjects* $P_1, P_2, \cdots, P_s$ *with the following properties:*

(1) *For each* $k \in \{1, 2, \cdots, s\}$, *either* a *or* b *induces an isomorphism of* P_k *onto* M_{i_k}.

(2) $M = (P_1 \oplus P_2 \oplus \cdots \oplus P_s) \oplus (\bigoplus_{i \in I'} M_i)$, *where we designate by* I' *the set* $I - \{i_1, i_2, \cdots, i_s\}$.

PROOF. Denote by $p_{i_1} : M \longrightarrow M_{i_1}$ the canonical projection of M onto M_{i_1} and by $\alpha_{i_1} : M_{i_1} \longrightarrow M$ the canonical injection of M_{i_1} into M. Clearly

$$p_{i_1} = p_{i_1}(a + b) = p_{i_1}a + p_{i_1}b.$$

$$1_{M_{i_1}} = p_{i_1}\alpha_{i_1} = p_{i_1}a\alpha_{i_1} + p_{i_1}b\alpha_{i_1}.$$

This implies, in view of the fact that the ring of endomorphisms of M_{i_1} is local, that either $p_{i_1}a\alpha_{i_1}$ or $p_{i_1}b\alpha_{i_1}$ is an isomorphism. Assume that

$p_{i_1} a \alpha_{i_1} : M_{i_1} \longrightarrow M_{i_1}$ is an isomorphism. Denote by P_1 the subobject $(a\alpha_{i_1})(M_{i_1})$. It follows evidently that P_1 is isomorphic with M_{i_1} and the restriction of p_{i_1} to P_1 is an isomorphism of P_1 with M_{i_1}. We may apply lemma 6.39. It follows that $M = P_1 \oplus (\bigoplus_{i \in I - i_1} M_i)$. We now repeat the argument for the index i_2. We thus obtain after a finite number of steps the objects $P_1, P_2, \cdots, P_s$ with the required properties.

LEMMA 6.43. *Assume that $M = \bigoplus_{i \in I} M_i$ and for each index $i \in I$ the ring of endomorphisms of M_i is a local ring. Under these conditions if $f: M \longrightarrow M$ is such that $f^2 = f$ then the following assertions are true:*

(a) *There exists at least one $i \in I$ such that f induces an isomorphism of M_i onto $f(M_i)$ and $f(M_i)$ is a direct factor of M.*

(b) *Any indecomposable direct factor of M is isomorphic with an M_i.*

PROOF. If we set $f' = 1 - f$ then we obviously have the following decomposition

$$M = f(M) \oplus f'(M).$$

Let J be an arbitrary finite subset of I. Then according to axiom AB5 we have:

$$f(M) = \sup_J [f(M) \cap (\bigoplus_{i \in J} M_i)].$$

Thus there exists a finite subset $\{i_1, i_2, \cdots, i_s\}$ of I such that

$$f(M) \cap (M_{i_1} \oplus M_{i_2} \oplus \cdots \oplus M_{i_s}) \neq 0.$$

We now apply lemma 6.42 with $a = f, b = f'$ and with the set of indices $i_1, i_2, \cdots, i_s$. We get the decomposition

$$M = (P_1 \oplus \cdots \oplus P_s) \oplus (\bigoplus_{i \in I} M_i)$$

where $I' = I - \{i_1, i_2, \cdots, i_s\}$. As on the other hand $f(M) \cap (M_{i_1} \oplus \cdots \oplus M_{i_s})$ belongs to the kernel of f', it follows that f' induces an isomorphism between $M_{i_1} \oplus \cdots \oplus M_{i_s}$ and $P_1 \oplus \cdots \oplus P_s$. Hence there exists at least one index i_k such that f induces an isomorphism between M_{i_k} and P_k, whence the assertion (a).

To prove the assertion (b), let P be an indecomposable direct summand of $M : M = P \oplus Q$. There exists an indempotent endomorphism $p: M \longrightarrow M$ such that $p(M) = P$. From the assertion (a) it follows therefore that there exists an index i such that p induces an isomorphism of M_i

onto $p(M_i) \subset P$. Since $p(M_i)$ is a direct factor of M and P is an indecomposable direct factor we infer that $p(M_i) = P$. Thus the assertion (b) is proved.

We are now in a position to state and prove the part concerning the uniqueness in the theorem of Krull–Remak–Schmidt.

THEOREM 6.44. (Theorem of Krull–Remak–Schmidt). *Assume that*

$$M = \bigoplus_{i \in I} M_i = \bigoplus_{j \in J} N_j$$

where the objects M_i, N_j are indecomposable and the rings of endomorphisms $\mathrm{Hom}_{\mathscr{C}}(M_i, M_i)$, $\mathrm{Hom}_{\mathscr{C}}(N_j, N_j)$ are local rings for any $i \in I$ and any $j \in J$. Under these conditions there exists a bijection $\pi : I \longrightarrow J$ such that M_i is isomorphic with $N_{\pi(i)}$ for any $i \in I$.

PROOF. By applying lemma 6.43 it results immediately that any object M_i is isomorphic with an object N_j and conversely. In the set of indices I we now introduce the following equivalence relation: We say that i_1 is equivalent with i_2 if M_{i_1} is isomorphic with M_{i_2}. An equivalence relation is introduced similarly in the set J. Let K, respectively L, be the quotient set obtained from the set I, respectively the set J, by this equivalence relation. From the above remarks it follows that K and L are equivalent. We shall therefore identify L with K.

Let now $k \in K$. Denote by $I(k)$ all the elements of the set I which belong to the class k, and by $J(k)$ all the elements of the set J which belong to the class k. It is sufficient to prove that the sets $I(k)$, $J(k)$ have the same cardinal for any $k \in K$. Let $c[I(k)]$, respectively $c[J(k)]$ be the cardinal of the set $I(k)$, respectively of the set $J(k)$. Since the sets I and J play a symmetric role, it is sufficient to prove that $c[I(k)] \geq c[J(k)]$. We distinguish two cases:

1. $c[I(k)]$ is finite. Let $j_1 \in J(k)$ and $p_{j_1} : M \longrightarrow M$ be such that $p_{j_1}(M) = N_{j_1}$, $p_{j_1}^2 = p_{j_1}$. By lemma 6.43 there exists $i_1 \in I(k)$ such that p_{j_1} induces an isomorphism of M_{i_1} onto N_{j_1} and by lemma 6.42 we have the following decomposition:

$$M = M_{i_1} \oplus (\bigoplus_{j \neq j_1} N_j). \qquad (**)$$

If j_1 is the only element of the set $J(k)$ then the relation $c[I(k)] \geq c[J(k)]$ is evidently proved. If $J(k)$ contains also the element j_2 then by considering the projection p_{j_2} in the decomposition (**) it follows that there exists an $i_2 \in I(k)$ such that p_{j_2} induces an isomorphism of M_{i_2} with N_{j_2}. All now reduces to prove that $i_2 \neq i_1$. But $p_{j_2}(M_{i_1}) = 0$, which shows that $i_2 \neq i_1$. By continuing in this manner the argument, we clearly achieve after a finite number of steps the proof of the relation $c[I(k)] \geq c[J(k)]$.

2. $c[I(k)]$ is infinite. Let $j \in J$ and $p_j : M \longrightarrow N_j$ be the canonical projection. For any $i \in I$ we have the relation

$$M_i = \sup_L (M_i \cap (\bigoplus_{j \in L} N_j))$$

where L runs through the finite subsets of J. Thus $M_i \cap \ker p_j$ is null only for a finite number of j's.

For a given i there exist only a finite number of j's such that p_j induces an isomorphism of M_i with N_j. Since $c[I(k)]$ is infinite it follows that $c[I(k)] \geq c[J(k)]$.

The theorem of Krull–Remak–Schmidt in the form of Atiyah.

The theorem of Krull–Remak–Schmidt given above regards the existence and uniqueness of decompositions into indecomposable objects of injective objects in Abelian categories subject to certain conditions. We propose now to prove a Krull–Remak–Schmidt type theorem which may be applied to any object of the considered category. The conditions we are going to impose to the category $\mathscr{C}$ are similar to the conditions of noetherian and artinian type in the theory of modules.

Definition. A double chain of an Abelian category $\mathscr{C}$ is a sequence of systems (A_n, i_n, p_n) with the following properties:

(i) A_n are objects of the category $\mathscr{C}$.
(ii) $i_n : A_n \longrightarrow A_{n-1}$ is a monomorphism for any $n \geq 1$.
(iii) $p_n : A_{n-1} \longrightarrow A_n$ is an epimorphism for any $n \geq 1$.

A double chain (A_n, i_n, p_n) is said to be stationary if there exists an integer N such that for any $n \geq N$, i_n and p_n are isomorphisms.

Our aim is to prove the following

THEOREM 6.45. (Krull–Remak–Schmidt). *If in the category $\mathscr{C}$ any double chain is stationary then any object M of $\mathscr{C}$ admits a decomposition $M = M_1 \oplus M_2 \oplus \cdots \oplus M_n$, where each M_i is indecomposable. If $M = N_1 \oplus \cdots \oplus N_m$ is another decomposition where each N_i is indecomposable then $n = m$ and there exists a permutation $\pi : \{1, 2, \cdots, n\} \longrightarrow \{1, 2, \cdots, n\}$ such that for any $i \in \{1, 2, \cdots, n\}$ M_i is isomorphic with $N_{\pi(i)}$.*

PROOF. Existence. Assume that we have an unlimited sequence of decompositions

$$M = A_1 \oplus B_1, \qquad A_1 = A_2 \oplus N_2, \cdots, \qquad A_{n-1} = A_n \oplus B_n, \cdots$$

and hence the morphisms

$$i_n : A_n \longrightarrow A_{n-1}$$

$$p_n : A_{n-1} \longrightarrow A_n.$$

Thus we obtain the double chain (A_n, i_n, p_n). But from the hypothesis it results that this double chain is stationary, which is in contradiction with our hypothesis.

The uniqueness of the decomposition will follow from theorem 6.44 if we prove that under our conditions if the object A is indecomposable then $\mathrm{Hom}_{\mathscr{C}}(A, A)$ is a local ring. This fact will follow immediately from the following lemmas:

LEMMA 6.46. *Let A be an arbitrary object of the category $\mathscr{C}$ and $\theta : A \longrightarrow A$. A double chain (A_n, i_n, p_n) can be associated to the morphism θ such that if this double chain is stationary then, for a sufficiently large n, A_n is a direct summand of A.*

PROOF. The construction of the double chain (A_n, i_n, p_n) is as follows: Let $(A_n, j_n) = \mathrm{im}\, \theta^n$. Here $j_n : A_n \longrightarrow A$ is the canonical injection of the object A_n into A. From the relation $\theta^n = \theta^{n-1} \circ \theta$ we deduce the relation $\mathrm{im}\, \theta^n \subset \mathrm{im}\, \theta^{n-1}$, in other words there exists a monomorphism $A_n \xrightarrow{i_n} A_{n-1}$ such that the diagram

$$\begin{array}{ccc} A_n & \xrightarrow{i_n} & A_{n-1} \\ & {\scriptstyle j_n} \searrow \quad \swarrow {\scriptstyle j_{n-1}} & \\ & A & \end{array}$$

is commutative: $j_{n-1} i_n = j_n$. Hence $j_n = i_1 i_2 \cdots i_n$. There exist also epimorphisms $p_{n-1} : A_{n-1} \longrightarrow A_n$ such that $j_n p_{n-1} = \theta j_{n-1}$. Thus we have defined the double chain (A_n, i_n, p_n).

Assume now that this double chain is stationary. Define the epimorphism $q_n : A \longrightarrow A_n$ by

$$q_n = p_n p_{n-1} \cdots p_1.$$

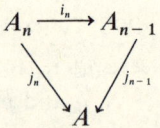

We show first that for a sufficiently large n $q_n j_n: A_n \longrightarrow A_n$ is an isomorphism. This follows from the relation

$$q_n j_n = i_{n+1} i_{n+2} \cdots i_{2n} p_{2n} \cdots p_{n+1}. \qquad (*)$$

To prove relation (**) we notice that

$$j_n q_n j_n = j_n(p_n p_{n-1} \cdots p_1) j_n = \theta j_{n-1} q_{n-1} j_n = \theta^n j_n$$
$$j_n(i_{n+1} i_{n+2} \cdots i_{2n} p_{2n} \cdots p_{n+1}) = j_{2n} p_{2n} \cdots p_{n+1} = \theta^n j_n.$$

Relation (*) now follows since j_n is a monomorphism. Let $\sigma_n = q_n j_n$. We have $\sigma_n^{-1} q_n j_n = 1_{A_n}$, whence it follows that A_n is a direct summand of A.

LEMMA 6.47. *If in the Abelian category $\mathscr{C}$ any double chain is stationary and A is indecomposable then any element $\theta \in \mathrm{Hom}_{\mathscr{C}}(A, A)$ either is an isomorphism or is nilpotent.*

PROOF. Using the notations and the results of lemma 6.46 it follows that for a sufficiently large n j_n and q_n are isomorphisms or are null. In the first case θ is an isomorphism, whereas in the second case $\theta^n = 0$.

LEMMA 6.48. *Under the conditions of lemma 6.47 if $\theta_1, \theta_2 \in \mathrm{Hom}_{\mathscr{C}}(A, A)$ and if $\theta_1 + \theta_2$ is an isomorphism then either θ_1 or θ_2 is an isomorphism.*

PROOF. We may assume $\theta_1 + \theta_2 = 1_A$. If neither θ_1 nor θ_2 is an isomorphism, then it follows from lemma 6.47 that $(\theta_1 + \theta_2)^{2n} = 0$, which is absurd. We have used here the relation $\theta_1\theta_2 = \theta_2\theta_1$ which results for instance from the relations:

$$\theta_1\theta_2 = \theta_1 - \theta_1^2 = \theta_2\theta_1.$$

Remark. One should not believe that in the Abelian categories in which any double chain is stationary any object is injective. Indeed, in the Abelian category of finite Abelian groups there exist objects which are not injective.

8. The Structure of Injective Objects in Locally Noetherian Categories

Let $\mathscr{C}$ be a Grothendieck category.

Definition. $\mathscr{C}$ is said to be *locally noetherian* if it possesses a family of noetherian generators.

From this definition it follows immediately that any object of $\mathscr{C}$ is the least upper bound of its noetherian subobjects. As examples of locally noetherian categories we can invoke the category $\mathscr{Ab}$ of Abelian groups and the category of graded unitary left-modules over a left noetherian ring Λ with a unit element. A more general example is the following. Let $\mathscr{C}$ be a small Abelian category in which any object is noetherian. Then the category Sex $(\mathscr{C}, \mathscr{Ab})$ of contravariant left exact functors from $\mathscr{C}$ to $\mathscr{Ab}$ is locally noetherian, and conversely, any locally noetherian category is of this form.

In what follows we shall prove a few results which will lead to a characterization of injective objects in locally-noetherian categories.

PROPOSITION 6.49. *Let $\mathscr{C}$ be a Grothendieck category, M a noetherian object of $\mathscr{C}$ and $(P_i)_{i \in I}$ a directed family of subobjects of P such that $P = \sup_{i \in I} P_i$. Then the canonical homomorphism*

$$\varphi : \sup_i \mathrm{Hom}_{\mathscr{C}}(M, P_i) \longrightarrow \mathrm{Hom}_{\mathscr{C}}(M, \sup_i P_i)$$

is an isomorphism.

PROOF. The fact that φ is a monomorphism is valid in general. It remains to show that φ is an epimorphism. Let $f: M \longrightarrow \sup_i P_i$ be a morphism in $\mathscr{C}$. $P_1' = \operatorname{im} f$ is a noetherian object, subobject of P. $(P_1' \cap P_i)_{i \in I}$ is a family of subobjects of P which has a maximal element, say $P_1' \cap P_{i_0}$. But then $P_1' = P_1' \cap P_{i_0}$. Indeed, for any $i \in I$ there exists $j(i) \in I$ such that $P_{j(i)} \supset P_{i_0}$, $P_{j(i)} \supset P_i$ and $P_{j(i)} \cap P_1' = P_{i_0} \cap P_1'$. But then $P_1' = \sup_i (P_1' \cap P_i) = \sup_i (P_{j(i)} \cap P_1') = P_{i_0} \cap P_1$. Thus $\operatorname{im} f \subset P_{i_0}$. The rest follows immediately.

COROLLARY 6.50. *In a locally-noetherian category $\mathscr{C}$, a direct sum of injective objects is an injective object.*

PROOF. Let $(Q_i)_{i \in I}$ be a family of injective objects of $\mathscr{C}$. Then the direct sum of any finite subfamily of this given family is an injective object. The object $Q = \bigoplus_{i \in I} Q_i$ is the least upper bound of its subobjects which are direct sums of finite subfamilies, denoted by (S_α), where α runs through the finite subsets of I. Let X be an object of $\mathscr{C}$ and $(X_j)_{j \in J}$ the family of noetherian subobjects of X. Then $X = \sup X_j$ and the family (X_j) is directed. If

$$0 \longrightarrow X' \longrightarrow X \xrightarrow{p} X'' \longrightarrow 0 \qquad (*)$$

is an exact sequence of $\mathscr{C}$, then $X'_j = X' \cap X_j (j \in J)$ is a family of noetherian subobjects of X' and $X' = \sup_j X'_j$. Also $(p(X_j))_{j \in J}$ is a family of noetherian subobjects of X'' and $X'' = \sup_j p(X_j)$. Moreover, the exact sequence (*) is the inductive limit of the directed family of exact sequences:

$$0 \longrightarrow X'_j \longrightarrow X_j \longrightarrow p(X_j) \longrightarrow 0.$$

But for any $j \in J$ we have according to proposition 6.49 the exact sequence:

$$0 \longrightarrow \mathrm{Hom}_{\mathscr{C}}(p(X_j), Q) \longrightarrow \mathrm{Hom}_{\mathscr{C}}(X_j, Q) \longrightarrow \mathrm{Hom}_{\mathscr{C}}(X'_j, Q) \longrightarrow 0.$$

Passing to the inductive limit we obtain the exact sequence

$$0 \longrightarrow \mathrm{Hom}_{\mathscr{C}}(X'', Q) \longrightarrow \mathrm{Hom}_{\mathscr{C}}(X, Q) \longrightarrow \mathrm{Hom}_{\mathscr{C}}(X', Q) \longrightarrow 0$$

which shows that Q is injective.

PROPOSITION 6.51. *Let $\mathscr{C}$ be a locally noetherian category and Q an injective object of $\mathscr{C}$. Then Q is a direct sum of a family $(Q_i)_{i \in I}$ of indecomposable injective objects and if $(Q'_j)_{j \in J}$ gives another representation of Q as a direct sum of indecomposable injective objects, then there exists a bijective map $\tau: I \longrightarrow J$ such that $Q_i \approx Q'_{\tau(i)}$ $(i \in I)$.*

PROOF. Q possesses non-null noetherian subobjects; the injective envelopes of these objects are—according to proposition 6.41—finite direct sums of indecomposable injective objects, which are subobjects of Q. Let then Q_0 be a maximal subobject of Q which is a direct sum of indecomposable injective objects. Q_0 is injective and therefore $Q = Q_0 \oplus Q_1$. If Q_1 is not 0, then it contains a non-null noetherian subobject Q'_1 and the injective envelope of Q'_1 which is of the form $R = \bigoplus_{1 \leq k \leq n} R_k$, with R_k indecomposable injective, is contained in Q_1. But $Q_0 \oplus R$ is a subobject of Q which contradicts the maximality of Q_0, if R is non-null.

The second part of the proposition results from proposition 6.38 and theorem 6.44.

Definition. Let $\mathscr{C}$ be a Grothendieck category and Q an indecomposable injective object of $\mathscr{C}$. We call *type* of Q the class of all objects of $\mathscr{C}$ which are isomorphic with Q. The class of types of C is called the *spectrum* of $\mathscr{C}$. Notation: Spec($\mathscr{C}$).

Propositions 6.50 and 6.51 show that if $\mathscr{C}$ is a locally noetherian category then there exists a one-to-one correspondence between the types of injective objects of $\mathscr{C}$ and the families of objects of Spec($\mathscr{C}$). Under this correspondence, to the injective envelopes of noetherian objects there correspond finite subsets of Spec($\mathscr{C}$) (cf. proposition 6.41).

Let $\mathscr{C}$ be a locally noetherian category and M a noetherian object of $\mathscr{C}$. Let $E(M)$ be the injective envelope of M and $E(M) = \bigoplus_{1 \leq i \leq n} Q_{s_i}$, where s_i belong to Spec($\mathscr{C}$) and Q_{s_i} are the corresponding indecomposable injective objects. We denote by L the subsets $\{s_i | 1 \leq i \leq n\}$ of Spec($\mathscr{C}$).

Proposition 6.52. *There exists a map f from L to the set of subobjects of $M : s_i \longrightarrow M_{s_i}$, which satisfies the following conditions:*

(a) *The injective envelope of M/M_{s_i} is a direct sum of indecomposable injective objects of type s_i.*

(b) $\cap M_{s_i} = 0$ *and this intersection is reduced, i.e. if we remove a member of it it ceases to be null.*

Proof. Let i be such that $1 \leq i \leq n$. We denote by J_i the subobject of $E(M)$ which is the direct sum of all subobjects Q_{s_j} which are isomorphic with Q_{s_i}. For distinct i, j we have that either $J_i \cap J_j = 0$ or $J_i = J_j$. Thus we obtain a family $J_{k_1}, \cdots, J_{k_r}$ of subobjects of $E(M)$, by making a suitable indexing. We denote by $M'_i = J_i \cap M$. The construction of the map f is obvious: $s_i \longrightarrow J_{k_l} \longrightarrow J_{k_l} \cap M \longrightarrow M_{s_i}$, $l \leq r$ (in view of the fact that any Q_{s_i} is a direct summand in a single J_{k_l}), where $M_{s_i} = \oplus M'_j$, such that $J_i \cap J_j = 0$. M_{s_i} are the desired objects. The verification of conditions (a) and (b) is immediate.

9. Applications to the Decomposition Theories

In the present paragraph we give a theorem which permits the description of all the decomposition theories given on a locally noetherian category.

Definition. Let $\mathscr{C}$ be a locally noetherian category and A a fixed object of $\mathscr{C}$. We denote by $\mathscr{P}(A)$ the set of subobjects of A (cf. proposition 5.35). We shall say that a *decomposition theory* has been given on $\mathscr{C}$, if a 'function' Γ has been given which associates to each object M of $\mathscr{C}$ a subset $\Gamma(M)$ of $\mathscr{P}(A)$ such that the following conditions are satisfied:

I. If $(M_i)_{i \in I}$ is a family of subobjects of M directed by the inclusion relation and if $M = \sup_{i \in I} M_i$, then $\Gamma(M) = \bigcup_{i \in I} \Gamma(M_i)$.

II. If the sequence

$$0 \longrightarrow M' \longrightarrow M \longrightarrow M'' \longrightarrow 0$$

is exact, then $\Gamma(M') \subset \Gamma(M) \subset \Gamma(M') \cup \Gamma(M'')$.

III. $\Gamma(M) = \varnothing$ if and only if $M = 0$.

IV. For any noetherian object M there exist the subobjects $M_1, M_2, \cdots, M_r$ such that

(i) $\Gamma(M/M_i) = \{P_i\}$, $i = 1, 2, \cdots, r$, where the P_i are subobjects of A.

(ii) $0 = M_1 \cap M_2 \cap \cdots \cap M_r$ and 0 is not the intersection of a subfamily of $(M_i)_{i=1,\cdots,r}$.

(iii) $\Gamma(M/M_i) \neq \Gamma(M/M_j)$ if $i \neq j$.

(iv) $\Gamma(M) = \{P_1, P_2, \cdots, P_r\}$.

We now give a theorem which characterizes completely the decomposition theories.

THEOREM 6.53. *To each function* $f: \mathrm{Spec}(\mathscr{C}) \longrightarrow \mathscr{P}(A)$ *we can associate in a unique manner a decomposition theory* Γ_f *such that*

$$\Gamma_f(X) = f(x)$$

for any injective indecomposable object X *of type* x. *The correspondence* $f \longrightarrow \Gamma_f$ *thus obtained between the functions* $f: \mathrm{Spec}(\mathscr{C}) \longrightarrow \mathscr{P}(A)$ *and the decomposition theories is a bijection.*

PROOF. Let $f: \mathrm{Spec}(\mathscr{C}) \longrightarrow \mathscr{P}(A)$ and M be an arbitrary object of the category $\mathscr{C}$. Let $E(M)$ be the injective envelope of M. According to proposition 6.51 there exists a decomposition of $E(M)$.

$$E(M) = \bigoplus_{i \in I} Q_i$$

where each Q_i is an injective indecomposable object. Thus M determines a subset $P(M)$ of $\mathrm{Spec}(\mathscr{C})$ and we may set

$$\Gamma_f(M) = f[P(M)].$$

We have to check that conditions I–IV are satisfied. Conditions I and III are clearly satisfied.

To verify condition II, let

$$0 \longrightarrow M' \longrightarrow M \longrightarrow M'' \longrightarrow 0$$

be an exact sequence. It is evident that $E(M') \subset E(M)$, hence $P(M') \subset P(M)$

and therefore $\Gamma_f(M') \subset \Gamma_f(M)$. It remains to show that $\Gamma_f(M) \subset \Gamma_f(M') \cup \Gamma_f(M'')$. This results from the remark that if $N = N_1 \oplus N_2$ then $\Gamma_f(N) = \Gamma_f(N_1) \cup \Gamma_f(N_2)$ and from the following

LEMMA. *If the sequence*

$$0 \longrightarrow M' \longrightarrow M \longrightarrow M'' \longrightarrow 0$$

is exact and if $Q = E(M')$ then there exists a monomorphism $\alpha : M \longrightarrow Q \oplus M''$.

PROOF. Let $\xi : M' \longrightarrow Q$ be the canonical injection. Then there exists a morphism $\eta : M \longrightarrow Q$ such that the diagram

$$\begin{array}{ccccccccc} 0 & \longrightarrow & M' & \longrightarrow & M & \longrightarrow & M'' & \longrightarrow & 0 \\ & & & \overset{\xi}{\searrow} & \downarrow{\scriptstyle \eta} & & & & \\ & & & & Q & & & & \end{array}$$

is commutative. This yields a morphism $\alpha : M \longrightarrow Q \oplus M''$ such that the diagram

$$\begin{array}{ccc} M & \longrightarrow & M'' \\ {\scriptstyle \eta}\downarrow & \overset{\alpha}{\searrow} & \uparrow \\ Q & \longleftarrow & M'' \oplus Q \end{array}$$

is commutative and it is obvious that $\ker \alpha = 0$.

Condition IV follows from proposition 6.52.

Let now Γ be a decomposition theory. We must show that there exists a function $f : \mathrm{Spec}(\mathscr{C}) \longrightarrow \mathscr{P}(A)$ such that $\Gamma_f = \Gamma$. It is sufficient to prove that if Q is an injective indecomposable object, then $\Gamma(Q)$ is reduced to a single element. To do this, let M be a noetherian subobject of Q. Since Q is indecomposable, M is coirreducible. Then condition IV implies that $\Gamma(M)$ is reduced to a single element since the only decomposition of 0 which satisfies condition (II) is $0 = 0$.

Let now N be another noetherian subobject of Q. We prove that $\Gamma(M) = \Gamma(N)$. Since Q is indecomposable, we may suppose that $N \subset M$. Then condition II implies that $\Gamma(N) \subset \Gamma(M)$, hence $\Gamma(N) = \Gamma(M)$. Now condition I shows that $\Gamma(Q)$ is reduced to a single element.

This concludes the proof of the theorem.

We now present as examples two classical decomposition theories within the framework of the above point of view.

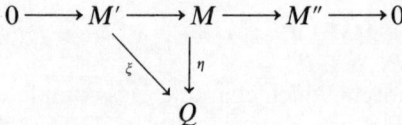

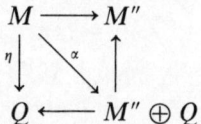

Example 1. *Primary decomposition theory* (*Lasker–Noether*). Let A be a commutative noetherian ring with a unity element. Let $\mathscr{C}$ denote the Abelian category of A-modules. For each $a \in A$ we designate by $\varphi_a : M \longrightarrow M$ the homothety $a \longrightarrow ax$ of the A-module M.

Definition. The module M is said to be *coprimary* if
(i) $M \neq 0$,
(ii) For each element $a \in A$, the homothety $\varphi_a : M \longrightarrow M$ is either injective or nilpotent.

LEMMA. *Let M be a noetherian coirreducible module. Then M is coprimary.*

PROOF. Assume there exists $a \in A$ such that φ_a is neither injective nor nilpotent. Then we can form the sequence of submodules of M

$$\ker \varphi_a \subset \ker \varphi_{a^2} \subset \cdots \subset \ker \varphi_{a^p} \subset \cdots.$$

Since M is noetherian there exists n such that

$$\ker \varphi_{a^n} = \ker \varphi_{a^{n+1}} = \cdots.$$

In particular, we have $\ker \varphi_{a^n} = \ker \varphi_{a^{2n}}$. Since φ_a is not an injection, $\ker \varphi_a \neq 0$ and therefore $\ker \varphi_{a^n} \neq 0$. Since φ_a is not nilpotent, $\operatorname{im} \varphi_{a^n} \neq 0$. But $\ker \varphi_{a^n} \cap \operatorname{im} \varphi_{a^n} = 0$. For, if $\xi \in \ker \varphi_{a^n} \cap \operatorname{im} \varphi_{a^n}$ then $\varphi_{a^n}(\xi) = 0$ and $\xi = \varphi_{a^n}(\eta)$. Hence $\varphi_{a^{2n}}(\eta) = 0$, i.e. $\eta \in \ker \varphi_{a^{2n}} = \ker \varphi_{a^n}$. This implies $\xi = \varphi_{a^n}(\eta) = 0$, which contradicts the fact that M is coirreducible.

Definition. The submodule N of M is said to be *primary* if M/N is coprimary. In this case the set of elements $a \in A$ such that $\varphi_a : M/N \longrightarrow M/N$ is nilpotent is a prime ideal of A which is called the primary radical of the submodule N of M and is designated by $\mathscr{R}_p(N, M)$.

Let now Q be an injective indecomposable object of $\mathscr{C}$ and M a noetherian subobject of Q. Then M is coirreducible. $\mathscr{R}_p(0, M)$ does not depend on M. We may set $f(q) = \mathscr{R}_p(0, M)$, where q denotes the type of Q. The decomposition theory thus obtained is nothing else than the primary theory.

Example 2. *The tertiary decomposition theory* (*Lesieur–Croisot*). Let A be a noetherian ring with a unity element. The bilateral ideal p of A is said to be *prime* if
(i) $p \neq A$,
(ii) $a \cdot A \cdot b \subset p$ implies $a \in p$ or $b \in p$.

Definition. An element $a \in A$ is said to be a semiannulator of the module M is there exists a submodule N of M such that $N \neq 0$ and $a \cdot N = \{0\}$.

Similarly a left-ideal $\mathfrak{a}$ of A is said to be a semiannulator of the module M if there exists a submodule N of M such that $N \neq 0$ and $\mathfrak{a} \cdot N = \{0\}$. A module M is said to be cotertiary if any element of A which is a semiannulator of M is a semiannulator of any non-trivial submodule of M.

LEMMA. *Let M be a coirreducible A-module. Then M is cotertiary.*

PROOF. Let $\lambda \in A$ and let N be a non-nul submodule of M such that $\lambda \cdot N = \{0\}$. If P is another submodule of M, we must prove that there exists a submodule P' of P such that $P' \neq 0$ and $\lambda \cdot P' = \{0\}$. Since M is irreducible we can take $P' = N \cap P$.

Definition. The submodule N of M is said to be tertiary if M/N is cotertiary. In this case the set $\mathscr{R}_t(N, M)$ of elements of A which are semiannulators of M/N is called the tertiary radical of the submodule N of M.

LEMMA. *If the module M is noetherian and coirreducible then $\mathscr{R}_t(0, M)$ is a bilateral prime ideal.*

PROOF. (a) $\mathscr{R}_t(0, M)$ is a bilateral ideal. It is evident that $\mathscr{R}_t(0, M)$ is a left-ideal of A. To prove that $\mathscr{R}_t(0, M)$ is a bilateral ideal, notice first that $\mathscr{R}_t(0, M)$ coincides with the set of semiannulators for all monogeneous submodules of M. Let now $a \in \mathscr{R}_t(0, M)$ and $\mu \in A$. We show that for any $b \in A$, $b \neq 0$, there exists a submodule of Ab which is annuled by $a\mu$. But, by hypothesis there exists $\xi \in A$ such that $\xi b \neq$ and $(a, \lambda)(\xi b) = 0$ for any $\lambda \in A$. We may therefore conclude that $(a, \mu)(\lambda \xi b) = 0$ for any $\lambda \in A$. Thus $a\mu \in R_t(0, M)$. Let $a_1, a_2 \in \mathscr{R}_t(0, M)$. We have to show that $a_1 + a_2 \in \mathscr{R}_t(0, M)$. To this end, let $b \in M$, $b \neq 0$ and $\xi_1 \in A$ such that $\xi_1 b \neq 0$ but $(a_1 \lambda) \cdot (\xi_1 b) = 0$ for any $\lambda \in A$. On the other hand, there exists $\xi_2 \in A$ such that $\xi_2 \xi_1 b \neq 0$ and $(a_2 \lambda)(\xi_2 \xi_1 b) = 0$ for any $\lambda \in A$. This implies $(a_1 + a_2) \times (\xi_2 \xi_1 b) = 0$ for any $\lambda \in A$ and therefore $a_1 + a_2 \in \mathscr{R}_t(0, M)$.

(b) $\mathscr{R}_t(0, M)$ is a prime ideal. To see this, let $\mathrm{Ann}(N)$ stand for the annulator of N, for each submodule N of M. The family $(\mathrm{Ann}(N))_{N \subset M, N \neq 0}$ possesses a maximal element $\mathrm{Ann}(N_0)$ which we denote by $A(M)$. Clearly $\mathscr{R}_t(0, M) \subset A(M)$. Conversely, if $\lambda \in A(M)$ then there exists a submodule N of M with $N \neq 0$ and $\lambda \cdot N = 0$. Since M is coirreducible, we may apply the preceding lemma to conclude that $\lambda \in \mathscr{R}_t(0, M)$. Thus $A(M) = \mathscr{R}_t(0, M)$.

Let now $a, b \in A$ such that $aAb \subset A(M)$. Assume $b \in A(M)$. We show that $a \in A(M)$. $b \notin A(M)$ implies $bN_0 \neq 0$, hence $AbN_0 \neq 0$. But the hypothesis $aAb \subset A(M)$ gives

$$a \in \operatorname{Ann}(AbN_0) = \operatorname{Ann}(N_0) = A(M) = \mathscr{R}_t(0, M),$$

by the maximality property of N_0. Indeed, $A(M)$ is the annulator of any submodule of N_0. This completes the proof of the lemma.

Let now Q be an injective indecomposable object of $\mathscr{C}$ and M a noetherian submodule of Q. Thus M is coirreducible and $\mathscr{R}_t(0, M)$ is a bilateral prime ideal. $\mathscr{R}_t(0, M)$ does not depend on M. We may set

$$f(q) = \mathscr{R}_t(0, M),$$

where q denotes the type of Q. The decomposition theory thus obtained is nothing else than the tertiary theory.

CHAPTER 7

Elements of Homological Algebra

1. Complexes. Homology. Cohomology

Definition. Let $\mathscr{C}$ be an Abelian category. A *complex* **K** with values in the category $\mathscr{C}$ consists of:

(i) a sequence $(K^n)_{n \in Z}$ of objects of the category $\mathscr{C}$.

(ii) a sequence $(d_{\mathbf{K}}^n)_{n \in Z}$ of morphisms $d_{\mathbf{K}}^n : K^n \longrightarrow K^{n+1}$ such that $d_{\mathbf{K}}^{n+1} d_{\mathbf{K}}^n = 0$ for any n.

If **L** and **K** are complexes with values in the category $\mathscr{C}$, a morphism from the complex **L** to the complex **K** is a sequence $\mathbf{f} = (f^n)_{n \in Z}$ where, for every n, f^n is a morphism $f^n : L^n \longrightarrow K^n$ such that the diagram:

$$\begin{array}{ccc} L^n & \xrightarrow{f^n} & K^n \\ {\scriptstyle d_{\mathbf{L}}^n}\downarrow & & \downarrow{\scriptstyle d_{\mathbf{K}}^n} \\ L^{n+1} & \xrightarrow[f^{n+1}]{} & K^{n+1} \end{array} \tag{1}$$

is commutative for any n, in other words, $d_{\mathbf{K}}^n f^n = f^{n+1} d_{\mathbf{L}}^n$.

Thus the class of all complexes with values in the category $\mathscr{C}$ forms a category which we designate by $\mathscr{K}(\mathscr{C})$.

Homology (cohomology) functors. Let n be a fixed integer and **K** a complex. We can associate to this complex the following objects of the category $\mathscr{C}$:

$$Z^n(\mathbf{K}) = \ker d_{\mathbf{K}}^n$$

$$B^n(\mathbf{K}) = \operatorname{im} d_{\mathbf{K}}^{n-1}.$$

Moreover, it is clear that if $\mathbf{f} = (f^n)_{n \in Z}$ is a morphism from the complex **L** to the complex **K**, then there exists a unique morphism $Z^n(\mathbf{f}) : Z^n(\mathbf{L}) \longrightarrow Z^n(\mathbf{K})$ such that the diagram

$$\begin{array}{ccccc} Z^n(\mathbf{L}) & \longrightarrow & L^n & \xrightarrow{d_{\mathbf{L}}^n} & L^{n+1} \\ {\scriptstyle Z^n(\mathbf{f})}\downarrow & & {\scriptstyle f^n}\downarrow & & \downarrow{\scriptstyle f^{n+1}} \\ Z^n(\mathbf{K}) & \longrightarrow & K^n & \xrightarrow[d_{\mathbf{K}}^n]{} & K^{n+1} \end{array}$$

is commutative. Thus we obtain a functor $Z^n: \mathcal{K}(\mathcal{C}) \longrightarrow \mathcal{C}$ and similarly a functor $B^n: \mathcal{K}(\mathcal{C}) \longrightarrow \mathcal{C}$. These functors are additive.

The object $Z^n(\mathbf{K})$ is called, by tradition, the object of the n-dimensional cycles of the complex $\mathbf{K}$, and $B^n(\mathbf{K})$ the object of the n-dimensional boundaries of the complex $\mathbf{K}$. The relation $d_K^n d_K^{n-1} = 0$ implies evidently that $B^n(\mathbf{K}) \subset Z^n(\mathbf{K})$ for any n. If we associate to each complex $\mathbf{K}$ the object $Z^n(\mathbf{K})/B^n(\mathbf{K}) = H^n(\mathbf{K})$ we obtain a new functor $H^n: \mathcal{K}(\mathcal{C}) \longrightarrow \mathcal{C}$, the n-dimensional homology functor. $H^n(\mathbf{K})$ is called the n-dimensional homology object of the complex $\mathbf{K}$.

We shall write sometimes $\mathbf{f}_* = H(\mathbf{f})$, if $\mathbf{f} = (f^n)_{n\in Z}$ is a morphism from the complex $\mathbf{L}$ to the complex $\mathbf{K}$.

Definition. Let $\mathbf{f} = (f^n)_{n\in Z}$, $\mathbf{g} = (g^n)_{n\in Z}$ be morphisms from the complex $\mathbf{L}$ to the complex $\mathbf{K}$. These morphisms are said to be *homotopic* if there exist morphisms $k^n: L^n \longrightarrow K^{n-1}$, $n \in Z$, such that the relation

$$g^{n+1} - f^{n+1} = d_K^n k^{n+1} + k^{n+2} d_L^{n+1} \tag{2}$$

holds for any $n \in Z$.

PROPOSITION 7.1. *If the morphisms* $\mathbf{f} = (f^n)_{n\in Z}$, $\mathbf{g} = (g^n)_{n\in Z}$ *from the complex* $\mathbf{L}$ *to the complex* $\mathbf{K}$ *are homotopic then* $H^n(\mathbf{f}) = H^n(\mathbf{g})$ *for any* $n \in Z$.

PROOF. Notice that $\mathbf{f} - \mathbf{g} = (f^n - g^n)_{n\in Z}$ is also a morphism from $\mathbf{L}$ to $\mathbf{K}$. Since the functor H^n is additive it follows that $H^n(\mathbf{f} - \mathbf{g}) = H^n(\mathbf{f}) - H^n(\mathbf{g})$. Hence it is sufficient to show that $H^n(\mathbf{f} - \mathbf{g}) = 0$. From the definition it results that $H^n(\mathbf{f} - \mathbf{g})$ is the unique morphism for which the diagram:

$$\begin{array}{ccccc} B^n(\mathbf{L}) & \longrightarrow & Z^n(\mathbf{L}) & \longrightarrow & H^n(\mathbf{L}) \\ \downarrow{\scriptstyle B^n(\mathbf{h})} & & \downarrow{\scriptstyle Z^n(\mathbf{h})} & & \downarrow{\scriptstyle H^n(\mathbf{h})} \\ B^n(\mathbf{K}) & \longrightarrow & Z^n(\mathbf{K}) & \longrightarrow & H^n(\mathbf{K}) \end{array} \tag{3}$$

is commutative, where we have set $\mathbf{h} = \mathbf{f} - \mathbf{g}$. But relation (2) implies that $Z^n(\mathbf{h})$ factorizes through $B^n(\mathbf{K})$, whence it follows that $H^n(\mathbf{h}) = 0$ since the morphism $O: H^n(\mathbf{L}) \longrightarrow H^n(\mathbf{K})$ satisfies the condition to make the diagram (3) commutative.

The complex $\mathbf{K}$ is said to be positive if $K^n = 0$ for any $n < 0$ and negative if $K^n = 0$ for any $n > 0$. A positive complex is also said to be a cochain complex, and its homology objects are called cohomology objects. A negative complex is also said to be a chain complex. For a chain complex $\mathbf{K}$ the following notations are traditionally used:

$$K^{-n} = K_n \qquad \text{for any } n > 0,$$
$$H^{-n}(\mathbf{K}) = H_n(\mathbf{K}) \qquad \text{for any } n > 0,$$
$$d_K^{-n} = d_n^K \qquad \text{for any } n > 0.$$

In other words a chain complex can be represented as follows:

$$\cdots \longrightarrow K_n \xrightarrow{d_n^K} K_{n-1} \xrightarrow{d_{n-1}^K} \cdots \xrightarrow{d_1^K} K_0 \longrightarrow 0 \longrightarrow \cdots.$$

Simplicial complexes. We shall now indicate a convenient procedure for constructing chain and cochain complexes.

Let Δ_n be the set $\{0, 1, 2, \cdots, n\}$. We define a category $\mathscr{S}$ as follows:

(i) the objects of the category $\mathscr{S}$ are the sets Δ_n for $n \geq 0$.

(ii) the morphisms from the object Δ_n to the object Δ_m are the increasing maps of the set Δ_n into the set Δ_m.

Consider in particular the following morphisms in the category $\mathscr{S}$ which will play an important role in the sequel:

$$d_n^i : \Delta_n \longrightarrow \Delta_{n+1} \qquad (0 \leq i \leq n+1)$$

defined by

$$d_n^i(p) = p \qquad \text{if } p < i,$$
$$d_n^i(p) = p + 1 \qquad \text{of } p \geq i,$$

and

$$s_n^i : \Delta_{n+1} \longrightarrow \Delta_n \qquad (0 \leq i \leq n)$$

defined by

$$s_n^i(p) = p \qquad \text{if } p \leq i,$$
$$s_n^i(p) = p - 1 \qquad \text{if } p > i.$$

A simple calculus concerning the composition of maps leads to the following relations:

(I) $\quad d_{n+1}^j d_n^i = d_{n+1}^i d_n^{j-1} \qquad (i < j)$

(II) $\quad s_n^j s_{n+1}^i = s_n^i s_{n+1}^{j+1} \qquad (i \leq j)$

(III) $\quad s_{n+1}^j d_{n+1}^i = d_n^i s_n^{j-1} \qquad (i < j)$

(IV) $\quad s_n^i d_n^i = s_n^i d_n^{i+1} = 1_{\Delta_n}$

(V) $\quad s_{n+1}^j d_{n+1}^i = d_n^{i-1} s_n^j \qquad (j + 1 < i).$

If $\mathscr{C}$ is an arbitrary category, a *simplicial object with values in* $\mathscr{C}$ is a contravariant functor $F:\mathscr{S}\longrightarrow\mathscr{C}$. If the functor F is covariant we obtain a *cosimplicial object with values in the category* $\mathscr{C}$.

PROPOSITION 7.2. *A chain complex is associated to every simplicial object with values in an Abelian category* $\mathscr{C}$. *A cochain complex is associated to every cosimplicial object with values in an Abelian category* $\mathscr{C}$.

PROOF. Consider for instance the case of cosimplicial objects. Let $F:\mathscr{S}\longrightarrow\mathscr{C}$ be a covariant functor, where $\mathscr{C}$ is an Abelian category. Set $K^n = F(\Delta_n)$. It follows that $F(d_n^i)$ are morphisms $F(d_n^i):K^n\longrightarrow K^{n+1}$ in the category $\mathscr{C}$. $\mathscr{C}$ being Abelian we can define for any n the morphism:

$$d_K^n : K^n \longrightarrow K^{n+1}$$

by the formula

$$d_K^n = \sum_{i=0}^{n+1} (-1)^i F(d_n^i).$$

We wish to prove that the sequence of objects $(K^n)_{n\geq 0}$ together with the sequence of morphisms $(d_K^n)_{n\geq 0}$ define a complex with values in the Abelian category $\mathscr{C}$. To achieve this, it is sufficient to show that

$$d_K^{n+1} d_K^n = 0.$$

But

$$\begin{aligned}
d_K^{n+1} d_K^n &= \left[\sum_{j=0}^{n+2} (-1)^j F(d_{n+1}^j)\right]\left[\sum_{i=0}^{n+1} (-1)^i F(d_n^i)\right] \\
&= \sum_{j=0}^{n+2}\sum_{i=0}^{n+1} (-1)^{i+j} F(d_{n+1}^j d_n^i) \\
&= \sum_{0\leq i<j\leq n+2} (-1)^{i+j} F(d_{n+1}^j d_n^i) + \sum_{0\leq j\leq i\leq n+1} (-1)^{i+j} F(d_{n+1}^j d_n^i) \\
&= \sum_{0\leq i<j\leq n+2} (-1)^{i+j} F(d_{n+1}^i d_n^{j-1}) \\
&\quad + \sum_{0\leq j\leq i\leq n+1} (-1)^{i+j} F(d_{n+1}^j d_n^i).
\end{aligned}$$

To write the last equality we have used relation (I). In the first sum we replace $j-1$ by i and i by j. Thus we get

$$d_K^{n+1} d_K^n = \sum_{0\leq j\leq i\leq n+1} (-1)^{i+j-1} F(d_{n+1}^j d_n^i) + \sum_{0\leq j\leq i\leq n+1} (-1)^{i+j} F(d_{n+1}^j d_n^i) = 0.$$

Examples. The standard complex of Grothendieck.

Let $\mathscr{C}$ be a category with direct sums and A an arbitrary object of $\mathscr{C}$. We denote by A^r the direct sum of the family of objects $(A_i)_{1 \leq i \leq r}$ where $A_i = A$ for any $1 \leq i \leq r$. We define a covariant functor $F: \mathscr{S} \longrightarrow \mathscr{C}$, in other words a cosimplicial object with values in the category $\mathscr{C}$ as follows:

We set by definition $F(\Delta_0) = A$ and $F(\Delta_n) = A^{n+1}$ for any $n \geq 1$. Let $\varphi \in \mathrm{Hom}_{\mathscr{S}}(\Lambda_r, \Delta_s)$. We have to define $F(\varphi): A^{r+1} \longrightarrow A^{s+1}$ such that

$$F(1_{\Delta_r}) = 1_{A^{r+1}}$$

$$F(\psi\varphi) = F(\psi)F(\varphi). \tag{4}$$

We denote by $i_0, i_1, \cdots, i_r$ the canonical injections of A into A^{r+1} and by $j_0, j_1, \cdots, j_s$ the canonical injections of A into A^{s+1}. To define the morphism $F(\varphi)$ we use the following construction: There exists $F(\varphi): A^{r+1} \longrightarrow A^{s+1}$ uniquely determined by the condition that the diagram

$$\begin{array}{ccc} A^{r+1} & \xrightarrow{F(\varphi)} & A^{s+1} \\ {\scriptstyle i_k}\nwarrow & & \nearrow{\scriptstyle j_{\varphi(k)}} \\ & A & \end{array}$$

be commutative for any $k \in \Delta_r$. With this definition of $F(\varphi)$ relations (4) are obviously fulfilled.

If $\mathscr{C}'$ is an Abelian category and $G: \mathscr{C} \longrightarrow \mathscr{C}'$ is a covariant functor, then we can associate to this functor in an obvious manner a cosimplicial object with values in $\mathscr{C}'$ and therefore by proposition 7.2 a cochain complex with values in $\mathscr{C}'$. This cochain complex is called the standard complex of Grothendieck associated to the functor G and the object A.

2. Resolutions

Definition. Let $\mathscr{C}$ be an Abelian category and A an object of $\mathscr{C}$. A right *resolution* of the object A is a positive complex $\mathbf{K}$ together with a morphism $\varepsilon: A \longrightarrow K^0$ (the augmentation morphism) such that the sequence

$$0 \longrightarrow A \xrightarrow{\varepsilon} K^0 \xrightarrow{d_k^0} K^1$$

is exact.

A resolution of the object A is said to be exact if $H^n(\mathbf{K}) = 0$ for any $n > 0$; it is said to be injective if the objects K^n are injective objects in the category $\mathscr{C}$.

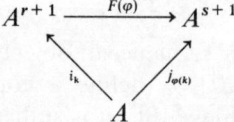

Dually one obtains the notions of a left resolution of an object, projective resolution, and so on.

In the sequel we shall confine our attention to the statements and proofs of the fundamental results concerning injective resolutions, leaving to the reader the formulation by duality of the corresponding results concerning projective resolutions.

PROPOSITION 7.3. *If the category $\mathscr{C}$ is with sufficiently many injectives then any object of $\mathscr{C}$ possesses at least one exact injective resolution.*

PROOF. Let A be an arbitrary object of the category $\mathscr{C}$ and $\varepsilon : A \longrightarrow K^0$ a monomorphism of A into an injective object K^0. Denote $L^0 = \operatorname{coker} \varepsilon$ and $\lambda^0 : K^0 \longrightarrow L^0$ the canonical morphism. Let now $\mu^0 : L^0 \longrightarrow K^1$ be a monomorphism of L^0 into an injective object K^1. Let $d_K^0 = \mu^0 \lambda^0$. The sequence

$$0 \longrightarrow A \xrightarrow{\varepsilon} K^0 \xrightarrow{d_K^0} K^1$$

is evidently exact. Let $L^1 = \operatorname{coker} d_K^0$, $\lambda^1 : K^1 \longrightarrow L^1$ be the canonical morphism, and $\mu^1 : L^1 \longrightarrow K^2$ be a monomorphism of L^1 into an injective object K^2. Denote $d_K^1 = \mu^1 \lambda^1$. By continuing this process we obtain an injective resolution of A.

LEMMA 7.4. *If in the diagram*

$$\begin{array}{ccccc} A_1 & \xrightarrow{u_1} & A_2 & \xrightarrow{u_2} & A_3 \\ & & \downarrow{\scriptstyle \varphi} & & \\ & & Q & & \end{array}$$

the sequence $A_1 \xrightarrow{u_1} A_2 \xrightarrow{u_2} A_3$ is exact, Q is injective and $\varphi u_1 = 0$, then there exists $\psi : A_3 \longrightarrow Q$ such that $\psi u_2 = \varphi$.

PROOF. If we consider the canonical factorization of the morphism u_2, we obtain the diagram:

$$\begin{array}{ccccccccc} \operatorname{im} u_1 & \xrightarrow{i} & A_2 & \xrightarrow{j_2} & \operatorname{coim} u_2 & \xrightarrow{\bar{u}_2} & \operatorname{im} u_2 & \xrightarrow{i_2} & A_3 \\ & & \downarrow{\scriptstyle \varphi} & & & & & & \\ & & Q & & & & & & \end{array}$$

where $\varphi i = 0$ and $\bar{u}_2$ is an isomorphism. Since $\operatorname{coim} u_2 = \operatorname{coker} i$ and $\varphi i = 0$, there exists $\theta : \operatorname{coim} u_2 \longrightarrow Q$ such that $\theta j_2 = \varphi$. Since $\bar{u}_2$ is an isomorphism, there obviously exists $\psi_1 : \operatorname{im} u_2 \longrightarrow Q$ such that $\psi_1 \bar{u}_2 j_2 = \varphi$. Now, because Q is injective and i_2 is an injection, there exists $\psi : A_3 \longrightarrow Q$ such that $\psi i_2 = \psi_1 \cdot \psi$ is evidently the required morphism.

PROPOSITION 7.5. *Let A and B be two objects of the category $\mathscr{C}$, $(\mathbf{K}, \varepsilon)$ an exact resolution of A and $(\mathbf{L}, \eta)$ an injective resolution of B. If $u : A \longrightarrow B$ is a morphism from A to B, then there exists at least one morphism $\mathbf{f} = (f^n)_{n \geq 0}$ from the complex $\mathbf{K}$ to the complex $\mathbf{L}$ such that the diagram*

$$\begin{array}{ccc} A & \xrightarrow{\varepsilon} & K^0 \\ {\scriptstyle u}\downarrow & & \downarrow{\scriptstyle f^0} \\ B & \xrightarrow[\eta]{} & L^0 \end{array} \tag{5}$$

is commutative. If $\mathbf{g} = (g^n)_{n \geq 0}$ is another morphism from the complex $\mathbf{K}$ to the complex $\mathbf{L}$ satisfying the same condition as $\mathbf{f}$, then $\mathbf{f}$ and $\mathbf{g}$ are homotopic.

PROOF. We define successively the morphisms $f^0, f^1, \cdots, f^n, \cdots$ such that the conditions stated in the proposition are fulfilled. The existence of a $f^0 : K^0 \longrightarrow L^0$ such that diagram (5) be commutative results simply from the fact that L^0 is injective and ε is a monomorphism. Here we do not use the fact that η is a monomorphism. To define the morphism $f^1 : K^1 \longrightarrow L^1$ consider the diagram

$$\begin{array}{ccccc} A & \xrightarrow{\varepsilon} & K^0 & \xrightarrow{d_K^0} & K^1 \\ & & {\scriptstyle d_L^0 f^0}\downarrow & & \\ & & L^1 & & \end{array}$$

and notice that we are under the conditions of lemma 7.4. Thus there exists $f^1 : K^1 \longrightarrow L^1$ such that $f^1 d_K^0 = d_L^0 f^0$. Using induction, assume that we have already constructed the morphisms $f^0, f^1, \cdots, f^{n-1}$ such that $f^i d_K^{i-1} = d_L^{i-1} f^{i-1}$ for any $0 \leq i \leq n-1$. In particular $f^{n-1} d_K^{n-2} = d_L^{n-2} f^{n-2}$. Thus we can form the diagram

$$\begin{array}{ccccc} K^{n-2} & \xrightarrow{d_K^{n-2}} & K^{n-1} & \xrightarrow{d_K^{n-1}} & K^n \\ & & \downarrow{\scriptstyle d_L^{n-1} f^{n-1}} & & \\ & & L^n & & \end{array}$$

We are again under the conditions of lemma 7.4. The only fact we have still to verify is that $d_L^{n-1}f^{n-1}d_K^{n-2} = 0$. But we have

$$d_L^{n-1}f^{n-1}d_K^{n-2} = (d_L^{n-1}d_L^{n-2})f^{n-2} = 0.$$

Hence there exists $f^n : K^n \longrightarrow L^n$ such that $f^n d_K^{n-1} = d_L^{n-1}f^{n-1}$. Thus the first part of the theorem is proved.

Let now $\mathbf{g} = (g^n)_{n\geq 0}$ be another morphism satisfying the same conditions as $\mathbf{f}$. To show that $\mathbf{f}$ and $\mathbf{g}$ are homotopic, we must construct morphisms $k^i : K^i \longrightarrow L^{i-1}$, $i \geq 1$, such that

$$f^i - g^i = d_L^{i-1}k^i + k^{i+1}d_K^i.$$

To define $k^1 : K^1 \longrightarrow L^0$, consider the diagram

$$\begin{array}{ccccc} A & \xrightarrow{\varepsilon} & K^0 & \xrightarrow{d_K^0} & K^1 \\ & & \Big\downarrow{\scriptstyle f^0 - g^0} & & \\ & & L^0 & & \end{array}$$

and notice that we are again under the conditions of lemma 7.4, since we have $(f^0 - g^0)\varepsilon = f^0\varepsilon - g^0\varepsilon = \eta u - \eta u = 0$. Thus there exists $k^1 : K^1 \longrightarrow L^0$ such that $f^0 - g^0 = k^1 d_K^0$. To define $k^2 : K^2 \longrightarrow L^1$ consider the diagram

$$\begin{array}{ccccc} K^0 & \xrightarrow{d_K^0} & K^1 & \xrightarrow{d_K^1} & K^2 \\ & & \Big\downarrow{\scriptstyle f^1 - g^1 - d_L^0 k^1} & & \\ & & L^1 & & \end{array}$$

and notice that we are again under the conditions of lemma 7.4. Indeed:

$$(f^1 - g^1 - d_L^0 k^1)d_K^0 = f^1 d_K^0 - g^1 d_K^0 - d_L^0 k^1 d_K^0 = d_L^0(f^0 - g^0)$$
$$- d_L^0(f^0 - g^0) = 0.$$

Thus there exists $k^2 : K^2 \longrightarrow L^1$ such that $f^1 - g^1 = d_L^0 k^1 + k^2 d_K^1$.

Assume now that we have constructed $k^1, k^2, \cdots, k^n$ such that

$$f^i - g^i = d_L^{i-1}k^i + k^{i+1}d_K^i \quad \text{for} \quad i + 1 \leq n.$$

In particular we have $f^{n-1} - g^{n-1} = d_L^{n-2}k^{n-1} + k^n d_K^{n-1}$. To construct k^{n+1} consider the diagram

$$\begin{array}{ccccc} K^{n-1} & \xrightarrow{d_K^{n-1}} & K^n & \xrightarrow{d_K^n} & K^{n+1} \\ & & \Big\downarrow{\scriptstyle f^n - g^n - d_L^{n-1}k^n} & & \\ & & L^n & & \end{array}$$

and we again verify that the conditions stated in lemma 7.4 are fulfilled, whence the existence of k^{n+1} with the desired property.

PROPOSITION 7.6. *Let*

$$0 \longrightarrow A' \xrightarrow{\psi} A \xrightarrow{\varphi} A'' \longrightarrow 0$$

be an exact sequence in the category $\mathscr{C}$, $(\mathbf{X}', \varepsilon')$ *an exact and injective resolution of* A', $(\mathbf{X}'', \varepsilon'')$ *an exact resolution of* A''. *Under these conditions there exists an exact resolution* $(\mathbf{X}, \varepsilon)$ *of* A *satisfying the following conditions*:

(a) *There exist the morphisms of complexes* $I' : \mathbf{X}' \longrightarrow \mathbf{X}$, $P'' : \mathbf{X} \longrightarrow \mathbf{X}''$ *such that the sequence of complexes and morphisms of complexes*

$$0 \longrightarrow \mathbf{X}' \xrightarrow{I'} \mathbf{X} \xrightarrow{P''} \mathbf{X}'' \longrightarrow 0$$

is exact.

(b) *For any* $n \geq 0$, $X_n = X'_n \oplus X''_n$. *We denote by* i'_n, i''_n *the canonical injections and by* p'_n, p''_n *the canonical projections.*

(c) *The diagram*

$$\begin{array}{ccccc} A' & \xrightarrow{\psi} & A & \xrightarrow{\varphi} & A'' \\ \downarrow{\scriptstyle \varepsilon'} & & \downarrow{\scriptstyle \varepsilon} & & \downarrow{\scriptstyle \varepsilon''} \\ X'_0 & \longrightarrow & X_0 & \longrightarrow & X''_0 \end{array}$$

is commutative.

(d) $I' = (i'_n)_{n \geq 0}$ *and* $P'' = (p''_n)_{n \geq 0}$.

An exact resolution $(\mathbf{X}, \varepsilon)$ *satisfying conditions* (a), (b), (c), (d) *is said to be a normal resolution.*

PROOF. We shall use for the proof the following lemma.

LEMMA 7.7. *If in the diagram*

$$\begin{array}{ccccccccc} & & 0 & & & & 0 & & \\ & & \downarrow & & & & \downarrow & & \\ 0 & \longrightarrow & A' & \xrightarrow{\psi} & A & \xrightarrow{\varphi} & A'' & \longrightarrow & 0 \\ & & \downarrow{\scriptstyle \varepsilon'} & & & & \downarrow{\scriptstyle \varepsilon''} & & \\ & & P' & & & & P'' & & \\ & & \downarrow & & & & \downarrow & & \\ & & L' & & & & L'' & & \\ & & \downarrow & & & & \downarrow & & \\ & & 0 & & & & 0 & & \end{array}$$

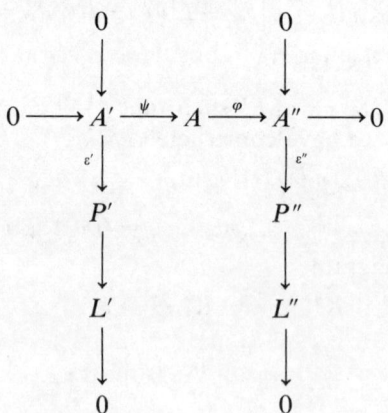

the row and columns are exact and P' is injective, then this diagram can be completed to obtain the commutative diagram

$$\begin{array}{ccccccccc}
 & & 0 & & 0 & & 0 & & \\
 & & \downarrow & & \downarrow & & \downarrow & & \\
0 & \longrightarrow & A' & \xrightarrow{\psi} & A & \xrightarrow{\varphi} & A'' & \longrightarrow & 0 \\
 & & \downarrow{\scriptstyle \varepsilon'} & & \downarrow{\scriptstyle \varepsilon} & & \downarrow{\scriptstyle \varepsilon''} & & \\
0 & \longrightarrow & P' & \xrightarrow[i']{} & P & \xrightarrow[p'']{} & P'' & \longrightarrow & 0 \\
 & & \downarrow & & \downarrow & & \downarrow & & \\
0 & \longrightarrow & L' & \longrightarrow & L & \longrightarrow & L'' & \longrightarrow & 0 \\
 & & \downarrow & & \downarrow & & \downarrow & & \\
 & & 0 & & 0 & & 0 & &
\end{array}$$

where all the rows and columns are exact and P is the direct sum of the objects P' and P'' having the canonical injections i', i'' and the canonical projections p', p''.

PROOF. Let $P = P' \oplus P''$, i', i'' be the canonical injections and p', p'' the canonical projections. To define the morphism ε consider first the morphism $\sigma_0 : A \longrightarrow P'$ such that $\sigma_0 \psi = \varepsilon'$. There exists such a σ_0 because P' is injective. We set $\varepsilon = i'\sigma_0 + i''\varepsilon''\varphi$. With this the upper part of the preceding diagram is commutative. We now show that ε is an injection. Let $u : X \longrightarrow A$ be such that $\varepsilon u = 0$. We have $p''\varepsilon u = 0$. Consequently $\varepsilon''\varphi u = 0$, hence $\varphi u = 0$. Thus there exists $u' : X \longrightarrow A'$ such that $u = \psi u'$. From the hypothesis it follows that

$$0 = \varepsilon u = \varepsilon \psi u' = (i'\sigma_0 + i''\varepsilon''\varphi)\psi u' = i'\sigma_0\psi u' = i'\varepsilon' u'.$$

Thus $u' = 0$ and therefore $u = 0$ since i' and ε' are monomorphisms. Hence ε is an injection.

To complete the diagram we set $L = \operatorname{coker} u$ and the morphisms indicated by arrows are canonical morphisms. The diagram thus obtained is commutative. The exactness follows from the '3×3 lemma' (proposition 5.25).

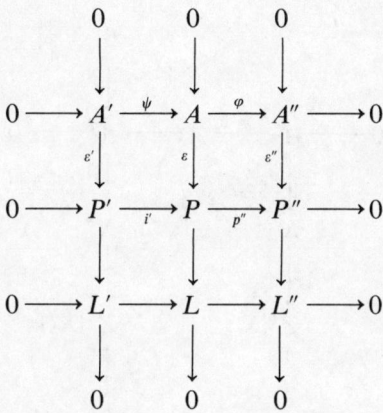

We are now in a position to prove proposition 7.6. By applying lemma 7.7 to the diagram

$$
\begin{array}{ccccccccc}
 & & 0 & & & & 0 & & \\
 & & \downarrow & & & & \downarrow & & \\
0 & \longrightarrow & A' & \xrightarrow{\psi} & A & \xrightarrow{\varphi} & A'' & \longrightarrow & 0 \\
 & & \downarrow{\scriptstyle \varepsilon'} & & & & \downarrow{\scriptstyle \varepsilon''} & & \\
 & & X'_0 & & & & X''_0 & & \\
 & & \downarrow & & & & \downarrow & & \\
 & & Z^1(\mathbf{X}') & & & & Z^1(\mathbf{X}'') & & \\
 & & \downarrow & & & & \downarrow & & \\
 & & 0 & & & & 0 & &
\end{array}
$$

we obtain the commutative diagram

$$
\begin{array}{ccccccccc}
 & & 0 & & 0 & & 0 & & \\
 & & \downarrow & & \downarrow & & \downarrow & & \\
0 & \longrightarrow & A' & \xrightarrow{\psi} & A & \xrightarrow{\varphi} & A'' & \longrightarrow & 0 \\
 & & \downarrow{\scriptstyle \varepsilon'} & & \downarrow{\scriptstyle \varepsilon} & & \downarrow{\scriptstyle \varepsilon''} & & \\
0 & \longrightarrow & X'_0 & \xrightarrow{i'_0} & X & \xrightarrow{p''_0} & X''_0 & \longrightarrow & 0 \\
 & & \downarrow{\scriptstyle \delta'_1} & & \downarrow{\scriptstyle \delta_1} & & \downarrow{\scriptstyle \delta''_1} & & \\
0 & \longrightarrow & Z^1(\mathbf{X}') & \longrightarrow & \operatorname{coker} \varepsilon & \longrightarrow & Z^1(\mathbf{X}'') & \longrightarrow & 0 \\
 & & \downarrow & & \downarrow & & \downarrow & & \\
 & & 0 & & 0 & & 0 & &
\end{array}
$$

where the rows and the columns are exact, $X = X'_0 \oplus X''_0$, and i'_0, i''_0 are the canonical injections and p'_0, p''_0 are the canonical projections.

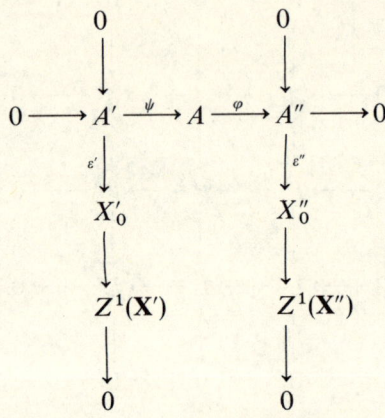

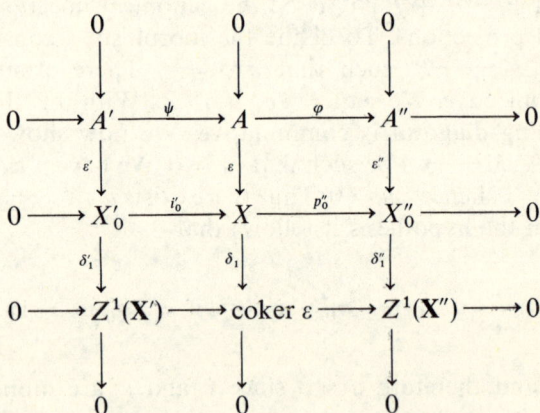

Now, if we apply lemma 7.7 to the diagram

$$
\begin{array}{ccccccccc}
 & & 0 & & & & 0 & & \\
 & & \downarrow & & & & \downarrow & & \\
0 & \longrightarrow & Z^1(\mathbf{X}') & \longrightarrow & \operatorname{coker} \varepsilon & \longrightarrow & Z^1(\mathbf{X}'') & \longrightarrow & 0 \\
 & & \downarrow{\scriptstyle \delta'_2} & & & & \downarrow{\scriptstyle \delta''_2} & & \\
 & & X'_1 & & & & X''_1 & & \\
 & & \downarrow & & & & \downarrow & & \\
 & & Z^2(\mathbf{X}') & & & & Z^2(\mathbf{X}'') & & \\
 & & \downarrow & & & & \downarrow & & \\
 & & 0 & & & & 0 & &
\end{array}
$$

we obtain the commutative diagram

$$
\begin{array}{ccccccccc}
 & & 0 & & 0 & & 0 & & \\
 & & \downarrow & & \downarrow & & \downarrow & & \\
0 & \longrightarrow & Z^1(\mathbf{X}') & \longrightarrow & \operatorname{coker} \varepsilon & \longrightarrow & Z^1(\mathbf{X}'') & \longrightarrow & 0 \\
 & & \downarrow{\scriptstyle \delta'_2} & & \downarrow{\scriptstyle \delta_2} & & \downarrow{\scriptstyle \delta''_2} & & \\
0 & \longrightarrow & X'_1 & \xrightarrow{i'_1} & X_1 & \xrightarrow{p''_1} & X''_1 & \longrightarrow & 0 \\
 & & \downarrow & & \downarrow & & \downarrow & & \\
0 & \longrightarrow & Z^2(\mathbf{X}') & \longrightarrow & \operatorname{coker} \delta_2 & \longrightarrow & Z^2(\mathbf{X}'') & \longrightarrow & 0 \\
 & & \downarrow & & \downarrow & & \downarrow & & \\
 & & 0 & & 0 & & 0 & &
\end{array}
$$

where $X_1 = X'_1 \oplus X''_1$, i'_1, i''_1 are the canonical injections, p'_1, p''_1 are the canonical projections and, in addition: $\delta'_2\delta'_1 = d^0_{X'}$, $\delta''_2\delta''_1 = d^0_{X''}$. We set $d^0_X = \delta_2\delta_1$. Continuing in the same manner we evidently obtain the exact resolution $(\mathbf{X}, \varepsilon)$ with the properties mentioned in the proposition.

To facilitate the proof of proposition 7.8, we now make a few general observations concerning normal resolutions. Let $(\mathbf{X}, \varepsilon)$ be such a resolution. Using the notations in proposition 7.6, we introduce the morphisms

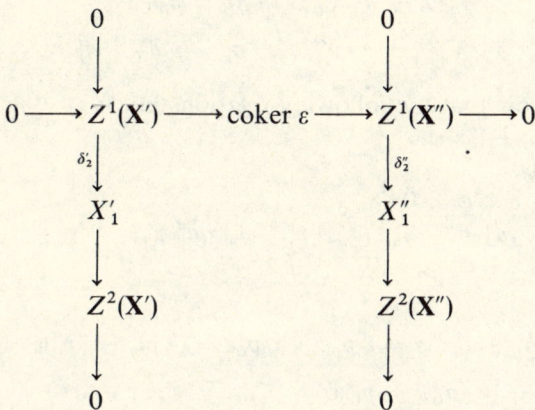

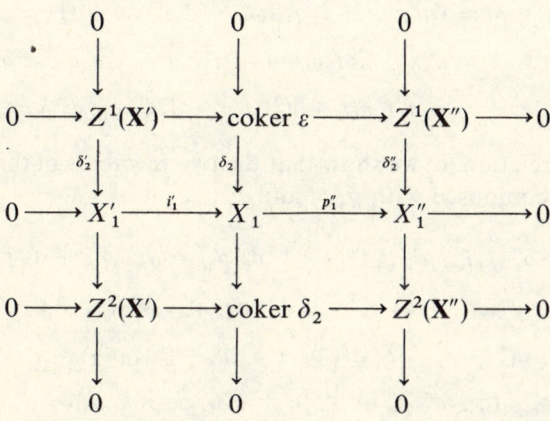

$\sigma_i, i \geq 0$ defined as follows:

$$\sigma_0 : A \longrightarrow X'_0, \qquad \sigma_0 = p'_0\varepsilon$$

$$\sigma_n : X''_{n-1} \longrightarrow X'_n, \qquad \sigma_n = p'_n d_X^{n-1} \cdot i''_{n-1}.$$

We now show that the following relations hold:

(a) $\varepsilon = i'_0\sigma_0 + i''_0\varepsilon''\varphi$,

(b) $\varepsilon' = \sigma_0\psi$,

(c) $d_{X'}^0\sigma_0 + \sigma_1\varepsilon''\varphi = 0$,

(d) $d_X^n = i''_{n+1}d_{X''}^n p''_n + i'_{n+1}\sigma_{n+1}p''_n + i'_{n+1}d_{X'}^n p'_n$,

(e) $\sigma_{n+1}d_{X''}^{n-1} + d_{X'}^n\sigma_n = 0$.

Indeed:

$$i'_0\sigma_0 + i''_0\varepsilon''\varphi = i'_0p'_0\varepsilon + i''_0p''_0\varepsilon = (i'_0p'_0 + i''_0p''_0)\varepsilon = \varepsilon.$$

$$\sigma_0\psi = p'_0\varepsilon\psi = p'_0i'_0\varepsilon' = \varepsilon'.$$

To prove relation (c) it is sufficient to prove that $i'_1(d'_0\sigma_0 + \sigma_1\varepsilon''\varphi) = 0$. But we have

$$\begin{aligned} i'_1(d'_0\sigma_0 + \sigma_1\varepsilon''\varphi) &= i'_1d'_0p'_0\varepsilon + i'_1p'_1d_X^0i''_0 = d_X^0i'_0p'_0\varepsilon + (1_{X_1} - i''_1p''_1)d_X^0i''_0\varepsilon''\varphi \\ &= d_X^0\varepsilon - d_X^0i''_0p''_0\varepsilon + d_X^0i''_0\varepsilon''\varphi - i''_1p''_1d_X^0i''_0\varepsilon''\varphi \\ &= -d_X^0i''_0p''_0\varepsilon + d_X^0i''_0p''_0\varepsilon - i''_1d_{X''}^0p''_0i''_0\varepsilon''\varphi = i''_1d_{X''}^0\varepsilon''\varphi = 0. \end{aligned}$$

To prove relation (d) we show that the two members of this relation are equal when composed with p'_{n+1} and p''_{n+1}.

$$\begin{aligned} p'_{n+1}(i''_{n+1}d_{X''}^n p''_n + i'_{n+1}\sigma_{n+1}p''_n + i'_{n+1}d_{X'}^n p'_n) &= \sigma_{n+1}p''_n + d_{X'}^n p'_n \\ = p'_{n+1}d_X^n i''_n p''_n + d_{X'}^n p'_n &= p'_{n+1}d_X^n - p'_{n+1}d_X^n i'_n p'_n + d_{X'}^n p'_n \\ = p'_{n+1}d_X^n - p'_{n+1}i'_{n+1}d_{X'}^n p'_n + d_{X'}^n p'_n &= p'_{n+1}d_X^n \end{aligned}$$

$$p''_{n+1}(i''_{n+1}d_{X''}^n p''_n + i'_{n+1}\sigma_{n+1}p''_n + i'_{n+1}d_{X'}^n p'_n) = d_{X''}^n p''_n = p''_{n+1}d_X^n.$$

To prove relation (a) we show that $i'_{n+1}(\sigma_{n+1}d_{X''}^{n-1} + d_{X'}^n\sigma_n) = 0$.

$$\begin{aligned} i'_{n+1}(\sigma_{n+1}d_{X''}^{n-1} + d_{X'}^n\sigma_n) &= i'_{n+1}p'_{n+1}d_X^n i''_n d_{X''}^{n-1} + i'_{n+1}d_{X'}^n p'_n d_X^{n-1} i''_{n-1} \\ &= d_X^n i''_n d_{X''}^{n-1} - i''_{n+1}p''_{n+1}d_X^n i''_n d_{X''}^{n-1} + d_X^n i'_n p'_n d_X^{n-1} i''_{n-1} \\ &= d_X^n i''_n d_{X''}^{n-1} - i''_{n+1}d_{X''}^n p''_n i''_n d_{X''}^{n-1} - d_X^n i''_n p''_n d_X^{n-1} i''_{n-1} \\ &= d_X^n i''_n d_{X''}^{n-1} - d_X^n i''_n d_{X''}^{n-1} p''_{n-1} i''_{n-1} = 0. \end{aligned}$$

Proposition 7.8. *Consider the commutative diagram*

$$\begin{array}{ccccccccc} 0 & \longrightarrow & A' & \xrightarrow{\psi} & A & \xrightarrow{\varphi} & A'' & \longrightarrow & 0 \\ && \downarrow{\scriptstyle h'} && \downarrow{\scriptstyle h} && \downarrow{\scriptstyle h''} && \\ 0 & \longrightarrow & B' & \xrightarrow{\xi} & B & \xrightarrow{\eta} & B'' & \longrightarrow & 0 \end{array}$$

where the rows are exact. Let $(\mathbf{X}', \varepsilon')$, $(\mathbf{X}, \varepsilon)$, $(\mathbf{X}'', \varepsilon'')$ *be exact resolutions of* A', A, A'', and $(\mathbf{W}', e')$, $(\mathbf{W}, e)$, $(\mathbf{W}'', e'')$ *be exact resolutions of* B', B, B''. *We assume that these resolutions satisfy the conditions of proposition 7.6 in other words* $(\mathbf{X}, \varepsilon)$ *and* $(\mathbf{W}, e)$ *are normal resolutions. In addition, we assume that the following conditions are satisfied*:

(a) $(\mathbf{W}'', e'')$ *is an injective resolution of* B''.

(b) *There exist morphisms of complexes*

$$\mathbf{f}' : \mathbf{X}' \longrightarrow \mathbf{W}', \qquad \mathbf{f}' = (f'_n)_{n \geq 0}, \qquad \mathbf{f}'' : \mathbf{X}'' \longrightarrow \mathbf{W}'', \qquad \mathbf{f}'' = (f''_n)_{n \geq 0}$$

such that the following diagrams are commutative:

$$\begin{array}{ccc} A' & \xrightarrow{h'} & B' \\ \downarrow{\scriptstyle \varepsilon'} && \downarrow{\scriptstyle e'} \\ X'_0 & \xrightarrow[f'_0]{} & W'_0 \end{array} \qquad\qquad \begin{array}{ccc} A'' & \xrightarrow{h''} & B'' \\ \downarrow{\scriptstyle \varepsilon''} && \downarrow{\scriptstyle e''} \\ X''_0 & \xrightarrow[f''_0]{} & W''_0. \end{array}$$

Under these conditions there exists a morphism of complexes $\mathbf{f} : \mathbf{X} \longrightarrow \mathbf{W}$, $\mathbf{f} = (f_n)_{n \geq 0}$ *such that the following diagrams are commutative*:

$$\begin{array}{ccc} A & \xrightarrow{h} & B \\ \downarrow{\scriptstyle \varepsilon} && \downarrow{\scriptstyle e} \\ X_0 & \xrightarrow[f_0]{} & W_0 \end{array} \quad \text{(i)} \qquad\qquad \begin{array}{ccccc} \mathbf{X}' & \xrightarrow{I'} & \mathbf{X} & \xrightarrow{P''} & \mathbf{X}'' \\ \downarrow{\scriptstyle \mathbf{f}'} && \downarrow{\scriptstyle \mathbf{f}} && \downarrow{\scriptstyle \mathbf{f}''} \\ \mathbf{W}' & \xrightarrow[J']{} & \mathbf{W} & \xrightarrow[\pi'']{} & \mathbf{W}''. \end{array} \quad \text{(ii)}$$

Proof. We denote by τ_i the morphisms similar to the morphisms σ_i for the resolutions $\mathbf{W}'$, $\mathbf{W}$, $\mathbf{W}''$. We are going to show that we can determine the morphisms f_n under the form:

$$f_n = j'_n q_n p''_n + j'_n f'_n p'_n + j''_n f''_n p''_n \tag{6}$$

where $q_n : X''_n \longrightarrow W'_n$ are morphisms to be determined. We first define $q_0 : X''_0 \longrightarrow W'_0$ such that diagram (i) is commutative. To this end, consider the diagram

$$\begin{array}{ccccc} A' & \xrightarrow{\psi} & A & \xrightarrow{\varepsilon''\varphi} & X''_0. \\ && \downarrow{\scriptstyle \tau_0 h - f'_0 \sigma_0} && \\ && W'_0 && \end{array}$$

We are under the conditions of lemma 7.4. For, by hypothesis W'_0 is injective and the sequence $A' \xrightarrow{\psi} A \xrightarrow{\varepsilon''\varphi} X''_0$ is exact. It remains to show that $(\tau_0 h - f'_0\sigma_0)\psi = 0$. But

$$(\tau_0 h - f'_0\sigma_0)\psi = \tau_0 h\psi - f'_0\sigma_0\psi = \tau_0\xi h' - f'_0\varepsilon' = e'h' - f'_0\varepsilon' = 0.$$

To write the last two equalities we have used relation (b) both for τ_0 and for σ_0, and evidently the commutativity of the first diagram in the proposition.

Then it follows from lemma 7.4 that there exists $q_0 : X''_0 \longrightarrow W'_0$ such that $q_0\varepsilon''\varphi = \tau_0 h - f'_0\sigma_0$. With this q_0, the morphism f_0 defined by relation (6) makes diagram (i) commutative, for

$$\begin{aligned} f_0\varepsilon &= (j'_0 q_0 p''_0 + j'_0 f'_0 p'_0 + j''_0 f''_0 p''_0)(i'_0\sigma_0 + i''_0\varepsilon''\varphi) \\ &= j'_0 f'_0\sigma_0 + j'_0 q_0\varepsilon''\varphi + j''_0 f''_0\varepsilon''\varphi. \\ eh &= (j'_0\tau_0 + j''_0 e''\eta)h = j'_0\tau_0 h + j''_0 e''\eta h = j'_0\tau_0 h + j''_0 e'' h''\varphi \\ &= j'_0\tau_0 h + j''_0 f''_0\varepsilon''\varphi. \end{aligned}$$

Now we will show by induction that the morphisms $q_n : X''_n \longrightarrow W''_n$, $n \geq 1$ can be determined such that the conditions

$$q_{n+1}d^n_{X''} - d^n_{W'}q_n = -f'_{n+1}\sigma_{n+1} + \tau_{n+1}f''_n \qquad (n \geq 0) \tag{7}$$

are satisfied.

For $n = 1$ consider the diagram

$$\begin{array}{ccccc} A & \xrightarrow{\varepsilon''\varphi} & X''_0 & \xrightarrow{d^0_{X''}} & X''_1. \\ & & \Big\downarrow{\scriptstyle d^0_{W'}q_0 - f'_1\sigma_1 + \tau_1 f''_0} & & \\ & & W'_1 & & \end{array}$$

We may again apply lemma 7.4, because

$$\begin{aligned} (d^0_{W'}q_0 - f'_1\sigma_1 + \tau_1 f''_0)\varepsilon''\varphi &= d^0_{W'}(\tau_0 h - f'_0\sigma_0) - f'_1\sigma_1\varepsilon''\varphi + \tau_1 f''_0\varepsilon''\varphi \\ &= -\tau_1 e'' h''\varphi - f'_1 d^0_{X'}\sigma_0 + f'_1 d^0_{X'}\sigma_0 - \tau_1 e'' h''\varphi = 0. \end{aligned}$$

Thus there exists $q_1 : X''_1 \longrightarrow W'_1$ such that $q_1 d^0_{X''} = \tau_1 f''_0 - f'_1\sigma_1 + d^0_{W'}q_0$.

Assume now that the morphisms $q_1, q_2, \cdots, q_n$ satisfying relation (7) have already been constructed. To define the morphism $q_{n+1}: X''_{n+1} \longrightarrow W'_{n+1}$, consider the diagram

$$\begin{array}{ccccc} X''_{n-1} & \xrightarrow{d^{n-1}_{X''}} & X''_n & \xrightarrow{d^n_{X''}} & X''_{n+1}. \\ & & \Big\downarrow{\scriptstyle \tau_{n+1}f''_n - f'_{n+1}\sigma_{n+1} + d^n_{W'}q_n} & & \\ & & W'_{n+1} & & \end{array}$$

We may apply lemma 7.4, since

$$\begin{aligned}(\tau_{n+1}f''_n - f'_{n+1}\sigma_{n+1} + d^n_{W'}q_n)d^{n-1}_{X''} &= (\tau_{n+1}f''_n - f'_{n+1}\sigma_{n+1})d^{n-1}_{X''} \\ &+ d^n_{W'}(d^{n-1}_{W'}q_{n-1} + \tau_n f''_{n-1} - f'_n\sigma_n) = \tau_{n+1}f''_n d^{n-1}_{X''} - f'_{n+1}\sigma_{n+1}d^{n-1}_{X''} \\ &+ d^n_{W'}\tau_n f''_{n-1} - d^n_{W'}f'_n\sigma_n = (\tau_{n+1}d^{n-1}_{W''} + d^n_{W'}\tau_n)f''_{n-1} \\ &- f'_{n+1}(\sigma_{n+1}d^{n-1}_{X''} + d^n_{X'}\sigma_n) = 0.\end{aligned}$$

Thus the existence of the family of morphisms q_n with property (7) is proved.

It remains to show that the family of morphisms $(f_n)_{n\geq 0}$ defined by (6) where the q_n's satisfy (7) constitute a morphism from the complex **X** to the complex **W** and that diagram (ii) is commutative. First, we show that $f_{n+1}d^n_X = d^n_W f_n$.

$$\begin{aligned}d^n_W f_n &= (j''_{n+1}d^n_{W''}\pi''_n + j'_{n+1}\tau_{n+1}\pi''_n + j'_{n+1}d^n_{W'}\pi'_n)\circ(j'_n q_n p''_n + j'_n f'_n p'_n \\ &+ j''_n f''_n p''_n) = j'_{n+1}d^n_{W'}q_n p''_n + j'_{n+1}d^n_{W'}f'_n p'_n + j''_{n+1}d^n_{W''}f''_n p''_n \\ &+ j'_{n+1}\tau_{n+1}f''_n p''_n = j'_{n+1}d^n_{W'}q_n p''_n + j'_{n+1}f'_{n+1}d^n_{X'}p'_n \\ &+ j''_{n+1}f''_{n+1}d^n_{X''}p''_n + j'_{n+1}\tau_{n+1}f''_n p''_n.\end{aligned}$$

$$\begin{aligned}f_{n+1}d^n_X &= (j'_{n+1}q_{n+1}p''_{n+1} + j'_{n+1}f'_{n+1}p'_{n+1} + j''_{n+1}f''_{n+1}p''_{n+1}) \\ &\circ (i''_{n+1}d^n_{X''}p''_n + i'_{n+1}\sigma_{n+1}p''_n + i'_{n+1}d^n_{X'}p'_n) = j'_{n+1}q_{n+1}d^n_{X''}p''_n \\ &+ j''_{n+1}f''_{n+1}d^n_{X''}p''_n + j_{n+1}f'_{n+1}\sigma_{n+1}p''_n + j'_{n+1}f'_{n+1}d^n_{X'}p'_n.\end{aligned}$$

From the comparison of these relations it follows that relation (7) implies $d^n_W f_n = f_{n+1}d^n_X$.

To verify the commutativity of diagram (ii) it is not necessary to impose any condition on the morphisms q_n and this verification is trivial. This concludes the proof.

3. Derived Functors

We are in a position to introduce one of the fundamental notions of homological algebra, namely the notion of derived functors.

Let $\mathscr{C}$ and $\mathscr{C}'$ be two Abelian categories and $F:\mathscr{C} \longrightarrow \mathscr{C}'$ an additive covariant functor. In addition, we assume that the category $\mathscr{C}$ is with sufficiently many injectives. Under these conditions any object of the category $\mathscr{C}$ possesses at least one exact and injective resolution. We suppose that for each object A of $\mathscr{C}$ we have chosen an exact and injective resolution $(\mathbf{R}_A, \varepsilon_A)$. (It should be pointed out that the subsequent construction depends on the choice of these resolutions). It is evident that $F(\mathbf{R}_A)$ is a complex in the category $\mathscr{C}'$.

We construct a sequence of covariant functors from the category $\mathscr{C}$ to the category $\mathscr{C}'$ $R^0F, R^1F, \cdots, R^qF, \cdots$ called the *right-derived functors* of the functor F and defined as follows:

$$(R^qF)(A) = H^q(F(\mathbf{R}_A)).$$

If $u:A \longrightarrow B$ is a morphism in the category $\mathscr{C}$ we must define the morphism $(R^qF)(u):(R^qF)(A) \longrightarrow (R^qF)(B)$ such that $(R^0F)(vu) = (R^qF)(v) \circ (R^qF)(u)$. According to proposition 7.5 there exists a morphism $\mathbf{f} = (f^n)_{n\geq 0}$ from the complex $\mathbf{R}_A$ to the complex $\mathbf{R}_B$ such that the diagram

$$\begin{array}{ccc} A & \xrightarrow{u} & B \\ {\scriptstyle \varepsilon_A}\downarrow & & \downarrow{\scriptstyle \varepsilon_B} \\ R_A^0 & \xrightarrow[f^0]{} & R_B^0 \end{array}$$

is commutative. Moreover, by the same proposition $\mathbf{f}$ is uniquely determined up to a homotopy. In other words, if $\mathbf{g} = (g^n)_{n\geq 0}$ is another morphism from $\mathbf{R}_A$ to $\mathbf{R}_B$ satisfying the same conditions as $\mathbf{f}$, then $\mathbf{f}$ and $\mathbf{g}$ are homotopic. $F(\mathbf{f}) = (F(f^n))_{n\geq 0}$, $F(\mathbf{g}) = (F(g^n))_{n\geq 0}$ are morphisms from the complex $F(\mathbf{R}_A)$ to the complex $F(\mathbf{R}_B)$ and it is evident that they are homotopic. Thus we can define starting from the morphism $\mathbf{f}$ a morphism from the complex $F(\mathbf{R}_A)$ to the complex $F(\mathbf{R}_B)$ and, by proposition 7.1, this morphism depends only on the morphism u. Consequently we set $(R^qF)(u) = H^q(F(\mathbf{f}))$. It is obvious that $(R^qF)(vu) = (R^qF)(v)\cdot(R^qF)(u)$. Hence $R^qF:\mathscr{C} \longrightarrow \mathscr{C}'$ is an additive covariant functor.

We now look closer into the manner in which the functors R^qF depend on the choice of the resolutions $\mathbf{R}_A$.

Thus, assume that we have chosen for each object A of the category $\mathscr{C}$ the exact and injective resolution $(\bar{\mathbf{R}}_A, \bar{\varepsilon}_A)$. We denote by $\bar{R}^0F, \bar{R}^1F, \cdots,$

$\bar{R}^qF, \cdots$ the right-derived functors of the functor F constructed by starting from the resolutions $\bar{\mathbf{R}}_A$. We have to compare the functors R^qF, $\bar{R}^qF$. We shall prove that there exists a canonical functorial isomorphism between these functors. To do this, we define the isomorphisms $f_A^q:(R^qF)(A) \longrightarrow (\bar{R}^qF)(A)$ such that the diagram

$$\begin{array}{ccc} (R^qF)(A) & \xrightarrow{f_A^q} & (\bar{R}^qF)(A) \\ \Big\downarrow{\scriptstyle (R^qF)(u)} & & \Big\downarrow{\scriptstyle (\bar{R}^qF)(u)} \\ (R^qF)(B) & \xrightarrow[f_B^q]{} & (\bar{R}^qF)(B) \end{array} \tag{8}$$

is commutative for any $u:A \longrightarrow B$.

To define the morphism f_A we use again proposition 7.5. Thus there exists a morphism $\mathbf{f} = (f^n)_{n \geq 0}$ from the complex $\mathbf{R}_A$ to the complex $\bar{\mathbf{R}}_A$ uniquely determined up to a homotopy such that the diagram

$$\begin{array}{ccc} & A & \\ {\scriptstyle \varepsilon_A}\swarrow & & \searrow{\scriptstyle \bar{\varepsilon}_A} \\ R_A^0 & \xrightarrow[f^0]{} & \bar{R}_A^0 \end{array}$$

is commutative. Set $f_A^q = H^q(F(\mathbf{f})):(R^qF)(A) \longrightarrow (\bar{R}^qF)(B)$. The commutativity of diagram (8) follows immediately by using proposition 7.5.

The above remark suggests the idea of choosing once for all a resolution for each object of the category $\mathscr{C}$. These resolutions will be said to be canonical.

The following two fundamental properties of right-derived functors of a covariant functor F follow from the definition.

PROPERTY 7.9. *If Q is an injective object of the category $\mathscr{C}$ then $(R^qF)(Q) = 0$ for any $q \geq 1$.*

PROPERTY 7.10. *If F is a left-exact functor then R^0F is isomorphic with F. If F is exact then $(R^qF)(A) = 0$ for any $q \geq 1$.*

PROOF. Property 7.9 follows from the remark that the sequence

$$0 \longrightarrow Q \xrightarrow{1_Q} Q \longrightarrow 0 \longrightarrow 0 \longrightarrow \cdots \longrightarrow 0 \longrightarrow \cdots.$$

is an exact and injective resolution of the object Q.

Property 7.10 results from the following remarks:

If $(\mathbf{R}_A, \varepsilon_A)$ is an exact and injective resolution of A and if F is left-exact then the sequence

$$0 \longrightarrow F(A) \xrightarrow{F(\varepsilon_A)} F(R_A^0) \xrightarrow{F(d_{\mathbf{R}_A}^0)} F(R_A^1)$$

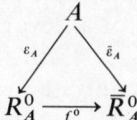

is exact and therefore

$$(R^0F)(A) = H^0(F(\mathbf{R}_A)) \approx F(A).$$

If in addition F is exact then the whole sequence

$$0 \longrightarrow F(A) \xrightarrow{F(\varepsilon_A)} F(R_A^0) \xrightarrow{F(d^0_{\mathbf{R}_A})} F(R_A^1) \xrightarrow{F(d^1_{\mathbf{R}_A})} F(R_A^2) \longrightarrow \cdots$$

is exact and therefore

$$(R^qF)(A) = H^q(F(\mathbf{R}_A)) = 0 \qquad (q \geq 1).$$

Besides these two properties and the general properties of additive functors, the derived functors have also other remarkable properties. However, to present them we need some other preliminary constructions and properties.

If instead of starting from exact injective resolution we start from exact projective resolutions then in the same manner as above we define the notion of left-derived functors of a covariant functor F and we denote them by $L^0F, L^1F, \cdots, L^nF, \cdots$.

Also, we can consider by duality the right- and left-derived functors of a contravariant functor. We leave to the reader the formulation of the properties dual to those obtained above.

Let $\mathbf{X}', \mathbf{X}, \mathbf{X}''$ be three complexes with values in the category $\mathscr{C}$. To simplify the writing we shall denote the elements and the differentials of these complexes as follows:

$$\mathbf{X}' = (\cdots \longrightarrow X'_i \xrightarrow{d'_i} X'_{i+1} \xrightarrow{d'_{i+1}} X'_{i+2} \longrightarrow \cdots)$$
$$\mathbf{X} = (\cdots \longrightarrow X_i \xrightarrow{d_i} X_{i+1} \xrightarrow{d_{i+1}} X_{i+2} \longrightarrow \cdots)$$
$$\mathbf{X}'' = (\cdots \longrightarrow X''_i \xrightarrow{d''_i} X''_{i+1} \xrightarrow{d''_{i+1}} X_{i+2} \longrightarrow \cdots).$$

We now assume that for each $i \in Z$, $X_i = X'_i \oplus X''_i$ and that the canonical injections $i'_j : X'_j \longrightarrow X'_j \oplus X''_j$, $j \in Z$ define a morphism $I' : \mathbf{X}' \longrightarrow \mathbf{X}$ from the complex $\mathbf{X}'$ to the complex $\mathbf{X}$. We also assume that the canonical projections $p''_j : X'_j \oplus X''_j \longrightarrow X''_j$, $j \in Z$ define a morphism $P'' : \mathbf{X} \longrightarrow \mathbf{X}''$ from the complex $\mathbf{X}$ to the complex $\mathbf{X}''$. Thus we obtain the exact sequence of complexes

$$0 \longrightarrow \mathbf{X}' \xrightarrow{I'} \mathbf{X} \xrightarrow{P''} \mathbf{X}'' \longrightarrow 0. \tag{E}$$

Our aim is to show that under these conditions there exists for any $i \in Z$ a morphism

$$\partial^i_E : H^i(\mathbf{X}'') \longrightarrow H^{i+1}(\mathbf{X}')$$

such that the sequence

$$\cdots \longrightarrow H^i(\mathbf{X}'') \xrightarrow{\partial_E^i} H^{i+1}(\mathbf{X}') \xrightarrow{H^{i+1}(I')} H^{i+1}(\mathbf{X}) \xrightarrow{H^{i+1}(P'')} H^{i+1}(\mathbf{X}'') \xrightarrow{\partial_E^{i+1}} \cdots$$

is exact.

To this end, we notice that for any i we have the commutative diagram

$$\begin{array}{ccccccccc} 0 & \longrightarrow & X'_i & \longrightarrow & X_i & \longrightarrow & X''_i & \longrightarrow & 0 \\ & & \downarrow{\scriptstyle d'_i} & & \downarrow{\scriptstyle d^i} & & \downarrow{\scriptstyle d''_i} & & \\ 0 & \longrightarrow & X'_{i+1} & \longrightarrow & X_{i+1} & \longrightarrow & X''_{i+1} & \longrightarrow & 0 \end{array}$$

which yields the exact sequences (see corollary 6.35).

$$0 \longrightarrow \ker d'_i \longrightarrow \ker d_i \longrightarrow \ker d''_i$$

$$\operatorname{coker} d'_i \longrightarrow \operatorname{coker} d_i \longrightarrow \operatorname{coker} d''_i \longrightarrow 0.$$

We remark that $\ker d_i = Z_i$ and $\operatorname{coker} d_i = X_{i+1}/B_{i+1}$. Moreover, d_{i+1} induces a morphism α_i from $\operatorname{coker} d_i$ to Z_{i+2} whose image is precisely B_{i+2}. Thus, we have the commutative diagram

$$\begin{array}{ccccccc} & & \operatorname{coker} d'_i & \longrightarrow & \operatorname{coker} d_i & \longrightarrow & \operatorname{coker} d''_i \longrightarrow 0 \\ & & \downarrow{\scriptstyle \alpha'_i} & & \downarrow{\scriptstyle \alpha_i} & & \downarrow{\scriptstyle \alpha''_i} \\ 0 & \longrightarrow & Z'_{i+2} & \longrightarrow & Z_{i+2} & \longrightarrow & Z''_{i+2}. \end{array}$$

To finish, it remains to remark that $\ker \alpha_i$ is isomorphic with $H_{i+1}(\mathbf{X})$, $\operatorname{coker} \alpha_i$ is isomorphic with $H_{i+2}(\mathbf{X})$, etc. and to apply corollary 6.22.

4. Other Properties of Derived Functors

The results obtained in the previous section enable us to complete the list of remarkable properties of derived functors.

PROPOSITION 7.11. *Let $\mathscr{C}$ be an Abelian category with sufficiently many injectives, $F:\mathscr{C} \longrightarrow \mathscr{C}'$ a left-exact covariant functor and*

$$0 \longrightarrow A' \xrightarrow{\alpha'} A \xrightarrow{\alpha''} A'' \longrightarrow 0 \qquad \text{(E)}$$

an exact sequence in $\mathscr{C}$. Then there exist for any $i \geq 0$ morphisms

$$\partial_E^i : (R^iF)(A'') \longrightarrow (R^{i+1}F)(A')$$

such that the sequence

$$(R^0F)(A') \xrightarrow{(R^0F)(\alpha')} (R^0F)(A) \xrightarrow{(R^0F)(\alpha'')} (R^0F)(A'') \xrightarrow{\partial_E^0} (R^1F)(A')$$
$$\longrightarrow \cdots \longrightarrow (R^iF)(A') \xrightarrow{(R^iF)(\alpha')} (R^iF)(A) \xrightarrow{(R^iF)(\alpha'')} (R^iF)(A'')$$
$$\xrightarrow{\partial_E^i} (R^{i+1}F)(A') \longrightarrow \cdots \qquad (14)$$

is exact.

If

$$0 \longrightarrow B' \xrightarrow{\beta'} B \xrightarrow{\beta''} B'' \longrightarrow 0 \qquad (F)$$

is another exact sequence in $\mathscr{C}$ *and the morphisms* $f' : A' \longrightarrow B', f : A \longrightarrow B,$ $f'' : A'' \longrightarrow B''$ *are such that the diagram*

$$\begin{array}{ccccccccc} 0 & \longrightarrow & A' & \xrightarrow{\alpha'} & A & \xrightarrow{\alpha''} & A'' & \longrightarrow & 0 \\ & & \downarrow{\scriptstyle f'} & & \downarrow{\scriptstyle f} & & \downarrow{\scriptstyle f''} & & \\ 0 & \longrightarrow & B' & \xrightarrow[\beta']{} & B & \xrightarrow[\beta'']{} & B'' & \longrightarrow & 0 \end{array}$$

is commutative, then the diagram

$$\begin{array}{ccc} (R^iF)(A'') & \xrightarrow{\partial_E^i} & (R^{i+1}F)(A') \\ \downarrow{\scriptstyle (R^iF)(f'')} & & \downarrow{\scriptstyle (R^iF)(f')} \\ (R^iF)(B'') & \xrightarrow[\partial_F^i]{} & (R^{i+1}F)(B') \end{array} \qquad (15)$$

is commutative for any $i \geq 0$.

PROOF. Let $(\mathbf{R}_{A'}, \varepsilon_{A'})$ be the canonical exact and injective resolution of A' and $(R_{A''}, \varepsilon_{A''})$ the canonical exact and injective resolution of A''. By choosing a normal resolution $(\mathbf{X}, \varepsilon)$ of A and by taking into account the considerations at the end of the preceding paragraph, we obtain the morphisms ∂_E^i such that the sequence (14) is exact. From the manner in which we have defined the morphisms ∂_E^i it would seem that they depend on the choice of the normal resolution $(\mathbf{X}, \varepsilon)$. In fact they do not depend on the particular choice of the normal resolution $(\mathbf{X}, \varepsilon)$ as follows immediately from relation (7) in the proof of proposition 7.8. This relation is applied here for the case where the morphism $\mathbf{f}'$ is the identity of the complex $\mathbf{R}_{A'}$ and $\mathbf{f}''$ is the identity of the complex $\mathbf{R}_{A''}$. The same relation considered this time in the general case implies the commutativity of diagram (15). This concludes the proof of the proposition.

5. Homology and Cohomology Functors

Closely related to the notion of derived functors of a functor are the notions of a homology functor and cohomology functor. In this connexion there are actually four notions, namely:

(1) Homology covariant functor,

(2) Homology contravariant functor,

(3) Cohomology covariant functor,

(4) Cohomology contravariant functor.

The notions (1) and (2) are dual to each other: the same holds for the notions (3) and (4). Also, the passage from the notion of a cohomology functor to that of a homology functor is immediate, so that, in developing the general theory it is sufficient for us to deal, for instance, with the case of cohomology covariant functors.

Definition. Let $\mathscr{C}$ and $\mathscr{C}'$ be two Abelian categories. A sequence $(T^i)_{i \geq 0}$ of covariant functors from the category $\mathscr{C}$ to the category $\mathscr{C}'$ is said to be a *cohomology covariant functor defined* on $\mathscr{C}$ and with values in $\mathscr{C}'$ if for any exact sequence of objects and morphisms in $\mathscr{C}$ of the form

$$0 \longrightarrow A' \xrightarrow{\alpha'} A \xrightarrow{\alpha''} A'' \longrightarrow 0 \tag{E}$$

and for any $i \geq 0$ there are given the morphisms

$$\partial_E^i : T^i(A'') \longrightarrow T^{i+1}(A')$$

such that the following conditions are fulfilled:

(1). The sequence

$$\begin{aligned} 0 \longrightarrow T^0(A') \xrightarrow{T^0(\alpha')} T^0(A) \xrightarrow{T^0(\alpha'')} T^0(A'') \xrightarrow{\partial_E^0} T^1(A') \longrightarrow \cdots \\ \longrightarrow T^i(A') \xrightarrow{T^i(\alpha')} T^i(A) \xrightarrow{T^i(\alpha'')} T^i(A'') \xrightarrow{\partial_E^i} T^{i+1}(A') \longrightarrow \cdots \end{aligned} \tag{16}$$

is exact.

(2) If

$$0 \longrightarrow B' \xrightarrow{\beta'} B \xrightarrow{\beta''} B'' \longrightarrow 0 \tag{F}$$

is another exact sequence in $\mathscr{C}$ and if the morphisms $f' : A \longrightarrow B'$, $f : A \longrightarrow B, f'' : A'' \longrightarrow B''$ are such that the diagram

$$\begin{array}{ccccccccc} 0 & \longrightarrow & A' & \xrightarrow{\alpha'} & A & \xrightarrow{\alpha''} & A'' & \longrightarrow & 0 \\ & & \downarrow{\scriptstyle f'} & & \downarrow{\scriptstyle f} & & \downarrow{\scriptstyle f''} & & \\ 0 & \longrightarrow & B' & \xrightarrow[\beta']{} & B & \xrightarrow[\beta'']{} & B'' & \longrightarrow & 0 \end{array}$$

is commutative, then the diagram

$$\begin{array}{ccc} T^i(A'') & \xrightarrow{\partial_E^i} & T^{i+1}(A') \\ {\scriptstyle T^i(f'')}\downarrow & & \downarrow{\scriptstyle T^{i+1}(f')} \\ T^i(B'') & \xrightarrow[\partial_F^i]{} & T^{i+1}(B') \end{array} \tag{17}$$

is commutative for any $i \geq 0$.

Consider now another cohomology covariant functor from the category $\mathscr{C}$ to the category $\mathscr{C}'$ defined by the sequence of functors $(S')_{i \geq 0}$ and the morphisms δ_E^i. We say that a functorial morphism has been given from the cohomology functor defined by the sequence $(T^i)_{i \geq 0}$ and the morphisms ∂_E^i to the cohomology functor defined by the sequence $(S^i)_{i \geq 0}$ and the morphisms ∂_E^i if there is given for any $i \geq 0$ a functorial morphism $\varphi^i : T^i \longrightarrow S^i$ such that for any exact sequence in $\mathscr{C}$

$$0 \longrightarrow A' \longrightarrow A \longrightarrow A'' \longrightarrow 0 \tag{E}$$

the diagram

$$\begin{array}{ccc} T^i(A'') & \xrightarrow{\partial_E^i} & T^{i+1}(A') \\ {\scriptstyle \varphi^i(A'')}\downarrow & & \downarrow{\scriptstyle \varphi^{i+1}(A')} \\ S^i(A'') & \xrightarrow[\delta_E^i]{} & S^{i+1}(A') \end{array}$$

is commutative.

A cohomology covariant functor (T^i, ∂_E^i) is said to be *universal* if for any cohomology covariant functor (S^i, δ_E^i) and for any functorial morphism $\varphi : T^0 \longrightarrow S^0$ there exists a unique morphism of the first cohomology functor to the second one, defined by a sequence of functorial morphisms $(\psi^i)_{i \geq 0}$, such that $\psi^0 = \varphi$.

If $F : \mathscr{C} \longrightarrow \mathscr{C}'$ is a left-exact covariant functor and (T^i, ∂_E^i) is a universal cohomology functor such that $T^0 = F$, then the universal cohomology functor (T^i, ∂_E^i) is uniquely determined (up to a canonical isomorphism) by this condition. In this case the functors T^i are said to be the satellites of the functor F and are denoted by S^iF.

For the notion of a cohomology contravariant functor one stipulates that to the sequence (E) there correspond the sequence of morphisms $\partial_E^i : T^i(A') \longrightarrow T^{i+1}(A'')$ such that the sequence

$$0 \longrightarrow T^0(A'') \xrightarrow{T^0(\alpha'')} T^0(A) \xrightarrow{T^0(\alpha')} T^0(A') \xrightarrow{\partial_E^0} T^1(A'') \longrightarrow \cdots$$
$$\longrightarrow T^i(A'') \longrightarrow T^i(A) \longrightarrow T^i(A') \xrightarrow{\partial_E^i} T^{i+1}(A'') \longrightarrow \cdots$$

be exact and, of course, that a commutativity condition of type (17) be satisfied.

For the notion of a homology covariant functor the difference consists of the fact that the morphisms ∂_E^i decrease the grades. In other words, $\partial_E^i : T^i(A'') \longrightarrow T^{i-1}(A')$ and instead of the sequence (16) we have the sequence

$$\cdots \longrightarrow T^i(A'') \xrightarrow{\partial_E^i} T^{i-1}(A') \xrightarrow{T^{i-1}(\alpha')} T^{i-1}(A) \xrightarrow{T^{i-1}(\alpha'')} T^{i-1}(A'') \longrightarrow \cdots$$
$$\longrightarrow T^0(A') \longrightarrow T^0(A) \longrightarrow T^0(A'') \longrightarrow 0.$$

It is now clear how one defines the notions of a homology contravariant functor.

Assume now that the category $\mathscr{C}$ is with sufficiently many injectives and $F : \mathscr{C} \longrightarrow \mathscr{C}'$ is covariant and left-exact. Then it is clear that the right-derived functors of F define a cohomology covariant functor. If F is right-exact then the left-derived functors of F define a homology covariant functor.

In fact, by means of derived functors we obtain even universal homology (or cohomology) functors, as results from the following:

PROPOSITION 7.12. *Assume that the category $\mathscr{C}$ is with sufficiently many injectives and that the cohomology functor (T^i, ∂_E^i) satisfies the condition:*

(a) *For any injective object Q of $\mathscr{C}$ and for any $i \geq 1$, $T^i(Q) = 0$.*

Under these conditions, the cohomology functor (T^i, ∂_E^i) is universal. Conversely, every universal cohomology functor (T^i, ∂_E^i) satisfies condition (a).

PROOF. Let (S^i, δ_E^i) be another cohomology covariant functor and let $\varphi : T^0 \longrightarrow S^0$ be a functorial morphism. We must show that there exists a morphism $(\psi^i)_{i \geq 0}$ from the cohomology functor (T^i, ∂_E^i) to the cohomology functor (S^i, δ_E^i). We set, of course, $\psi^0 = \varphi$. The definition of ψ^1 is as follows: We take an arbitrary object A of the category $\mathscr{C}$ and imbed it into an injective object Q. Thus we obtain the following exact sequence

$$0 \longrightarrow A \longrightarrow Q \longrightarrow Q/A \longrightarrow 0 \qquad \text{(E)}$$

which yields a morphism $\psi_Q^1(A) : T^1(A) \longrightarrow S^1(A)$ such that the diagram

$$\begin{array}{ccc} T^0(Q/A) & \xrightarrow{\partial_E^0} & T^1(A) \\ \downarrow{\scriptstyle \varphi(Q/A)} & & \downarrow{\scriptstyle \psi_Q^1(A)} \\ S^0(Q/A) & \xrightarrow[\delta_E^0]{} & S^1(A) \end{array}$$

is commutative. Our immediate goal is to prove that $\psi_Q^1(A)$ does not depend on the choice of the injective object Q into which we imbed A, which will justify us in writing $\psi^1(A)$ rather than $\psi_Q^1(A)$.

To this end, consider besides the sequence (E) the following sequence:

$$0 \longrightarrow A \longrightarrow Q_1 \longrightarrow Q1/A \longrightarrow 0. \qquad (E_1)$$

Then there exist $\alpha : Q \longrightarrow Q_1$, $\beta : Q/A \longrightarrow Q_1/A$ such that the diagram

$$\begin{array}{ccccccccc} 0 & \longrightarrow & A & \longrightarrow & Q & \longrightarrow & Q/A & \longrightarrow & 0 \\ & & \downarrow{\scriptstyle 1_A} & & \downarrow{\scriptstyle \alpha} & & \downarrow{\scriptstyle \beta} & & \\ 0 & \longrightarrow & A & \longrightarrow & Q_1 & \longrightarrow & Q_1/A & \longrightarrow & 0 \end{array}$$

is commutative. We have to prove that $\psi^1_Q(A) = \psi^1_{Q_1}(A)$. Since ∂^0_E is a surjection it is sufficient to check that $\psi^1_Q(A)\partial^0_E = \psi^1_{Q_1}(A)\partial^0_E$. To do this we use the commutativity of the following diagrams:

$$\begin{array}{ccc} T^0(Q/A) & \xrightarrow{\partial^0_E} & T^1(A) \\ {\scriptstyle T^0(\beta)}\downarrow & \nearrow{\scriptstyle \partial^0_{E_1}} & \\ T^0(Q_1/A) & & \end{array} \qquad \begin{array}{ccc} S^0(Q/A) & \xrightarrow{\delta^0_E} & S^1(A) \\ {\scriptstyle S^0(\beta)}\downarrow & \nearrow{\scriptstyle \delta^0_{E_1}} & \\ S^0(Q_1/A) & & \end{array}$$

$$\begin{array}{ccc} T^0(Q/A) & \xrightarrow{\varphi(Q/A)} & S^0(Q/A) \\ {\scriptstyle T^0(\beta)}\downarrow & & \downarrow{\scriptstyle S^0(\beta)} \\ T^0(Q_1/A) & \xrightarrow[\varphi(Q_1/A)]{} & S^0(Q_1/A) \end{array} \qquad \begin{array}{ccc} T^0(Q_1/A) & \xrightarrow{\partial^0_{E_1}} & T^1(A) \\ {\scriptstyle \varphi(Q_1/A)}\downarrow & & \downarrow{\scriptstyle \psi^1_{Q_1}(A)} \\ S^0(Q_1/A) & \xrightarrow[\delta^0_{E_1}]{} & S^1(A). \end{array}$$

We have

$$\begin{aligned} \psi^1_Q(A)\partial^0_E &= \delta^0_E\varphi(Q/A) = \delta^0_{E_1}S^0(\beta)\varphi(Q/A) = \delta^0_{E_1}\varphi(Q_1/A)T^0(\beta) \\ &= \psi^1_{Q_1}(A)\partial^0_{E_1}T^0(\beta) = \psi^1_{Q_1}(A)\partial^0_E. \end{aligned}$$

We now prove that the morphisms $\psi^1(A)$ define a functorial morphism from T^1 to S^1. Let $u : A \longrightarrow B$. We have to check the commutativity of the diagram

$$\begin{array}{ccc} T^1(A) & \xrightarrow{T^1(u)} & T^1(B) \\ {\scriptstyle \psi^1(A)}\downarrow & & \downarrow{\scriptstyle \psi^1(B)} \\ S^1(A) & \xrightarrow[S^1(u)]{} & S^1(B). \end{array} \qquad (18)$$

Resuming the definition of the morphisms $\psi^1(A)$, $\psi^1(B)$ we consider the commutative diagram

$$\begin{array}{ccccccccc} 0 & \longrightarrow & A & \longrightarrow & Q & \longrightarrow & Q/A & \longrightarrow & 0 \\ & & \downarrow{\scriptstyle u} & & \downarrow{\scriptstyle u_1} & & \downarrow{\scriptstyle u_2} & & \\ 0 & \longrightarrow & B & \longrightarrow & Q_1 & \longrightarrow & Q_1/B & \longrightarrow & 0 \end{array}$$

where Q and Q_1 are injective objects. To verify the commutativity of diagram (18) we use the commutativity of the following diagrams

$$\begin{array}{ccc} T^0(Q/A) & \xrightarrow{\partial_E^0} & T^1(A) \\ {\scriptstyle T^0(u_2)}\downarrow & & \downarrow{\scriptstyle T^1(u)} \\ T^0(Q_1/A) & \xrightarrow[\partial_F^0]{} & T^1(B) \end{array} \qquad \begin{array}{ccc} T^0(Q/A) & \xrightarrow{\partial_E^0} & T^1(A) \\ {\scriptstyle \varphi(Q/A)}\downarrow & & \downarrow{\scriptstyle \psi^1(A)} \\ S^0(Q/A) & \xrightarrow[\delta_E^0]{} & S^1(A) \end{array}$$

$$\begin{array}{ccc} S^0(Q/A) & \xrightarrow{\delta_E^0} & S^1(A) \\ {\scriptstyle S^0(u_2)}\downarrow & & \downarrow{\scriptstyle S^1(u)} \\ S^0(Q_1/A) & \xrightarrow[\delta_F^0]{} & S^1(B) \end{array} \qquad \begin{array}{ccc} T^0(Q_1/B) & \xrightarrow{\partial_F^0} & T^1(B) \\ {\scriptstyle \varphi(Q_1/B)}\downarrow & & \downarrow{\scriptstyle \psi^1(B)} \\ S^0(Q_1/B) & \xrightarrow[\delta_F^0]{} & S^1(B) \end{array}$$

$$\begin{array}{ccc} T^0(Q/A) & \xrightarrow{T^0(u_2)} & T^0(Q_1/B) \\ {\scriptstyle \varphi(Q/A)}\downarrow & & \downarrow{\scriptstyle \varphi(Q_1/B)} \\ S^0(Q/A) & \xrightarrow[S^0(u_2)]{} & S^0(Q_1/B). \end{array}$$

Since ∂_E^0 is a surjection it is sufficient to prove that $\psi^1(B)T^1(u)\partial_E^0 = S^1(u)\psi^1(A)\partial_E^0$.

$$\begin{aligned} \psi^1(B)T^1(u)\partial_E^0 &= \psi^1(B)\partial_F^0 T^0(u_2) = \delta_F^0\varphi(Q_1/B)T^0(u_2) \\ &= \delta_F^0 S^0(u_2)\varphi(Q/A) = S^1(u)\delta_E^0\varphi(Q/A) = S^1(u)\varphi^1(A)\partial_E^0. \end{aligned}$$

It remains to show that if

$$0 \longrightarrow A' \longrightarrow A \longrightarrow A'' \longrightarrow 0 \tag{U}$$

is an exact sequence, then the diagram

$$\begin{array}{ccc} T^0(A'') & \xrightarrow{\partial_U^0} & T^1(A') \\ {\scriptstyle \varphi(A'')}\downarrow & & \downarrow{\scriptstyle \psi^1(A')} \\ S^0(A'') & \xrightarrow[\delta_U^0]{} & S^1(A') \end{array} \tag{19}$$

is commutative. To do this we consider the exact sequence

$$0 \longrightarrow A' \longrightarrow Q \longrightarrow Q/A' \longrightarrow 0 \tag{V}$$

where Q is injective, which we insert into the commutative diagram

$$\begin{array}{ccccccccc} 0 & \longrightarrow & A' & \longrightarrow & A & \longrightarrow & A'' & \longrightarrow & 0 \\ & & {\scriptstyle 1_{A'}}\downarrow & & {\scriptstyle u}\downarrow & & {\scriptstyle v}\downarrow & & \\ 0 & \longrightarrow & A' & \longrightarrow & Q & \longrightarrow & Q/A' & \longrightarrow & 0. \end{array}$$

We use the commutativity of the following diagrams:

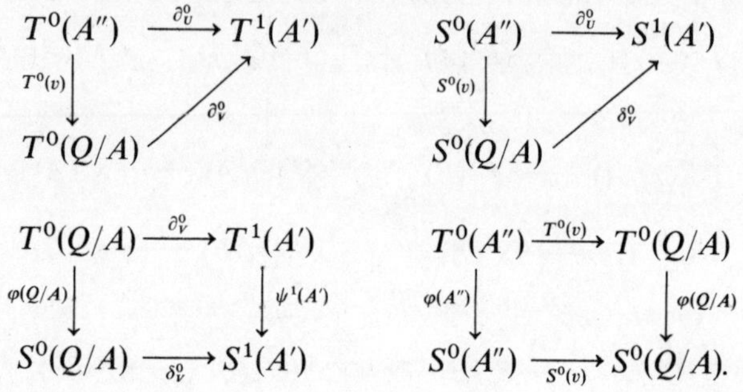

$$
\begin{array}{ccc}
T^0(A'') \xrightarrow{\partial_U^0} T^1(A') & \quad & S^0(A'') \xrightarrow{\partial_U^0} S^1(A') \\
{\scriptstyle T^0(v)}\downarrow \nearrow {\scriptstyle \partial_V^0} & & {\scriptstyle S^0(v)}\downarrow \nearrow {\scriptstyle \delta_V^0} \\
T^0(Q/A) & & S^0(Q/A)
\end{array}
$$

$$
\begin{array}{ccc}
T^0(Q/A) \xrightarrow{\partial_V^0} T^1(A') & \quad & T^0(A'') \xrightarrow{T^0(v)} T^0(Q/A) \\
{\scriptstyle \varphi(Q/A)}\downarrow \quad \downarrow {\scriptstyle \psi^1(A')} & & {\scriptstyle \varphi(A'')}\downarrow \quad \downarrow {\scriptstyle \varphi(Q/A)} \\
S^0(Q/A) \xrightarrow{\delta_V^0} S^1(A') & & S^0(A'') \xrightarrow{S^0(v)} S^0(Q/A).
\end{array}
$$

We obtain

$$\psi^1(A')\partial_U^0 = \psi^1(A')\partial_V^0 T^0(v) = \delta_V^0 \varphi(Q/A) T^0(v) = \delta_V^0 S^0(v)\varphi(A'')$$
$$= \delta_U^0 \varphi(A''),$$

i.e. precisely the commutativity of diagram (19).

By continuing in the same manner it is possible to determine the functorial morphisms $\psi^2, \psi^3, \ldots, \psi^n, \ldots$. Thus the first part of the proposition is completely proved. To prove the second part, we argue as follows: Under the conditions of our proposition we can construct the right-derived functors of the functor T^0; we thus obtain according to the first part of the proposition, a universal cohomology functor $(R^i T^0, \partial_E^i)$, which is canonically isomorphic with the functor (T^i, ∂_E^i) since $R^0 T^0 = T^0$. Since $(R^i T^0)(Q) = 0$ for any injective object Q and any $i \geq 1$, the same follows for the functors T^i.

Clearly a statement similar to proposition 7.12 holds for the case of cohomology contravariant functors as well as for the case of homology functors.

COROLLARY 7.13. *If the category $\mathscr{C}$ is with sufficiently many injectives and if $F: \mathscr{C} \longrightarrow \mathscr{C}'$ is a left-exact covariant functor, then there exists a universal cohomology covariant functor (T^i, ∂_E^i) such that $T^0 = F$; moreover, (T^i, ∂_E^i) is uniquely determined up to a canonical isomorphism.*

COROLLARY 7.14. *If the category $\mathscr{C}$ is with sufficiently many projectives and if $F: \mathscr{C} \longrightarrow \mathscr{C}'$ is a right-exact covariant functor, then there exists a universal homology covariant functor (T^i, ∂_E^i) such that $T^0 = F$; moreover, (T^i, ∂_E^i) is uniquely determined up to a canonical isomorphism.*

COROLLARY 7.15. *If the category $\mathscr{C}$ is with sufficiently many projectives and if $F:\mathscr{C} \longrightarrow \mathscr{C}'$ is a right-exact contravariant functor, then there exists a universal cohomology contravariant functor (T^i, ∂^i_E) such that $T^0 = F$; moreover, (T^i, ∂^i_E) is uniquely determined up to a canonical isomorphism.*

COROLLARY 7.16. *If the category $\mathscr{C}$ is with sufficiently many injectives and if $F:\mathscr{C} \longrightarrow \mathscr{C}'$ is a left-exact contravariant functor, then there exists a universal homology contravariant functor (T^i, ∂^i_E) such that $T^0 = F$; moreover, (T^i, ∂^i_E) is uniquely determined up to a canonical isomorphism.*

6. Other Properties of Homology and Cohomology Functors

To be precise we shall consider the case of cohomology contravariant functors.

PROPOSITION 7.17. *Let (T^i, ∂^i_E) be a cohomology contravariant functor from the category $\mathscr{C}$ to the category $\mathscr{C}'$. For any exact sequence of the form*

$$0 \longrightarrow B \xrightarrow{\beta} X_n \xrightarrow{\xi_n} \cdots \longrightarrow X_0 \xrightarrow{\xi_0} A \longrightarrow 0 \qquad \text{(S)}$$

and for any $i \geq 0$ there exists a morphism $D^i_S : T^i(B) \longrightarrow T^{i+n+1}(A)$ such that the following conditions are satisfied:

(a) *If we have another exact sequence of the form*

$$0 \longrightarrow B_1 \xrightarrow{\beta_1} Y_n \xrightarrow{\eta_n} \cdots \longrightarrow Y_0 \xrightarrow{\eta_0} A_1 \longrightarrow 0 \qquad \text{(U)}$$

and the morphisms $f:A \longrightarrow A_1$, $g:B \longrightarrow B_1$, $f_i:X_i \longrightarrow Y_i$, $i = 0, 1, \cdots, n$ such that the diagram:

$$\begin{array}{ccccccccccc}
0 & \longrightarrow & B & \xrightarrow{\beta} & X_n & \xrightarrow{\xi_n} \cdots \longrightarrow & X_0 & \xrightarrow{\xi_0} & A & \longrightarrow & 0 \\
 & & \downarrow{\scriptstyle g} & & \downarrow{\scriptstyle f_n} & & \downarrow{\scriptstyle f_0} & & \downarrow{\scriptstyle f} & & \\
0 & \longrightarrow & B_1 & \xrightarrow[\beta_1]{} & Y_n & \xrightarrow[\eta_n]{} \cdots \longrightarrow & Y_0 & \xrightarrow[\eta_0]{} & A_1 & \longrightarrow & 0
\end{array}$$

is commutative, then the diagram

$$\begin{array}{ccc}
T^i(B) & \xrightarrow{D^i_S} & T^{i+n+1}(A) \\
\uparrow{\scriptstyle T^i(g)} & & \uparrow{\scriptstyle T^{i+n+1}(f)} \\
T^i(B_1) & \xrightarrow[D^i_U]{} & T^{i+n+1}(A_1)
\end{array}$$

is commutative for any $i \geq 0$.

(b) *If in the sequence (S) the objects $X_i, i = 0, 1, \cdots, n$ are such that $T^j(X_i) = 0$ for any $j \geq 1$, then the morphisms D^i_S are isomorphisms for any*

$i \geq 1$ and in addition the sequence

$$T^0(X_n) \longrightarrow T^0(B) \xrightarrow{D_S^0} T^{n+1}(A) \longrightarrow 0$$

is exact.

(c) *If (R^i, δ_E^i) is another cohomology contravariant functor from $\mathscr{C}$ to $\mathscr{C}'$ and if $\varphi = (\varphi^i)_{i \geq 0}$ is a morphism from the cohomology functor (T^i, ∂_E^i) to the cohomology functor (R^i, δ_E^i), then for any exact sequence (S) and any $i \geq 0$ the diagram*

$$\begin{array}{ccc} T^i(B) & \xrightarrow{D_S^i} & T^{i+n+1}(A) \\ \downarrow{\scriptstyle \varphi^i(B)} & & \downarrow{\scriptstyle \varphi^{i+n+1}(A)} \\ R^i(B) & \xrightarrow[\Delta_S^i]{} & R^{i+n+1}(A) \end{array}$$

is commutative (Δ_S^i are the morphisms analogous to the morphisms D_S^i for the case of the functor (R^i, δ_E^i)).

PROOF. We reason by induction with respect to n. If $n = 0$, we set $D_S^i = \partial_S^i$ and conditions (a), (b) are obviously fulfilled.

For an arbitrary n, we decompose the sequence (S) into the following two exact sequences:

$$0 \longrightarrow B \xrightarrow{\beta} X_n \xrightarrow{\xi_n'} \operatorname{im} \xi_n \longrightarrow 0 \qquad (S_1)$$

$$0 \longrightarrow \operatorname{im} \xi_n \xrightarrow[\xi_n'']{} X_{n-1} \xrightarrow[\xi_{n-1}]{} \cdots \longrightarrow X_0 \xrightarrow[\xi_0]{} A \longrightarrow 0. \qquad (S_2)$$

By the induction hypothesis, the morphisms $D_{S_2}^i : T^i(\operatorname{im} \xi_n) \longrightarrow T^{i+n}(A)$ are defined and fulfil the required conditions. Then we set by definition $D_S^i = D_{S_2}^{i+1} \partial_{S_1}^i$. It remains to check conditions (a) (b).

For condition (a), we first remark that the diagrams

$$\begin{array}{ccccccccc} 0 & \longrightarrow & B & \xrightarrow{\beta} & X_n & \xrightarrow{\xi_n'} & \operatorname{im} \xi_n & \longrightarrow & 0 \\ & & \downarrow{\scriptstyle g} & & \downarrow{\scriptstyle f_n} & & \downarrow{\scriptstyle f'} & & \\ 0 & \longrightarrow & B_1 & \xrightarrow[\beta_1]{} & Y_n & \longrightarrow & \operatorname{im} \eta_n & \longrightarrow & 0 \end{array}$$

$$\begin{array}{ccccccccccccc} 0 & \longrightarrow & \operatorname{im} \xi_n & \longrightarrow & X_{n-1} & \longrightarrow & \cdots & \longrightarrow & X_0 & \longrightarrow & A & \longrightarrow & 0 \\ & & \downarrow{\scriptstyle f'} & & \downarrow{\scriptstyle f_{n-1}} & & & & \downarrow{\scriptstyle f_0} & & \downarrow{\scriptstyle f} & & \\ 0 & \longrightarrow & \operatorname{im} \eta_n & \longrightarrow & Y_{n-1} & \longrightarrow & \cdots & \longrightarrow & Y_0 & \longrightarrow & A_1 & \longrightarrow & 0 \end{array}$$

are commutative. This implies the commutativity of the diagram

$$\begin{array}{ccccc} T^i(B) & \xrightarrow{\partial^i_{S_1}} & T^{i+1}(\operatorname{im}\xi_n) & \xrightarrow{D^{i+1}_{S_2}} & T^{i+1+n}(A) \\ \uparrow{\scriptstyle T^i(g)} & & \uparrow{\scriptstyle T^{i+1}(f')} & & \uparrow{\scriptstyle T^{i+1+n}(f)} \\ T^i(B_1) & \xrightarrow[\partial^i_{U_1}]{} & T^{i+1}(\operatorname{im}\eta_n) & \xrightarrow[D^{i+1}_{U_2}]{} & T^{i+1+n}(A_1) \end{array}$$

whence we deduce immediately the commutativity of the diagram in the proposition.

Condition (b) is evidently satisfied for $n = 0$. By induction it follows immediately that D^i_S are isomorphisms for $i > 0$. For the case $i = 0$ we have the exact sequence

$$T^0(X_n) \longrightarrow T^0(B) \xrightarrow{\partial^0_{S_1}} T^1(\operatorname{im}\xi_n) \longrightarrow 0$$

as well as the isomorphism $D^1_{S_2}: T^1(\operatorname{im}\xi_n) \longrightarrow T^{n+1}(A)$, whence clearly the exact sequence in the proposition.

Condition (c) is immediate.

Definition. If (T^i, ∂^i_E) is a homology or a cohomology functor, the object A is said to be *T-acyclic* if $T^i(A) = 0$ for any $i \geq 1$.

The following proposition shows that if (T^i, ∂^i_E) is a homology or a cohomology functor, then we may use T-acyclic exact resolutions instead of projective or injective exact resolutions to determine the objects $T^i(A)$. Since we are primarily concerned with the applications to the theory of sheaves, we shall state it for the case of cohomology covariant functors.

PROPOSITION 7.18. *Let (T^i, ∂^i_E) be a universal cohomology covariant functor and let A be an arbitrary object. Let $(\mathbf{X}, \varepsilon)$ be an exact resolution of A such that X_i is T-acyclic for any i. Under these assumptions there exists an isomorphism $T^i(A) \approx H^i(T^0(\mathbf{X}))$ for any i.*

PROOF. Denote by $Z^i(\mathbf{X})$ the i-dimensional cycles of the complex $\mathbf{X}$. We have the following exact sequence:

$$0 \longrightarrow A \xrightarrow{\varepsilon} X_0 \xrightarrow{\zeta_0} \cdots \longrightarrow X_{i-1} \xrightarrow{\zeta'_i} Z^i(\mathbf{X}) \longrightarrow 0. \qquad (\xi_i)$$

By proposition 7.17, we obtain the exact sequence

$$T^0(X_{i-1}) \xrightarrow{T^0(\zeta'_i)} T^0(Z_i(\mathbf{X})) \xrightarrow{D^0_{\xi_i}} T^i(A) \longrightarrow 0.$$

In other words, $T^i(A) \approx T^0(Z^i(\mathbf{X}))/\operatorname{im} T^0(\zeta'_i)$. But since T^0 is left-exact $T^0(Z^i(\mathbf{X})) = Z^i(T^0(\mathbf{X}))$. On the other hand, if we denote by ζ''_i the canonical injection of $Z^i(\mathbf{X})$ into X_i, it follows that $T^0(\zeta''_i)$ is an injection. Since $\zeta_i = \zeta''_i\zeta'_i$ and therefore $T^0(\zeta_i) = T^0(\zeta''_i)T^0(\zeta'_i)$, we have $\operatorname{im} T^0(\zeta_i) = \operatorname{im} T^0(\zeta'_i)$, whence the relation in the proposition.

The following proposition offers a simple procedure for defining certain classes of T-acyclic objects.

PROPOSITION 7.19. *Let (T^i, ∂^i_E) be a universal cohomology covariant functor from the category $\mathscr{C}$ to the category $\mathscr{C}'$. Assume that $\mathscr{C}$ is with sufficiently many injectives. Let $\mathscr{M}$ be a class of objects of $\mathscr{C}$ which satisfies the following conditions:*

1. *For any object A of $\mathscr{C}$ there exists at least one monomorphism of A into an object of $\mathscr{M}$.*

2. *Any object of $\mathscr{C}$ isomorphic with a direct summand of an object M of $\mathscr{M}$ is an object of $\mathscr{M}$.*

3. *If*

$$0 \longrightarrow M' \longrightarrow M \longrightarrow M'' \longrightarrow 0$$

is an exact sequence in $\mathscr{C}$ and if $M, M' \in \mathscr{M}$ then $M'' \in \mathscr{M}$ and the sequence

$$0 \longrightarrow T^0(M') \longrightarrow T^0(M) \longrightarrow T^0(M'') \longrightarrow 0$$

is exact.

Under these assumptions any injective object of $\mathscr{C}$ belongs to the class $\mathscr{M}$ and any object of $\mathscr{M}$ is T-acyclic.

PROOF. Let Q be an injective object of $\mathscr{C}$ and $\alpha : Q \longrightarrow M$ a monomorphism of Q into an object of the class $\mathscr{M}$. Q is a direct summand of M, hence, by condition (2), Q belongs to the class $\mathscr{M}$.

Let now M be an arbitrary object of $\mathscr{M}$ and

$$0 \longrightarrow M \xrightarrow{\mu} L^0 \xrightarrow{\lambda^0} L^1 \xrightarrow{\lambda^1} L^2 \longrightarrow \cdots \longrightarrow L^i \xrightarrow{\lambda^i} \cdots$$

be an exact and injective resolution of M. To prove that M is T-acyclic it is sufficient to prove that the sequence

$$0 \longrightarrow T^0(M) \longrightarrow T^0(L^0) \longrightarrow T^0(L^1) \longrightarrow \cdots \tag{20}$$

is exact.

We show by induction that $Z^1(\mathbf{L})$ is an object of $\mathscr{M}$ and the sequence (20) is exact at $T^0(L^i)$.

The case when $i = 0$ is clear. Since, by induction, $Z^{i-1}(\mathbf{L}) \in \mathscr{M}$, it follows from the exact sequence

$$0 \longrightarrow Z^{i-1}(\mathbf{L}) \longrightarrow L^{i-1} \longrightarrow Z^i(\mathbf{L}) \longrightarrow 0$$

that $Z^i(\mathbf{L}) \in \mathscr{M}$ and the sequence

$$0 \longrightarrow T^0(Z^{i-1}(\mathbf{L})) \longrightarrow T^0(L^{i-1}) \longrightarrow T^0(Z^i(\mathbf{L})) \longrightarrow 0$$

is also exact. Similarly, the sequences

$$0 \longrightarrow T^0(Z^i(\mathbf{L})) \longrightarrow T^0(L^i) \longrightarrow T^0(Z^{i+1}(\mathbf{L})) \longrightarrow 0$$

$$0 \longrightarrow T^0(Z^{i+1}(\mathbf{L})) \longrightarrow T^0(L^{i+1}) \longrightarrow T^0(Z^{i+2}(\mathbf{L})) \longrightarrow 0$$

are exact, whence clearly the exactness of the sequence (20) at $T^0(L^i)$.

7. Examples of Homology and Cohomology Functors

Example 1. *The functors* $\mathrm{Ext}^i_{\mathscr{C}}$.

Let $\mathscr{C}$ be an Abelian category with sufficiently many injectives and let A be a fixed object of $\mathscr{C}$. Consider the functor $\mathrm{Hom}_{\mathscr{C}}(A, \quad):\mathscr{C} \longrightarrow \mathscr{Ab}$. This functor is covariant and left-exact. According to corollary 7.8 there exists a unique universal cohomology functor (T^i_A, ∂^i_E) such that $T^0_A = \mathrm{Hom}_{\mathscr{C}}(A, \quad)$. In fact, T^i_A are the right-derived functors of the given functor. We write for each object B of the category $\mathscr{C}$ $T^i_A(B) = \mathrm{Ext}^i_{\mathscr{C}}(A, B)$.

Assume now that we have fixed an element B in the category $\mathscr{C}$ and let $u:A_1 \longrightarrow A$ be a morphism. We define a morphism

$$\mathrm{Ext}^i_{\mathscr{C}}(u, B):\mathrm{Ext}^i_{\mathscr{C}}(A, B) \longrightarrow \mathrm{Ext}^i_{\mathscr{C}}(A_1, B)$$

as follows: Consider the exact and injective canonical resolution $(\mathbf{R}_B, \varepsilon_B)$ of B, where

$$\mathbf{R}_B = (X_0 \xrightarrow{d_0} X_1 \xrightarrow{d_1} \cdots \longrightarrow X_n \xrightarrow{d_n} \cdots).$$

Under these conditions, $(\mathrm{Hom}_{\mathscr{C}}(A, X_n), \mathrm{Hom}_{\mathscr{C}}(A, d_n))$, $(\mathrm{Hom}_{\mathscr{C}}(a_1, X_n), \mathrm{Hom}_{\mathscr{C}}(A_1, d_n))$ are complexes with values in the category of Abelian groups and it is evident that u induces a well-defined morphism from the first complex to the second one, hence a well-defined morphism from $\mathrm{Ext}^i_{\mathscr{C}}(A, B)$ to $\mathrm{Ext}^i_{\mathscr{C}}(A_1, B)$ for any i. We denote this morphism by $\mathrm{Ext}^i_{\mathscr{C}}(u, B)$.

We have thus defined a sequence of contravariant functors from the category $\mathscr{C}$ to the category $\mathscr{Ab}$. This sequence depends obviously on the object B. We now show that we can define for any exact sequence

$$0 \longrightarrow A' \longrightarrow A \longrightarrow A'' \longrightarrow 0$$

the morphisms $\delta^i_E:\mathrm{Ext}^i_{\mathscr{C}}(A', B) \longrightarrow \mathrm{Ext}^{i+1}_{\mathscr{C}}(A'', B)$ and that we obtain a cohomology contravariant functor. Since the resolution $\mathbf{R}_B$ is injective, the sequence (E) generates an exact sequence of complexes

$$0 \longrightarrow \mathrm{Hom}_{\mathscr{C}}(A'', \mathbf{R}_B) \longrightarrow \mathrm{Hom}_{\mathscr{C}}(A, \mathbf{R}_B) \longrightarrow \mathrm{Hom}_{\mathscr{C}}(A', \mathbf{R}_B) \longrightarrow 0$$

in the category of Abelian groups. This obviously yields the morphisms δ_E^i with the desired properties. In particular we have the exact sequence:

$$0 \longrightarrow \mathrm{Hom}_{\mathscr{C}}(A'', B) \longrightarrow \mathrm{Hom}_{\mathscr{C}}(A, B) \longrightarrow \mathrm{Hom}_{\mathscr{C}}(A', B)$$
$$\xrightarrow{\delta_E^0} \mathrm{Ext}^1_{\mathscr{C}}(A'', B) \longrightarrow \mathrm{Ext}^2_{\mathscr{C}}(A, B) \longrightarrow \cdots \longrightarrow \mathrm{Ext}^i_{\mathscr{C}}(A'', B)$$
$$\longrightarrow \mathrm{Ext}^i_{\mathscr{C}}(A, B) \longrightarrow \mathrm{Ext}^i_{\mathscr{C}}(A', B) \xrightarrow{\delta_E^i} \mathrm{Ext}^{i+1}_{\mathscr{C}}(A'', B) \longrightarrow \cdots.$$

If A is a projective object of $\mathscr{C}$ then the functor $\mathrm{Hom}_{\mathscr{C}}(A, \ )$ is exact and therefore $\mathrm{Ext}^i_{\mathscr{C}}(A, B) = 0$ for any object B of $\mathscr{C}$.

It follows that if $\mathscr{C}$ is with sufficiently many projectives then the sequence of contravariant functors $\mathrm{Ext}^i_{\mathscr{C}}(A, B)$ for a fixed object B, together with the morphisms δ_E^i coincides with the universal cohomology contravariant functor associated to the functor $\mathrm{Hom}_{\mathscr{C}}(\ , B): \mathscr{C} \longrightarrow \mathscr{Ab}$.

As an example of a category which satisfies the above conditions we can take the category of right or left Λ-modules over the ring Λ with a unity element. The fact that the categories ${}_{\Lambda}\mathscr{C}, \mathscr{C}_{\Lambda}$ are with sufficiently many injectives results from the general theory of injective objects since each of these categories possesses generators and satisfies axiom AB5. In view of the importance of these categories we have exposed in chapter 6 (proposition 6.5) a special proof for this case.

Assume now that the ring Λ is commutative. In this case we can introduce in a natural way a Λ-module structure on $\mathrm{Ext}^i_{\Lambda}(A, B)$ as follows: Any element $\lambda \in \Lambda$ induces an endomorphism of the Λ-module B as follows: $b \longrightarrow \lambda b$. This defines a functorial morphism of the functor $A \longrightarrow \mathrm{Hom}_{\Lambda}(A, B)$ into itself and hence a morphism of the cohomology functor defined by the sequence $\mathrm{Ext}^i_{\Lambda}(\ , B)$ into itself. It is easy to check that a Λ-module structure is thus introduced on the Abelian groups $\mathrm{Ext}^i_{\Lambda}(A, B)$. Also, it is proved immediately that the structure induced by starting from the endomorphisms of A defined by the elements of Λ coincides with the preceding one.

Example 2. The functors Tor_i^{Λ}

Let Λ be a ring with a unity element and ${}_{\Lambda}\mathscr{C}$ the category of left Λ-modules. The category ${}_{\Lambda}\mathscr{C}$ is with sufficiently many projectives. Let A be a fixed right Λ-module and $F_A: {}_{\Lambda}\mathscr{C} \longrightarrow \mathscr{Ab}$ the functor which associates with every left Λ-module B the abelian group $A \otimes_{\Lambda} B$. This is a right-exact covariant functor. By corollary 7.14 there exists a unique universal homology covariant functor (T_A^i, ∂_E^i) such that $T_A^0 = F_A$. We shall use the notation $T_A^i(B) = \mathrm{Tor}_i^\Lambda(A, B)$. Of course the functors T_A^i are the left-derived functors of the functor F_A.

Example 3. Cohomology of topological spaces.

Let X be an arbitrary topological space (No separation axiom is assumed!) If $\mathscr{C}$ is an Abelian category with a generator and satisfying axiom AB5, we denote by $\mathscr{C}(X)$ the Abelian category of presheaves on X with values in $\mathscr{C}$. Obviously, there exists an exact covariant functor $\gamma:\mathscr{C}(X) \longrightarrow \mathscr{C}$ which assigns to each presheaf $\mathscr{F}$ its value $\mathscr{F}(X)$. We denote by $\tilde{\mathscr{C}}(X)$ the category of sheaves on X with values in the category $\mathscr{C}$. $\tilde{\mathscr{C}}(X)$ is a full subcategory of $\mathscr{C}(X)$ and γ induces a left-exact functor $\Gamma:\tilde{\mathscr{C}}(X) \longrightarrow \mathscr{C}$. This is the functor of sections over the whole space X. As we have shown, the category $\tilde{\mathscr{C}}(X)$ is with sufficiently many injectives. Thus there exists a universal cohomology covariant functor (H^i, ∂_E^i) such that $H^0 = \Gamma$. This functor is said to be the functor of the 'good cohomology' of the space X.

The value of the functor H^i for the sheaf $\mathscr{F}$ is designated usually by $H^i(X, \mathscr{F})$ and is called the i-dimensional cohomology object of the space X with values in the sheaf $\mathscr{F}$. $H^i(X, \mathscr{F})$ is an object of the category $\mathscr{C}$. In practice one considers frequently the particular case when $\mathscr{C}$ is the category of right or left Λ-modules over the ring Λ. In this case $H^i(X, \mathscr{F})$ is a Λ-module.

For the special case when $\mathscr{C} = {}_\Lambda\mathscr{C}$ we shall give a proof independent of the general theory of injective objects and of localization for the fact that the category $\tilde{\mathscr{C}}(X)$ is with sufficiently many injectives. We shall lean heavily on the fact that ${}_\Lambda\mathscr{C}$ is with sufficiently many injectives. Denote for each sheaf $\mathscr{F}$ (in the present case of Λ-modules) by $\mathscr{F}_x$ the fiber of the sheaf $\mathscr{F}$ at the point $x \in X$. $\mathscr{F}_x$ is a Λ-module. If $s \in \mathscr{F}(U)$ we denote by $s(x)$ the element associated to x in $\mathscr{F}_x$ (of course, $x \in U$). Since the category of Λ-modules is with sufficiently many injectives, the Λ-module $\mathscr{F}_x$ can be imbedded into an injective Λ-module Q_x. Consider now the sheaf $\mathscr{Q}$ defined by:

$$\mathscr{Q}(U) = \prod_{x \in U} Q_x,$$

the restrictions being defined in an obvious manner. There exists a monomorphism of the sheaf $\mathscr{F}$ into the sheaf $\mathscr{Q}$ which associates to each $s \in \mathscr{F}(U)$ the family $(s(x))_{x \in U}$ in $\mathscr{Q}(U)$. It remains to show that $\mathscr{Q}$ is an injective object of the category ${}_\Lambda\tilde{\mathscr{C}}(X)$. To this end, we define for each $x \in X$ the sheaf $\mathscr{L}^x$ as follows:

$$\mathscr{L}^x(U) = \mathscr{Q}_x \quad \text{if} \quad x \in U$$

$$\mathscr{L}^x(U) = \{0\} \quad \text{if} \quad x \notin U\,;$$

the restrictions are defined in an obvious manner. It is immediately proved (without using the fact that the Λ-modules $\mathscr{Q}_x$ are injective) that $\mathscr{Q} = \prod_{x \in X} \mathscr{L}^x$. From this we deduce that for any sheaf $\mathscr{M}$ of Λ-modules we have the canonical isomorphisms

$$\mathrm{Hom}_{\Lambda\tilde{\mathscr{C}}(X)}(\mathscr{M}, \mathscr{Q}) \approx \prod_{x \in X} \mathrm{Hom}_{\Lambda}(\mathscr{M}_x, Q_x).$$

By using now the fact that each $\mathscr{Q}_x$ is an injective Λ-module it follows that the sheaf $\mathscr{Q}$ is injective.

8. The Homological Dimension of Abelian Categories

Let $\mathscr{C}$ be an Abelian category and A an object of $\mathscr{C}$. We say that the projective dimension of A is smaller than n if A possesses at least one exact projective resolution:

$$\longrightarrow X_i \longrightarrow X_{i-1} \longrightarrow \cdots \longrightarrow X_0 \longrightarrow A \longrightarrow 0$$

such that $X_i = 0$ for $i > n$. In this case we write $\mathrm{dp}\, A \leq n$. If there exists no such n then we say that the projective dimension of A is infinite.

By dualization we obtain the notion of injective dimension of an object A of the category $\mathscr{C}$; we denote it by $\mathrm{di}\, A$.

PROPOSITION 7.20. *If the category $\mathscr{C}$ is with sufficiently many projectives, then the following assertions are equivalent*:

(a) $\mathrm{dp}\, A \leq n$

(b) $\mathrm{Ext}^{n+1}_{\mathscr{C}}(A, C) = 0$ *for any object C of the category $\mathscr{C}$.*

(c) $\mathrm{Ext}^{n}_{\mathscr{C}}(A, \quad)$ *is a right-exact functor.*

(d) *If in the exact sequence*

$$0 \longrightarrow X_n \longrightarrow X_{n-1} \longrightarrow \cdots \longrightarrow X_0 \longrightarrow A \longrightarrow 0 \qquad \text{(S)}$$

the objects X_k $(k < n)$ are projective then the object X_n is also projective.

PROOF. The implications (a) $\Rightarrow$ (b) and (b) $\Rightarrow$ (c) are evident.

(c) $\Rightarrow$ (d). If $\gamma : C \longrightarrow C''$ is a morphism, then the diagram

$$\begin{array}{ccccccc} \mathrm{Hom}_{\mathscr{C}}(X_{n-1}, C) & \longrightarrow & \mathrm{Hom}_{\mathscr{C}}(X_n, C) & \xrightarrow{D_S^0} & \mathrm{Ext}^n_{\mathscr{C}}(A, C) & \longrightarrow & 0 \\ \downarrow \xi & & \downarrow \eta & & \downarrow \zeta & & \\ \mathrm{Hom}_{\mathscr{C}}(X_{n-1}, C'') & \longrightarrow & \mathrm{Hom}_{\mathscr{C}}(X_n, C'') & \xrightarrow[D_S^0]{} & \mathrm{Ext}^n_{\mathscr{C}}(A, C'') & \longrightarrow & 0 \end{array}$$

is commutative, where the vertical morphisms are induced by the morphism γ. The rows of this diagram are exact according to proposition 7.17. Assume now that γ is an epimorphism. It is sufficient to prove that η is also an epimorphism. But ξ is an epimorphism since X_{n-1} is by hypothesis projective. ζ is an epimorphism since $\mathrm{Ext}^n_{\mathscr{C}}(A,\ \ )$ is right-exact by (c). It follows then that η is also an epimorphism.

(d) $\Rightarrow$ (a). Since $\mathscr{C}$ is with sufficiently many projectives there exists an exact sequence of the form

$$0 \longrightarrow X_n \longrightarrow X_{n-1} \longrightarrow \cdots \longrightarrow X_1 \longrightarrow X_0 \longrightarrow A \longrightarrow 0$$

such that X_i $(i < n)$ are projective. But this implies by virtue of (d) that X_n is projective, hence $\mathrm{dp}\, A \leq n$.

Corollary 7.21. *In order that the object A be projective it is necessary and sufficient that* $\mathrm{Ext}^1_{\mathscr{C}}(A, Y) = 0$ *for any object Y of the category $\mathscr{C}$.*

By dualization of proposition 7.20 and corollary 7.21 we obtain:

Proposition 7.22. *If the category $\mathscr{C}$ is with sufficiently many injectives then the following assertions are equivalent:*

(a) $\mathrm{di}\, A \leq n$.

(b) $\mathrm{Ext}^{n+1}_{\mathscr{C}}(C, A) = 0$ *for any object C of the category $\mathscr{C}$.*

(c) $\mathrm{Ext}^n_{\mathscr{C}}(\ \ , A)$ *is a right-exact functor.*

(d) *If in the exact sequence*

$$0 \longrightarrow A \longrightarrow Y^0 \longrightarrow \cdots \longrightarrow Y^{n-1} \longrightarrow Y^n \longrightarrow 0$$

the objects Y_k $(k < n)$ are injective then the object Y_n is also injective.

Corollary 7.23. *In order that the object A be injective it is necessary and sufficient that* $\mathrm{Ext}^1_{\mathscr{C}}(Y, A) = 0$ *for any object Y of the category $\mathscr{C}$.*

Definition. We say that the homological dimension of the category $\mathscr{C}$ is smaller than n and we write $\mathrm{dh}\, \mathscr{C} \leq n$ if $\mathrm{dp}\, A \leq n$ for any object A of the category $\mathscr{C}$.

If Λ is a ring with a unity element and if $\mathrm{dh}\, {}_{\Lambda}\mathscr{C} \leq n$ then we say that the left global homological dimension of the ring Λ is smaller than n and we write lgl dim $\Lambda \leq n$. Similar definition for the right global homological dimension of the ring Λ.

PROPOSITION 7.24. *If the category $\mathscr{C}$ is with sufficiently many injectives and with sufficiently many projectives then the following assertions are equivalent*

(a) dh $\mathscr{C} \leq n$.

(b) dp $A \leq n$ *for any object A of the category $\mathscr{C}$.*

(c) di $A \leq n$ *for any object A of the category $\mathscr{C}$.*

(d) $\text{Ext}^k_{\mathscr{C}}(X, Y) = 0$ *for any $k > n$ and any pair (X, Y) of objects of $\mathscr{C}$.*

(e) $\text{Ext}^{n+1}_{\mathscr{C}}(X, Y) = 0$ *for any pair (X, Y) of objects of $\mathscr{C}$.*

PROOF. The equivalence between (a) and (b) follows from the definition of dh $\mathscr{C}$. The equivalence between (b) and (c) follows from propositions 7.20 and 7.22(a) and 7.22(b). The equivalence between (c) and (d) follows from proposition 7.22(a) and 7.22(b), and the equivalence between (d) and (e) is evident.

PROPOSITION 7.25. *If the category $\mathscr{C}$ is with sufficiently many injectives and if $\mathscr{C}$ possesses a generator Λ then the following assertions are equivalent:*

(a) di $A \leq n$.

(b) $\text{Ext}^i_{\mathscr{C}}(\Lambda/a, A) = 0$, *for any $i \geq n + 1$ and for any subobject a of the generator Λ.*

PROOF. The implication (a) $\Rightarrow$ (b) is clear. To prove the implication (b) $\Rightarrow$ (a) we use induction with respect to n. Consider first the case $n = 0$. We must show that the relation $\text{Ext}^1_{\mathscr{C}}(\Lambda/a, A) = 0$ for any subobject a of the generator Λ implies that A is injective. To this end it is sufficient to prove that the morphism

$$\text{Hom}_{\mathscr{C}}(\Lambda, A) \longrightarrow \text{Hom}_{\mathscr{C}}(a, A)$$

induced by the inclusion of a into Λ is an epimorphism. But this follows immediately by considering the exact sequence:

$$0 \longrightarrow a \longrightarrow \Lambda \longrightarrow \Lambda/a \longrightarrow 0$$

and the associated exact sequence

$$0 \longrightarrow \text{Hom}_{\mathscr{C}}(\Lambda/a, A) \longrightarrow \text{Hom}_{\mathscr{C}}(\Lambda, A) \longrightarrow \text{Hom}_{\mathscr{C}}(a, A) \longrightarrow \text{Ext}^1_{\mathscr{C}}(\Lambda/a, A)$$

which by the hypothesis yields precisely the desired result.

Assume now that $n \geq 1$. We can imbed A into an injective object Q

and we obtain the exact sequence

$$0 \longrightarrow A \longrightarrow Q \longrightarrow Q/A \longrightarrow 0.$$

From this it follows that $\mathrm{Ext}^i_{\mathscr{C}}(X, Q/A) \approx \mathrm{Ext}^{i+1}_{\mathscr{C}}(X, A)$ for any object X of $\mathscr{C}$. In particular $\mathrm{Ext}^i_{\mathscr{C}}(\Lambda/a, Q/A) = 0$ for any $i \geq n$ and any subobject a of the generator Λ. By induction, we have $\mathrm{di}\,(Q/A) \leq n - 1$. But clearly $\mathrm{di}\,A = (\mathrm{di}\,(Q/A) + 1)$, whence the desired result.

COROLLARY 7.26. *If* Λ *is a ring with a unity element then* $\mathrm{lgl\,dim}\,\Lambda = \sup \mathrm{dp}\,M$, *where* M *runs through the class of cyclic* Λ*-modules.*

PROOF. For each pair (M, N) of Λ-modules we denote by $\mathrm{d}\,(M, N)$ the upper bound of the numbers p for which $\mathrm{Ext}^p_\Lambda(M, N) \neq 0$. By proposition 7.24 we have

$$\mathrm{lgl\,dim}\,\Lambda = \sup_{M,N} \mathrm{d}\,(M, N) = \sup_N(\sup_M \mathrm{d}\,(M, N)) = \sup_N(\mathrm{di}\,N). \qquad (21)$$

On the other hand, we may apply proposition 7.25 to the category ${}_\Lambda\mathscr{C}$. We obtain

$$\mathrm{di}\,N = \sup_{M'} \mathrm{d}\,(M', N) \qquad (22)$$

where M' runs through the class of cyclic Λ-modules, since Λ is a generator of the category ${}_\Lambda\mathscr{C}$. From relations (21), (22) we infer

$$\mathrm{lgl\,dim}\,\Lambda = \sup_N(\sup_{M'} \mathrm{d}\,(M', N)) = \sup_{M'}(\sup_N \mathrm{d}\,(M', N)) = \sup_{M'} \mathrm{dp}\,M'.$$

9. Minimal Projective Resolutions

The notion of a projective envelope is dual to the notion of an injective envelope. We shall use the following definition, which is evidently equivalent to the one given before.

Definition. The epimorphism $P \xrightarrow{\varphi} Q$ is said to be a *projective envelope* of the object Q if the following conditions are satisfied:

(a) P is a projective object.

(b) If P' is a subobject of P such that the restriction of φ to P' is also an epimorphism, then $P' = P$.

It may happen that a category $\mathscr{C}$ is with sufficiently many projectives and nevertheless there exist objects of $\mathscr{C}$ which do not possess any projective envelope.

Example. Let $\mathscr{C} = \mathscr{Ab}$. The object Z_2 does not possess any projective envelope.

We shall indicate below examples of categories in which each object possesses at least one projective envelope.

PROPOSITION 7.27. *If $\varphi : P \longrightarrow Q$ is a projective envelope of Q, if $\varphi_1 : P_1 \longrightarrow Q$ is an epimorphism and P_1 is projective then there exists an epimorphism $\chi : P_1 \longrightarrow P$ such that the diagram*

$$\begin{array}{ccc} P & \xrightarrow{\varphi} & Q \\ {\scriptstyle \chi}\uparrow & \nearrow{\scriptstyle \varphi_1} & \\ P_1 & & \end{array} \tag{23}$$

is commutative.

PROOF. The existence of χ such that diagram (23) is commutative results from the fact that P_1 is projective. Since φ_1 is an epimorphism $\operatorname{im} \chi = P$. Otherwise the subobject $\operatorname{im} \chi$ of P would contradict condition (b).

PROPOSITION 7.28. *If $P_1 \xrightarrow{\varphi_1} Q$, $P_2 \xrightarrow{\varphi_2} Q$ are two projective envelopes of the object Q, then there exists an isomorphism $\chi : P_1 \longrightarrow P_2$ such that the diagram*

$$\begin{array}{ccc} P_1 & & \\ {\scriptstyle \chi}\downarrow & \searrow{\scriptstyle \varphi_1} & \\ & & Q \\ & \nearrow{\scriptstyle \varphi_2} & \\ P_2 & & \end{array} \tag{24}$$

is commutative.

PROOF. By proposition 7.27 there exists an epimorphism χ such that diagram (24) is commutative. We prove that χ is an isomorphism. The sequence

$$0 \longrightarrow \ker \chi \longrightarrow P_1 \xrightarrow{\chi} P_2 \longrightarrow 0$$

is exact. Since P_2 is projective, it follows that $P_1 = P_2 \oplus \ker \chi$ P_1 being a projective envelope of Q it follows that $\ker \chi = 0$, hence χ is injective.

COROLLARY 7.29. *If $\alpha : Q_1 \longrightarrow Q_2$ is an isomorphism, $\varphi_1 : P_1 \longrightarrow Q_1$ a projective envelope of Q_1 and $\varphi_2 : P_2 \longrightarrow Q_2$ a projective envelope of Q_2, then there exists an isomorphism $\chi : P_1 \longrightarrow P_2$ such that the diagram*

$$\begin{array}{ccc} P_1 & \xrightarrow{\varphi_1} & Q_1 \\ \chi \downarrow & & \downarrow \alpha \\ P_2 & \xrightarrow[\varphi_2]{} & Q_2 \end{array}$$

is commutative.

PROOF. It is sufficient to notice that $\alpha\varphi_1 : P_1 \longrightarrow Q_2$ is a projective envelope of Q_2.

Remark. The isomorphism χ is not in general uniquely determined by α.

Definition. A resolution

$$\cdots \xrightarrow{d_{n-1}} X_n \xrightarrow{d_n} X_{n-1} \xrightarrow{d_{n-1}} \cdots \xrightarrow{d_2} X_1 \xrightarrow{d_1} X_0 \xrightarrow{\varepsilon} Q \longrightarrow 0$$

is said to be a *minimal projective resolution* if:

(a) $X_0 \xrightarrow{\varepsilon} Q$ is a projective envelope of Q.

(b) The canonical surjections

$$X_n \longrightarrow \operatorname{im} d_n \qquad (n \geq 1)$$

are projective envelopes.

PROPOSITION 7.30. *Two minimal projective resolutions of the same object are isomorphic.*

The proof is immediate by using proposition 7.28.

Clearly there may exist objects which do not possess any minimal projective resolution. Also, it is evident that if any object possesses a projective envelope, then any object possesses at least one projective resolution.

Examples of categories in which any object possesses at least one projective envelope.

1. Let Λ be a local ring, not necessarily commutative, but noetherian and let $\mathfrak{m}$ be the maximal ideal of Λ. As usually ${}_\Lambda\mathscr{C}$ stands for the category of left Λ-modules. We denote by ${}_\Lambda\mathscr{C}'$ the full subcategory of the category ${}_\Lambda\mathscr{C}$ whose objects are the finitely generated Λ-modules.

PROPOSITION 7.31. *In the category ${}_\Lambda\mathscr{C}'$ each object possesses at least one projective envelope.*

PROOF. Let M be a finitely generated Λ-module and $k = \Lambda/\mathfrak{m}$. k is a field. $M/\mathfrak{m}M$ has a natural structure of finite dimensional vector space over k. Let $e_1, e_2, \cdots, e_n$ be a base of it. Let $y_1, y_2, \cdots, y_n \in M$ be such that $e_i = y_i \bmod \mathfrak{m}M$, $i = 1, 2, \cdots, n$.

The family $y_1, y_2, \cdots, y_n$ is a system of generators for M (Lemma of Nakayama).

Let $\Lambda^{(n)}$ be the free Λ-module generated by the family of elements $x_1, x_2, \cdots, x^n$ and let $\varphi : \Lambda^{(n)} \longrightarrow M$ be the homomorphism defined by

$$\varphi(x_i) = y_i, \qquad i = 1, 2, \cdots, n.$$

It is obvious that φ is an epimorphism and that $\Lambda^{(n)}$ is projective.

Let now Γ be a submodule of $\Lambda^{(n)}$ such that $\varphi(\Gamma) = M$. We show that $\Gamma = \Lambda^{(n)}$. Notice first that in the commutative diagram

$$\begin{array}{ccc} \Lambda^{(n)} & \xrightarrow{\varphi} & M \\ \downarrow & & \downarrow \\ \Lambda^{(n)}/\mathfrak{m}\Lambda^{(n)} & \xrightarrow[\bar{\varphi}]{} & M/\mathfrak{m}M \end{array}$$

$\bar{\varphi}$ is an isomorphism. It follows that in the commutative diagram

$$\begin{array}{ccc} \Gamma & \xrightarrow{\gamma} & \Lambda^{(n)} \\ \downarrow & & \downarrow \\ \Gamma/\mathfrak{m}\Gamma & \xrightarrow[\bar{\gamma}]{} & \Lambda^{(n)}/\mathfrak{m}\Lambda^{(n)} \end{array}$$

$\bar{\gamma}$ is an isomorphism.

In particular the composition $\Gamma \xrightarrow{\gamma} \Lambda^{(n)} \longrightarrow \Lambda^{(n)}/\mathfrak{m}\Lambda^{(n)}$ is a surjection. Hence

$$\mathfrak{m}\Lambda^{(n)} + \Gamma = \Lambda^{(n)},$$

and therefore, by the Lemma of Nakayama, $\Gamma = \Lambda^{(n)}$.

The proof of proposition 7.31 together with proposition 7.30 lead immediately to the following

PROPOSITION 7.32. *In order that the morphism $\varphi : P \longrightarrow Q$ in the category ${}_{\Lambda}\mathscr{C}'$ be a projective envelope of Q the following conditions are necessary and sufficient:*

(a) *P is projective.*

(b) *The homomorphism $\bar{\varphi} : P/\mathfrak{m}P \longrightarrow Q/\mathfrak{m}Q$ is an isomorphism.*

PROPOSITION 7.33. *Let*

$$\mathbf{X} = \cdots \longrightarrow X_n \xrightarrow{d_n} X_{n-1} \longrightarrow \cdots \longrightarrow X_1 \xrightarrow{d_1} X_0 \xrightarrow{\varepsilon} Q \longrightarrow 0$$

be a minimal projective resolution of a finitely generated left Λ-module Q.

Then the differentials of the complex

$$\mathbf{X}/\mathfrak{m}\mathbf{X} = \cdots X_n/\mathfrak{m}X_n \xrightarrow{d_n} X_{n-1}/\mathfrak{m}X_{n-1} \longrightarrow \cdots$$
$$\longrightarrow X_1/\mathfrak{m}X_1 \xrightarrow{d_1} X_0/\mathfrak{m}X_0$$

are zero.

PROOF. The hypothesis implies that the morphism

$$d_n : X_n \longrightarrow X_{n-1}$$

can be factorized into an essential epimorphism $f: X_n \longrightarrow Z_{n-1}$ and an injection $g: Z_{n-1} \longrightarrow X_{n-1}$. It is sufficient to prove that the morphism

$$\bar{g}: Z_{n-1}/\mathfrak{m}Z_{n-1} \longrightarrow X_{n-1}/\mathfrak{m}X_{n-1}$$

is zero. However, the composition

$$Z_{n-1}/\mathfrak{m}Z_{n-1} \longrightarrow X_{n-1}/\mathfrak{m}X_{n-1} \longrightarrow Z_{n-2}/\mathfrak{m}Z_{n-2}$$

is zero. Since $X_{n-1} \longrightarrow Z_{n-2}$ is a projective envelope it follows that the last morphism in the preceding sequence is an isomorphism, by proposition 7.32. But this implies $\bar{g} = 0$, which completes the proof.

Let Λ be a local ring (noetherian), m its maximal ideal and $k = \Lambda/m$.

PROPOSITION 7.34. *Let*

$$X = \cdots \longrightarrow X_n \xrightarrow{d_n} X_{n-1} \longrightarrow \cdots \longrightarrow X_1 \xrightarrow{d_1} X_0 \xrightarrow{\varepsilon} Q \longrightarrow 0$$

be a minimal projective resolution of the finitely generated left Λ-module Q. We have the isomorphisms:

I) $\mathrm{Tor}_i^\Lambda(k, Q) \simeq X_i/\mathfrak{m}X_i$

II) $\mathrm{Ext}^i_\Lambda(Q, k) \simeq \mathrm{Hom}_k(\mathrm{Tor}_i^\Lambda(k, Q), k)$.

PROOF. $\mathrm{Tor}_i^\Lambda(k, Q) \simeq H_i(X \otimes_\Lambda k) = H_i(\mathbf{X}/\mathfrak{m}\mathbf{X})$

Since by proposition 7.33 the differentials of the complex $X/\mathfrak{m}X$ are zero, it results that

$$H_i(X/\mathfrak{m}X) \simeq X_i/\mathfrak{m}X_i$$

$$\mathrm{Ext}^i_\Lambda(Q, k) \simeq H^i(\mathrm{Hom}_\Lambda(\mathbf{X}, k)) \simeq H^i(\mathrm{Hom}_\Lambda(\mathbf{X}/\mathfrak{m}\mathbf{X}, k))$$
$$= \mathrm{Hom}(X_i/\mathfrak{m}X_i, k) \simeq \mathrm{Hom}_k(X_i/\mathfrak{m}X_i, k) \simeq \mathrm{Hom}_k(\mathrm{Tor}_i^\Lambda(k, Q), k).$$

Definition. Let M be a left A-module and n an integer ≥ 0. We shall say that the weak homological dimension of M is less than n and we shall write $\mathrm{wdh}_A(M) \le n$, if for any right A-module N we have

$$\mathrm{Tor}(N, M) = 0 \text{ for any } m > n.$$

PROPOSITION 7.35. *If Λ is a local ring and M is a finitely generated left Λ-module, then we have*

$$\mathrm{wdh}_\Lambda(M) = \mathrm{dp}(M).$$

PROOF. Let

$$\cdots \longrightarrow P_n \xrightarrow{d_n} \cdots \longrightarrow P_1 \xrightarrow{d_1} P \xrightarrow{\varepsilon} M \longrightarrow 0$$

be a minimal projective resolution of M and suppose that $\mathrm{Tor}_{n+i}(L, M) = 0$ for every $i > 0$ and every right Λ-module L. We shall prove that $P_{n+i} = 0$ for every $i > 0$, i.e. $\mathrm{dp}(M) \le n$. But we have the relations (proposition 7.34)

$$0 = \mathrm{Tor}_{n+i}(k, M) \simeq P_{n+i}/\mathfrak{m}P_{n+i} \qquad (i > 0).$$

It follows that $P_{n+i} = \mathfrak{m}P_{n+i}$, hence (Lemma of Nakayama) $P_{n+i} = 0$. Thus we have proved the relation $\mathrm{dp}(M) \le \mathrm{wdh}(M)$. The converse inequality is obvious.

10. Relative Homological Algebra

We consider the following classical example:

Let $\varphi:\Lambda \longrightarrow \Gamma$ be a homomorphism of rings with unity elements. φ defines a functor $\varphi_*:{}_\Gamma\mathscr{C} \longrightarrow {}_\Lambda\mathscr{C}$. According to chapter 6, φ_* possesses a right-adjoint functor and a left-adjoint functor. We call the monomorphism $i:X \longrightarrow Y$ in ${}_\Gamma\mathscr{C}$ *φ-split* if $\varphi_*(i)$ defines $\varphi_*(X)$ as a direct summand of $\varphi_*(Y)$. We denote by $\mathscr{T}$ the class of all φ-split monomorphisms in ${}_\Gamma\mathscr{C}$. Let also $\mathscr{P}(\mathscr{T})$ be the class of all epimorphisms $p:X \longrightarrow Y$ in ${}_\Gamma\mathscr{C}$ such that the kernel of p is in $\mathscr{T}$. It can be readily seen that $\mathscr{T}$ and $\mathscr{P}(\mathscr{T})$ verify the following properties:

(i) $\mathscr{T} = \mathscr{T}(\mathscr{P}(\mathscr{T}))$.
(ii) If $poi = 1_X$, then $i \in \mathscr{T}$ (hence also $p \in \mathscr{P}(\mathscr{T})$).
(iii) If $\alpha, \beta \in \mathscr{T}$ and $\alpha\beta$ is defined, then $\alpha\beta \in \mathscr{T}$.
(iii′) If $\gamma, \delta \in \mathscr{P}(\mathscr{T})$ and $\gamma\delta$ is defined, then $\gamma\delta \in \mathscr{P}(\mathscr{T})$.
(iv) If α, β are monomorphisms, then $\alpha\beta \in \mathscr{T}$ implies $\beta \in \mathscr{T}$.
(iv′) If γ, δ are epimorphisms and $\gamma\delta \in \mathscr{P}(\mathscr{T})$, then $\gamma \in \mathscr{P}(\mathscr{T})$.

In this manner we can speak about the following notions which will be said to be relative to the class $\mathscr{T}$ or to the triple $(\varphi, \Lambda, \Gamma)$: An object P

of ${}_{\Gamma}\mathscr{C}$ is said to be $\mathscr{T}$-projective (i.e. projective relatively to $\mathscr{T}$) if for any diagram

$$\begin{array}{ccccc} & & P & & \\ & & \downarrow{\scriptstyle g} & & \\ A & \xrightarrow{p} & A'' & \longrightarrow & 0 \end{array}$$

where the row is exact and $p \in \mathscr{P}(\mathscr{T})$ there exists $f : P \longrightarrow A$ such that $pf = g$. Thus we can construct the derived functors of a functor by using $\mathscr{T}$-projective ($\mathscr{T}$-injective) resolutions.

Another method is given in [9], chapter 9, sections 6 and 8. We do not give the details, but content ourselves to show that the relative homological algebra can be reduced to the 'absolute' one.

To do this, we make the following considerations.

Let $\mathscr{C}$ be a small Abelian category. Assume that in $\mathscr{C}$ a family $\mathscr{T}$ of monomorphisms has been given. We associate to the family $\mathscr{T}$ the family $\mathscr{P} = \mathscr{P}(\mathscr{T})$ consisting of the epimorphisms formed by the cokernels of the monomorphisms belonging to $\mathscr{T}$. We denote by $\mathscr{M}$ the family of morphisms u of $\mathscr{C}$ such that $u = u''u'$, where $u' \in \mathscr{P}$, $u'' \in \mathscr{T}$. We suppose that the following conditions are satisfied

(i) $\mathscr{T} = \mathscr{T}(\mathscr{P}(\mathscr{T}))$.
(ii) if $pi = 1_A$, then $i \in \mathscr{T}$.
(iii) if $\alpha, \beta \in \mathscr{T}$ and $\alpha\beta$ is defined, then $\alpha\beta \in \mathscr{T}$.
(iii′) if $\gamma, \delta \in \mathscr{P}(\mathscr{T})$ and $\gamma\delta$ is defined, then $\gamma\delta \in \mathscr{P}(\mathscr{T})$.
(iv) if α, β are monomorphisms and $\alpha\beta \in \mathscr{T}$, then $\beta \in \mathscr{T}$.
(iv′) if γ, δ are epimorphisms and $\gamma\delta \in \mathscr{P}(\mathscr{T})$, then $\gamma \in \mathscr{P}(\mathscr{T})$.

By means of the class $\mathscr{T}$ we define a Grothendieck topology τ on $\mathscr{C}$ which is 'less fine' than the canonical topology (chapter 5, section 10) as follows:

We shall say that a morphism $p : A \longrightarrow X$ in $\mathscr{C}$ is covering if $p \in \mathscr{P}(\mathscr{T})$. We denote by (A, p) a covering in this topology.

Using the properties of $\mathscr{T}$, the reader can verify that, by defining for any object X of $\mathscr{C}$ Cov(X) as being the family of all couples (A, p) where p is covering, we get a Grothendieck topology τ on $\mathscr{C}$ (cf. chapter 2, section 3).

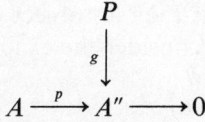

Denote by $\mathscr{C}^{\wedge}$ the category Fonct$(\mathscr{C}^0, \mathscr{Ab})$ consisting of all contravariant additive functors from $\mathscr{C}$ to $\mathscr{Ab}$ and by $h : \mathscr{C} \longrightarrow \mathscr{C}^{\wedge}$ the canonical functor (chapter 5, section 10). Also, denote by $\overline{\mathscr{C}}$ the category of sheaves relatively to the Grothendieck topology τ, and by $S:\overline{\mathscr{C}} \longrightarrow \mathscr{C}^{\wedge}$ the inclusion functor. By reproducing the construction of chapter 5, section 10, we obtain an exact functor $R:\mathscr{C}^{\wedge} \longrightarrow \overline{\mathscr{C}}$ which is a right-adjoint of S. This is done as follows: Let F be an object of $\mathscr{C}^{\wedge}$ and X an object of $\mathscr{C}$. For any $(A, p) \in \mathrm{Cov}(X)$ we consider the exact sequence

$$0 \longrightarrow \ker p \longrightarrow A \xrightarrow{\;p\;} X \longrightarrow 0,$$

whence we obtain the commutative diagram

$$\begin{array}{ccccccc} 0 \longrightarrow & F(A, p) & \longrightarrow & F(A) & \longrightarrow & F(\ker p) \\ & {\scriptstyle F(X,p)}\nwarrow & & \nearrow{\scriptstyle F(p)} & & \\ & & F(X) & & & \end{array}$$

where the row is exact. The assignment $X \longrightarrow \varinjlim_{(A,p)\in \mathrm{Cov}(X)} F(A, p)$ defines an additive left-exact functor $H:\mathscr{C}^{\wedge} \longrightarrow \mathscr{C}^{\wedge}$.

Now, the propositions 5.36, 5.37, can be adapted to the present situation with the necessary alterations.

The functor $R \circ h$ is not exact, but we have the following

PROPOSITION 7.34. *If the sequence*

$$0 \longrightarrow A' \xrightarrow{\;\alpha'\;} A \xrightarrow{\;\alpha''\;} A'' \longrightarrow 0$$

is exact in $\mathscr{C}$ *and* $\alpha'' \in \mathscr{P}$, *then the sequence*

$$0 \longrightarrow R \circ h(A') \xrightarrow{\;R \circ h(\alpha')\;} R \circ h(A) \xrightarrow{\;R \circ h(\alpha'')\;} R \circ h(A'') \longrightarrow 0$$

is exact in $\overline{\mathscr{C}}$.

PROOF. We first notice that any left-exact functor from $\mathscr{C}$ to $\mathscr{Ab}$ is a sheaf in the topology τ, which is immediate by the definition of $\mathrm{Cov}(X)$. Thus it remains to prove that $(R \circ h(A''), R \circ h(\alpha''))$ is the cokernel of $R \circ h(\alpha')$. To do this, let F be an arbitrary object of $\overline{\mathscr{C}}$ and $\varphi : R \circ h(A) \longrightarrow F$ a morphism such that $\varphi \circ Rh(\alpha') = 0$.

There exists a $\xi \in F(A)$ such that for any object X of $\mathscr{C}$ and morphism $u: X \longrightarrow A$ we have

$$\varphi(X)(u) = (F(u))(\xi).$$

For, $Rh(A)$ is isomorphic—according to the above remark—with h_A

and ξ is given by theorem 1.6. Thus for any morphism $v: X \longrightarrow A'$ we have $(F(\alpha' v))(\xi) = 0$.

From the commutative diagram

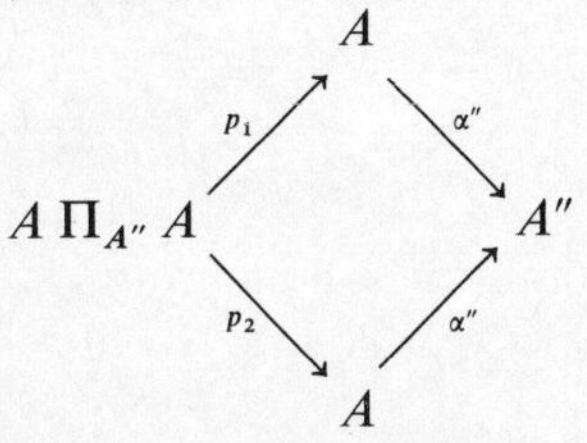

which is constructed in a canonical way, we obtain the exact sequence

$$0 \longrightarrow F(A'') \longrightarrow F(A) \xrightarrow{F(p_1) - F(p_2)} F(A \,\Pi_{A''}\, A),$$

since F is a sheaf.

We claim that $F(p_1)(\xi) = F(p_2)(\xi)$. For, from the construction of the fibered product, we have the exact sequence

$$0 \longrightarrow A \,\Pi_{A''}\, A \xrightarrow{\mu} A \,\Pi\, A \xrightarrow{\alpha''(\pi_1 - \pi_2)} A''$$

where π_1 and π_2 are the canonical projections, $p_i = \pi_i \mu$ $(i = 1, 2)$. This implies that $p_1 - p_2$ can be factorized through α', because $\alpha''(p_1 - p_2) = 0$. Hence $p_1 - p_2 = \alpha' v$, where $v: X \longrightarrow A'$. Thus there exists $\xi'' \in F(A'')$ such that $(F(\alpha''))(\xi'') = \xi$.

We deduce, since $Rh(A'')$ is representable, that ξ'' defines a functorial morphism $\psi: Rh(A'') \longrightarrow F$ such that $\psi \circ Rh(\alpha'') = \varphi$. From the construction it follows that ψ is unique.

Remark. It can be shown that any $\mathscr{T}$-projective object of the category $\mathscr{C}$ becomes projective in the category of sheaves.

BIBLIOGRAPHY

S. A. Amitsur

1. Derived functors in Abelian categories. *J. Math. Mech.*, **10** (1961), 971–994.

E. F. Assmus, Jr.

1. On the homology of local rings. *Ill. J. Math.*, **3** (1959), 187–199.

M. F. Atiyah

1. On the Krull–Schmidt theorem with application to sheaves. *Bull. Soc. Math. France*, **84** (1956), 307–317.

2. Characters and cohomology of finite groups. *Pub. Math.* no. 9, *Inst. Hautes Etudes, Paris.*

M. Auslander

1. On the dimension of modules and algebras. III. *Nagoya Math. J.*, **9** (1955), 67–77.

M. Auslander and D. A. Buchsbaum

1. Homological dimension in Noetherian rings. *Proc. NAS USA*, **42,** 9 (1955), 36–38.

2. Homological dimension in local rings. *Trans. AMS*, **85** (1957), 390–405.

3. Homological dimension in Noetherian rings II, *Trans. AMS*, **88** (1968), 194–206.

4. Codimension and multiplicity. *Ann. Math.*, **68** (1958), 625–657.

5. Unique factorization in regular local rings. *Proc. NAS USA*, **45** (1959), 733–734.

H. Bass

1. Finitistic dimension and a homological generalization of semi-primary rings. *Trans. AMS*, **95** (1960), 466–488.

C. Bănică and N. Popescu

1. Sur l'exactitude des foncteurs, *Bull. Soc. Math. Phys. R.P.R.* (1963), 3–8.

I. Berstein and P. J. Hilton

1. Category and generalized Hopf invariants. *Ill. J. Math.*, **4** (1960), 437–451.

2. Homomorphism of homotopy structures. *Top. Géom. Dif.* (Sém. Ehresmann) (1963), Paris, 24 p.

D. A. Buchsbaum

1. Exact categories and duality. *Trans. AMS*, **80** (1955), 1–34.

2. A note on homology in categories. *Ann. Math.*, **69** (1958), 66–74.

3. Satellites and universal functors. *Ann. Math.*, **71** (1960), 199–209.

I. Bucur

1. Fonctions définies sur le spectre d'une catégorie et théorie de décompositions. *Rev. Roum. Math. Pure Appl.*, **IX**, 6 (1964), 583–588.

2. Sur un problème posé par Buchsbaum. *Rev. Roum. Math. Pure Appl.*, **X**, 1 (1965).

L. Budach

1. *Quotientenfunktoren und Erweiterungstheorie*. VEB Deutscher Verlag der Wissenschaftem, Berlin, 1967.

H. Cartan and S. Eilenberg

1. *Homological Algebra*, Princeton (1956).

2. Foundations of fibre bundles. *Symp. Intern. Top. Alg. Mexico* (1958), 16–23.

P. Dedecker

1. Introduction aux structures locales. *Coll. Géom. Dif. Glob. Bruxelles* (1957), 103–136.

R. Deheuvels

1. Homologie des ensembles ordonnées et des espaces topologiques. *Bull. Soc. Math. France*, **90**, 2 (1962), 261–321.

J. Dixmier

1. Homologie des anneaux de Lie. *Ann. Sci. Ecole Norm. Supp.*, **74** (1957), 25–83.

A. Dold

1. Homology of symmetric products and other functors of complexes. *Ann. Math.*, **68** (1958), 58–80.

2. Universelle Koeffizienten. *Math. Z.*, **80** (1962), 63–88.

A. Dold and D. Puppe

1. Homologie nicht-additiver Funktoren, Anwendungen. *Ann. Inst. Fourier*, **11** (1961), 201–312.

B. Eckmann

1. Der Cohomologie-Ring einer beliebigen Gruppe. *Comment. Math. Helv.*, **18** (1945–1946), 232–282.

2. Zur Cohomologietheorie von Räumen und Gruppen. *Proc. Int. Congr. Math, Amsterdam*, **III** (1954), 170–177.

B. Eckmann and P. J. Hilton

1. On the homology and homotopy decomposition of continuous maps. *Proc. NAS, USA*, **45** (1959), 372–375.

2. Homotopy groups of maps and exact sequences. *Comment. Math. Helv.*, **34** (1960), 271–304.

3. Operators and cooperators in homotopy theory. *Math. Ann.*, **14** (1960), 1–21.

4. Structure maps in group theory. *Fund. Math.*, **50** (1961/1962), 207–221.

5. Group-like structures in general categories, I. *Math. Ann.*, **145** (1961/1962), 227–244.

6. Group-like structures in general categories, II. Equalizers, limits, lengths. *Math. Ann.*, **151** (1963), 150–186.

7. Group-like structures in general categories, III. Primitive categories. *Math. Ann.*, **150** (1963), 165–187.

B. Eckmann and A. Schopf

1. Uber injektive Moduln. *Arch. Math.*, **4** (1953), 75–78.

C. Ehresmann

1. Structures locales. *Ann. di Matem. IV*, **36** (1954), 133–142.

2. Gattungen von Lokalen Strukturen. *Jahresb. D.M.V.*, **60**, 2, (1957), 49–77.

3. Catégories topologiques et catégories différentiables. *Coll. Géom. Dif. Glob. Bruxelles* (1958), 137–150.

4. Catégorie des foncteurs types. *Rev. Un. Mat. Arg.*, **20** (1960), 194–209.

5. Catégories inductives et pseudogroupes. *Ann. Inst. Fourier*, **X** (1960), 317–332.

6. Espèces de structures locales, élargissements de catégories. *Top. et Geom. Dif.* (Sém. Ehresmann) III (1961), Paris, 73 p.

7. Structures feuilletées. *Proc. 5th Canad. Math. Cong.* (1961), 109–172.

8. Catégories structurées, I et II: *Ann. Ec. Norm. Sup.*, **80** (1963), 349–426; III: *Top. et Geom. Dif.*, **V** (1963), 21 p.

9. Catégories structurées quotient. *Top. et Géom. Dif.*, **V** (1963), Paris, 5 p.

10. Structures quotient. *Comm. Math. Helv.*, **38** (1963), 219–283.

11. Groupoides sous-inductifs. *Ann. Inst. Fourier*, **13**, 2 (1963), 1–60.

12. Sous-structures et catégories ordonnées. *Fund. Math.*, **14** (1964), 211–228.

13. Catégories ordonnées, Holonomie et Cohomologie. *Ann. Inst. Fourier*, **14**, 1 (1964), 205–268.

14. Complétion des catégories ordonnées. *Ann. Inst. Fourier*, **14**, 2 (1964), 89–144.

15. Catégories et structures. *Top. et Géom. Dif.*, **V** (1964), 31 p.
16. Prolongements des catégories différentiables. *Top. et Géom.. Dif.*, **VI** (1964), 8 p.
17. *Catégories et structures.* Dunod, Paris, 1965.
18. *Introduction to the theory on structured categories.* Univ. of Kansas, Dept. of Math., 1966.
19. Sur l'existence de structures libres et de foncteurs adjoints. *Top. et Géom. Dif., Paris*, 1966.

S. Eilenberg
1. Singular homology theory. *Ann. Math.*, **45** (1944), 407–447.
2. Homology of spaces with operators, I. *Trans. AMS*, **61** (1947), 378–417.
3. Homological dimension and syzygies. *Ann. Math.*, **64** (1956), 328–336.
4. Errata thereto: *Ann. Math.*, **65** (1957), 593.
5. Abstract description of some basic functors. *J. Indian Math. Soc.*, **24** (1960), 231–234.

S. Eilenberg and S. MacLane
1. Group extensions and homology. *Ann. Math.*, **43** (1942), 757–831.
2. Relations between homology and homotopy groups. *Proc. NAS USA*, **29** (1943), 155–158.
3. General theory of natural equivalences. *Trans. AMS*, **58** (1945), 231–294.
4. Relations between homology and homotopy groups of spaces. *Ann. Math.*, **46** (1945), 480–509.
5. Cohomology theory in abstract groups, I. *Ann. Math.*, **48** (1947), 51–78.
6. Cohomology theory in abstract groups, II. Groups extensions with a nonabelian kernel. *Ann. Math.*, **48** (1947), 326–341.
7. Cohomology and Galois theory, I. Normality of algebras and Teichmüller's cocycle. *Trans. AMS*, **64** (1948), 1–20.
8. Homology of spaces with operators, II. *Trans. AMS*, **65** (1949), 49–99.
9. Relations between homology and homotopy groups of spaces, II. *Ann. Math.*, **51** (1950), 514–533.
10. Cohomology theory of Abelian groups and homotopy theory, I. *Proc. NAS USA*, **36** (1950), 443–447.
11. Homology theories for multiplicative systems. *Trans. AMS*, **71** (1951), 294–330.
12. Acyclic models. *Am. J. Math.*, **75** (1953), 189–199.
13. On the groups H(π, n), I. *Ann. Math.*, **58** (1953), 55–106.
14. On the groups H(π, n), II. Methods of computation. *Ann. Math.*, **60** (1954), 49–139.
15. On the groups H(π, n), III. Operations and obstructions. *Ann. Math.*, **60** (1954), 513–557.
16. On the homology theory of Abelian groups. *Can. J. Math.*, **7** (1955), 43–55.

S. Eilenberg and J. C. Moore
1. Limits and spectral sequences. *Topology*, **1** (1962), 1–23.
2. Adjoint functors and triples. *Ill. J. Math.*, **9** (1965), 381–398.

S. Eilenberg and N. Steenrod
1. *Foundations of algebraic topology*. Princeton, 1952.

S. Eilenberg and J. A. Zilber
1. Semi-simplicial complexes and singular homology. *Ann. Math.*, **51** (1950), 499–513.

P. Freyd
1. *Functor theory*. Dissertation, Princeton Univ., 1960.
2. *Abelian categories*. Harper and Row, New York, 1964.

P. Gabriel
1. Des catégories abéliennes. *Bull. Soc. Math., France*, **90** (1962), 323–448.

P. Gabriel and N. Popescu
1. Caractérisations des catégories abéliennes avec générateurs et limites inductives exactes. *C. R. Acad. Sc. Paris*, **258**, 4188–4190 (1964).
P. Gabriel and M. Zisman
1. *Séminaire Homotopique*. Strasbourg (1964).
J. Giraud
1. Analysis situs. *Sem. Bourbaki* (1962–63), **257**, 11 p.
2. *Cohomologie non-abélienne de degré 2* (*thèse*). Fac. des Sciences de l'Univ. de Paris.
R. Godement
Théorie des faisceaux. Hermann, Paris (1958).
J. W. Gray
1. Extensions of sheaves of algebras. *Ill. J. Math.*, **5** (1961), 159–174.
2. Extensions of sheaves of associative algebras by non-trivial kernels. *Pac. J. Math.*, **11** (1961), 909–917.
A. Grothendieck
1. Sur quelques points d'algèbre homologique. *Tohoku Math. J. 2*, **9** (1957), 119–221.
2. Techniques de descente et théorèmes d'existence en géométrie algébrique. *Sem. Bourbaki*: I, **190** (1959–60), 29 p; II, **195** (1959–60), 22 p; III, **212** (1960–61), 20 p; IV, **221** (1960–61), 28 p; V, **232** (1961–62), 19 p.
3. Techniques de construction en géométrie analytique. *Sém. Cartan*, **13** (1960–1961); I, 33 p; II, 14 p; III, 11 p; IV, 28 p; V, 15 p; VI, 13 p; VII, 27 p; VIII, 10 p; IX, 20 p; X, 20 p.
4. *Fondements de la géométrie algébrique*. Paris (1961).
A. Grothendieck and J. Dieudonné
1. *Elements de géométrie algébrique, I–IV*. Publ. Inst. des Hautes Etud. Sci.

A. Heller
1. Homological algebra in abelian categories. *Ann. Math.*, **68** (1958), 484–525.
P. J. Hilton
1. Note on free and direct products in general categories. *Bull. Soc. Math. Belg.*, **13** (1961), 38–49.
2. The fundamental group as a functor. *Bull. Soc. Math. Belg.*, **14** (1962), 153–177.
P. J. Hilton and W. Ledermann
1. Homology and ringoids, I. *Proc. Camb. Phil. Soc.*, **54** (1958), 152–167.
2. On the Jordan–Hölder theorem in homological monoids. *Proc. London Math. Soc.*, **14** (1960), 321–334.
P. J. Huber
1. Homotopy theory in general categories. *Math. Ann.*, **144** (1961), 361–385.
2. Standard constructions in abelian categories. *Math. Ann.*, **146** (1962), 321–325.

D. M. Kan
1. On the homotopy relation for c.s.s. maps. *Col. Soc. Mat. Mec. 2*, **2** (1957), 75–81.
2. On c.s.s. categories, id. 82–94.
3. Adjoint functors. *Trans. A.M.S.*, **294** (1958), 294–329.
4. Functors involving c.s.s. complexes. *Trans. A.M.S.*, **87** (1958), 330–346.
5. On monoids and their duals. *Bol. Soc. Mat. Mec.*, **3** (1958), 52–61.
I. Kaplansky
1. On the dimension of modules and algebras, X. *Nagoya Math. J.*, **13** (1958), 85–88.
J. L. Kelley–E. Pitcher
1. Exact homomorphism sequences in homology theory. *Ann. Math.*, **48** (1947), 682–709.
H. Kleisli
1. Homotopy theory in abelian categories. *Can. J. of Math.*, **14**, 1 (1962), 139–169.

А. Г. Курощ

1. Прямое разложение в алгебраических категориях, Труды Моск. Мат. Об–в, **8** (1959): 3/9–421.

А. Г. Курощ and А. Х. Ливмиц and Е. Г. Щульгейфер

1. Основы теорий категорий У.М.Н. XV, **6** (1960), 3–52.

O. A. Laudal

1. Sur la limite projective et la théorie de la dimension. *Top. et Géom. Dif.* (*Sém. Ehresmann*) *III*, (1961), 23 p.

W. V. Lawvere

1. Functorial semantics of algebraic theories. *Proc. Nat. Acad. Sci.*, **50** (1963), 869–872.

2. An elementary theory of the category of sets. *Proc. Nat. Acad. Sci.* (1964), 1506–1511.

S. Lubkin

1. Imbedding of abelian categories. *Trans. A.M.S.*, **97** (1960), 410–417.

S. MacLane

1. Duality for groups. *Bull. A.N.S.*, **56** (1950), 485–516.

2. Slide and torsion products for modules. *Rend. Del. Sem. Math.*, *Torino*, **15** (1955/1956), 281–309.

3. Homologie des anneaux et des modules. *Colloque de Topologie algébrique*, Louvain (1956), 55–80.

4. Extension and obstructions for rings. *Ill. J. Math.*, **2** (1958), 316–345.

5. Locally small categories and the foundations of set theory. *Symp. Warsaw* (1959), 25–83.

6. Group extensions by primary abelian groups. *Trans. AMS*, **95** (1960), 1–16.

7. Triple torsion products and multiple Künneth formulas. *Math. Ann.*, **140** (1960), 51–64.

8. An algebra of additive relations. *Proc. Nat. Acad. Sc., USA*, **47** (1961), 1043–1051.

9. *Homology*, Springer-Berlin, 1963.

10. *Natural associativity and commutativity*. Rice University Studies 49 (1963), p. 28–46.

11. Categorical algebra. *Bull. Amer. Mat. Soc.*, **71** (1965), 40–106.

S. MacLane–G. Birkhoff

1. *Algebra*, Macmillan Company, 1967.

J. M. Maranda

1. Some remarks on limits in categories. *Canad. Math. Bull.* **5** (1962), 133–146.

E. Matlis

1. Injective modules over Noetherian rings. *Pac. J. Math.*, **8** (1958), 511–528.

2. Applications of duality. *Proc. AMS*, **10** (1959), 659–662.

3. Modules with descending chain condition. *Trans. AMS*, **97** (1960), 495–508.

B. Mitchell

1. The full embedding theorem. *Am. J. Math.*, **86** (1964), 619–637.

2. *Theory of categories*. New York, Academic Press, 1965.

C. Năstăsescu and N. Popescu

1. Quelques observations sur les topos abéliens. *Rev. Roum. Math. Pure et Appl.*, **12** (1967), 553–563.

2. Sur la structure des objects de certaines catégories abéliennes. *C. R. Acad. Sci., Paris*, **262** (1966), 1295–1297.

D. G. Northcott

1. *An introduction to homological algebra*. Cambridge Univ. Press, 1960.

F. Oort

1. Natural maps of extension functors. *Need. Akad. Proc.*, **66** (1963), 559–566.

2. Yoneda extensions in abelian categories. *Math. Ann.*, **153** (1964), 227–235.

Proceedings of the Conference on Categorical Algebra, La Jolla, 1965 (Edited by S. Eilenberg, D. K. Harison, S. MacLane, H. Rohrl).

D. Puppe

1. Korrespondenzen in abelschen Kategorien. *Math. Ann.*, **148** (1962), 1–30.

J. A. Riley

1. Axiomatic primary and tertiary decomposition theory. *Amer. Math. Soc.*, **105** (1962), 177–201.

H. Röhrl

1. Über Satelliten Halbexacter Funktoren. *Math. Zeitsch.*, **79** (1962), 193–228.

D. S. Rim

1. Modules over finite groups. *Ann. Math.*, **69** (1959), 700–712.

J. Roos

1. Sur les foncteurs dérivés de lim. Applications. *C. R. Acad. Sci., Paris*, **252** (1961), 3702–3704.

2. Locally distributive spectral categories and strongly regular rings. Midwest category seminar, lecture notes, 1967 (Springer).

3. Sur la condition AB6 et ses variantes dans les catégories abéliennes. *C. R. Acad. Sci., Paris*, **264** (1967), 991–994.

4. Sur les catégories spectrales localement distributives. *C. R. Acad. Sci., Paris*, **265** (1967), 14–17.

5. Sur la structure des catégories spectrales et les coordonnées de von Neumann des treillis modulaires et complémentés. *C. R. Acad. Sci., Paris*, **265** (1967), 42–45.

Z. Semadeni

1. Projectivity, injectivity and duality. *Rozprawy Mate.*, 35 Varsovie (1963), 47 p.

2. Free and direct objects. *Bull. Amer. Math. Soc.*, **69** (1963), 63–65.

U. Shukla

1. Cohomologie des algèbres associatives. *Ann. Sci. Ecole Norm Sup.*, **78** (1961), 163–209.

R. G. Swan

1. A simple proof of the cup product reduction theorem. *Proc. NAS, USA*, **46** (1960), 114–117.

2. Induced representations and projective modules. *Ann. Math.*, **71** (1960), 552–578.

J. A. Zilber

1. *Categories in homotopy theory*. Dissertation, Harvard University, Cambridge, Mass., 1963.

INDEX